Abiotic Stress-Induced Secondary Metabolites Regulating Plant Metabolism

Abiotic Stress-Induced Secondary Metabolites Regulating Plant Metabolism

Guest Editor

Runqiang Yang

MDPI

Basel • Beijing • Wuhan • Barcelona • Belgrade • Novi Sad • Cluj • Manchester

Guest Editor
Runqiang Yang
College of Food Science and Technology
Nanjing Agricultural University
Nanjing
China

Editorial Office
MDPI AG
Grosspeteranlage 5
4052 Basel, Switzerland

This is a reprint of the Special Issue, published open access by the journal *Plants* (ISSN 2223-7747), freely accessible at: https://www.mdpi.com/journal/plants/special_issues/6G0F5S70PT.

For citation purposes, cite each article independently as indicated on the article page online and as indicated below:

Lastname, A.A.; Lastname, B.B. Article Title. *Journal Name* **Year**, *Volume Number*, Page Range.

ISBN 978-3-7258-5549-0 (Hbk)
ISBN 978-3-7258-5550-6 (PDF)
https://doi.org/10.3390/books978-3-7258-5550-6

Contents

plants

MDPI

Article

GABA Application Enhances Drought Stress Tolerance in Wheat Seedlings (*Triticum aestivum* L.)

Qiuyan Zhao, Yan Ma *, Xianqing Huang, Lianjun Song, Ning Li, Mingwu Qiao, Tiange Li, Dan Hai and Yongxia Cheng

Henan Engineering Technology Research Center of Food Processing and Circulation Safety Control, College of Food Science and Technology, Henan Agricultural University, Zhengzhou 450002, China
* Correspondence: mayan201509@163.com; Tel./Fax: +86-0371-63558150

Abstract: In this study, the effects of γ-aminobutyric acid (GABA) on physio-biochemical metabolism, phenolic acid accumulation, and antioxidant system enhancement in germinated wheat under drought stress was investigated. The results showed that exogenous GABA reduced the oxidative damage in wheat seedlings caused by drought stress and enhanced the content of phenolics, with 1.0 mM being the most effective concentration. Six phenolic acids were detected in bound form, including *p*-hydroxybenzoic acid, vanillic acid, syringic acid, *p*-coumaric acid, ferulic acid, and sinapic acid. However, only syringic acid and *p*-coumaric acid were found in free form. A total of 1.0 mM of GABA enhanced the content of total phenolic acids by 28% and 22%, respectively, compared with that of drought stress, on day four and day six of germination. The activities of phenylalanine ammonia lyase (PAL), cinnamic acid 4-hydroxylase (C4H) and 4-coumarate coenzyme A ligase (4CL) were activated by drought stress plus GABA treatment. Antioxidant enzyme activities were also induced. These results indicate that GABA treatment may be an effective way to relieve drought stress as it activates the antioxidant system of plants by inducing the accumulation of phenolics and the increase in antioxidant enzyme activity.

Keywords: GABA; drought stress; wheat seedlings; phenolic acids; antioxidant system

1. Introduction

Drought is considered an important abiotic stress with major impacts on plant growth and productivity [1]. Drought stress causes physiological, biochemical, and morphological changes in plants. In addition, respiration, photosynthesis, enzyme activities, redox homeostasis, and chloroplast metabolism are also affected by drought [2]. Plants have evolved different defense systems to avoid oxidative damage caused by drought stress, including overproduction of antioxidant metabolites that prevent the spread of the oxidative chain reaction. In this case, polyphenols such as phenolic acids, flavonoids, and anthocyanins play an important role in reducing the adverse effects of stress [3].

Wheat (*Triticum aestivum* L.), as one of the earliest cultivated gramineae plants in the world, is an important grain crop in many countries. Wheat grains are usually ground into flour to make food for satiation. In recent years, germinated grains have attracted much attention due to their unique function and nutritional value. Studies have shown that wheat sprouts are good sources of minerals (potassium, magnesium, and calcium), vitamins (vitamins C and E), enzymes (superoxide dismutase and catalase), amino acids, bioflavonoids, phenolic acids (ferulic acid and vanillic acid), etc. [4]. Among the bioactive phytochemicals, phenolics were cited as the most important contributors to the antioxidant capacity of cereal grains and have many physiological functions in the human body [5]. At present, most studies on germinated wheat focus on the regulation of germination conditions [6] and the change in the processing properties of malt flour [7]. In recent years, more and more studies have focused on the content and composition of phenolics in wheat germination from different countries during germination [8–10].

Phenolics are secondary metabolites of plants, which are mainly synthesized by the phenylpropane metabolic pathway, and phenylalanine ammonia lyase (PAL), cinnamic acid 4-hydroxylase (C4H), and 4-coumarate CoA ligase (4CL) are the key enzymes involved in its biosynthesis. The biosynthesis of phenolics is a complex network of chemical reactions and an endogenously regulated process during plant growth and development [11], or which can be stimulated by exogenous factors such as moisture [12,13]. Environmental stresses induce the accumulation of phenolic compounds to protect tissues from damage caused by free-radical-induced oxidative stress [14]. When wheat and other grains germinated under stress conditions such as drought, stress signals were recognized and amplified by cell membrane receptors, related endogenous enzymes were activated, and the metabolic direction of substances changed. Finally, the composition and content of phenolics changed [15]. At present, studies on the key enzymes of phenolic biosynthesis in wheat seedlings mostly focus on changes in PAL enzyme activity [8,16]. There are few studies on C4H and 4CL during wheat germination under stress, and the study on the relationship between phenolics and the key enzymes of phenolic biosynthesis during wheat germination under stress needs to be strengthened.

γ-Aminobutyric acid (GABA) is a four-carbon non-protein amino acid and is widespread in nature. Previous studies have shown that GABA accumulated rapidly in various stresses, which implied that GABA might play a dual role as both a metabolite and a legitimate plant-signaling molecule [17,18]. The application of GABA under different abiotic stresses has been proven to alleviate stress by regulating the metabolism of sugar and proline [19,20], alleviating the effects of photosynthesis and mitochondrial activity inhibition and maintaining the integrity of chloroplasts under stress [19,20], increasing the activity of guaiacol peroxidase, catalase, superoxide dismutase, ascorbate peroxidase, and other antioxidant enzymes [19,20], decreasing Na^+ concentration and increasing K^+ concentration [21], increasing amino acid and organic acid accumulation [22], promoting the production of polyamines (PAs) and the inhibition of their metabolism [23], and increasing the content of unsaturated fatty acids (γ -linolenic acid) and reducing the content of saturated fatty acids [24]. In wheat, studies have shown that GABA increased salt tolerance by promoting nitrogen and carbon assimilation [25], improving photosynthesis, and enhancing the activities of antioxidant enzymes [26], as well as altering signal transduction pathways [27]. While under drought stress, GABA reduced the activity loss or degeneration of some enzymes and proteins, so as to maintain the balance of cell metabolism [28]. The mechanism of GABA is different for plants to resist different stresses, such as salt stress and drought stress. Additionally, studies have shown that GABA may respond to stress by inducing phenolic accumulation and enhancing antioxidant activity in barley seedlings [29] and soybean sprouts [30] under NaCl stress. To our knowledge, the regulatory role of GABA in the biosynthesis of phenolics induced by drought stress in germinated wheat has not been reported in the literature.

In this study, the effects of GABA on phenolic acids and antioxidant systems in wheat seedlings under drought-stress conditions were investigated. This would provide the basis for further research on the mechanism of GABA in enhancing plant stress resistance and bring new ideas for the enrichment of phenolics.

2. Results

2.1. Effects of GABA on Length and Weight of Wheat Seedlings under Drought Stress

Changes in the length and weight of wheat seedlings under different treatments were presented in Figure 1. With the prolonging of germination time, the length of wheat seedlings increased significantly (Figure 1A). The response of seedling fresh weight to different germination stages was similar to that of seedling length, and it increased significantly during germination (Figure 1C). However, a slight decrease in freeze-dried weight was observed (Figure 1D). Drought stress resulted in a significant reduction in seedling length during the whole growth process (Figure 1A), but root length only decreased at the initial stage of stress and had no significant change at the later stage (Figure 1B). The fresh weight of wheat seedlings also declined markedly under drought stress (Figure 1C). Meanwhile, PEG6000 treatment caused a considerable increase in freeze-dried weight (Figure 1D). Exogenous GABA treatment significantly alleviated the growth inhibition of wheat seedlings under drought stress, and the seedling length were enhanced by 2.69%, 19.53%, 34.27%, 19.67% and 16.68% on the 6th day, respectively.

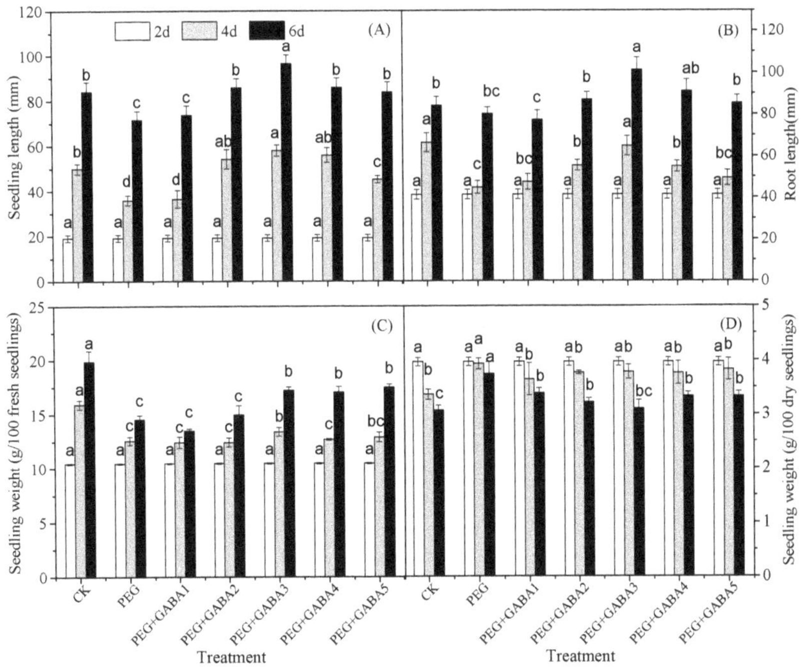

Figure 1. Effects of GABA on length and weight of wheat seedlings under drought stress. (**A**) seedling length, (**B**) root length, (**C**) fresh weight of wheat seedlings, (**D**) dry weight of wheat seedlings. The length and weight were determined on day 2, 4, and 6. Twenty wheat seedlings were selected and measured 20 times, and the average value was taken as the length of wheat seedlings and roots. One hundred wheat seedlings were selected as the sampling group, and a total of three groups were selected, whose average value was taken as the fresh weight and dry weight of wheat seedlings. Different letters following values indicate significant differences between treatments at the same germination time ($p < 0.05$). CK means the wheat was sprayed with distilled water; PEG means 100 g/L PEG 6000 treatment; PEG + GABA1 means 100 g/L PEG 6000 + 0.25 mM GABA treatment; PEG + GABA2 means 100 g/L PEG 6000 + 0.5 mM GABA treatment; PEG + GABA3 means 100 g/L PEG 6000 + 1.0 mM GABA treatment; PEG + GABA4 means 100 g/L PEG 6000 + 2.0 mM GABA treatment; PEG + GABA5 means 100 g/L PEG 6000 + 4.0 mM GABA treatment.

2.2. *Effects of GABA on the MDA Content and Electrolyte Leakage of Wheat Seedlings under Drought Stress*

As shown in Figure 2, germination resulted in a significant increase in the MDA level and electrolyte leakage. The largest increase in MDA level occurred from 4 d to 6 d of germination, and the largest increase in electrolyte leakage occurred from 2 d to 4 d of germination. Drought stress led to pathophysiological stress and damage of germinated wheat, which caused a noticeable improvement in MDA level and electrolyte leakage of 83% and 33%, respectively, on day 6 of germination, compared with the control. Application of exogenous GABA significantly alleviated the growth-inhibiting effects of drought stress. The maximum reduction in electrolyte leakage and MDA level of wheat treated with drought stress plus 1.0 mM GABA was 23% and 26%, respectively, after germination for 6 d, compared with that of drought stress.

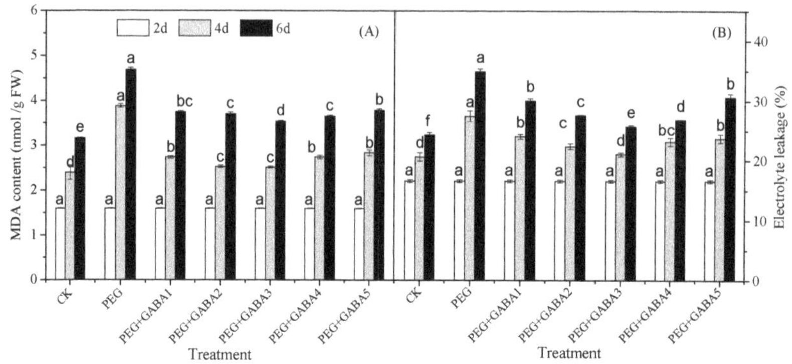

Figure 2. Effects of GABA on MDA content (**A**) and the electrolyte leakage (**B**) of wheat seedlings under drought stress. Different letters following values indicate significant differences between treatments at the same germination time ($p < 0.05$). CK, PEG, PEG+GABA1, PEG+GABA2, PEG+GABA3, PEG+GABA4, and PEG+GABA5 indicate distilled water treatment, 100 g/L PEG 6000 treatment, 100 /L PEG 6000 + 0.25 mM GABA treatment, 100 g/L PEG 6000 + 0.5 mM GABA treatment, 100 g/L PEG 6000 + 1.0 mM GABA treatment, 100 g/L PEG 6000 + 2.0 mM GABA treatment, and 100 g/L PEG 6000 + 4.0 mM GABA treatment, respectively.

2.3. Effects of GABA on the Total Phenolic Content of Wheat Seedlings under Drought Stress

Results showed that the total phenolic content on day 4 and day 6 of germination induced by drought stress increased by 64% and 29%, respectively (Figure 3). Exogenous GABA significantly increased total phenolic content under drought stress, which were 1.05, 1.10, 1.21, 1.15, and 1.11 times higher than that under only drought stress, respectively, under 0.25 mM, 0.5 mM, 1.0 mM, 2.0 mM, and 4.0 mM GABA at the 6 d of germination. The addition of 1.0 mM GABA had the most significant effect, resulting in an increase in the total phenolic content of 116% and 121% relative to that of drought stress on day 4 and day 6 of germination.

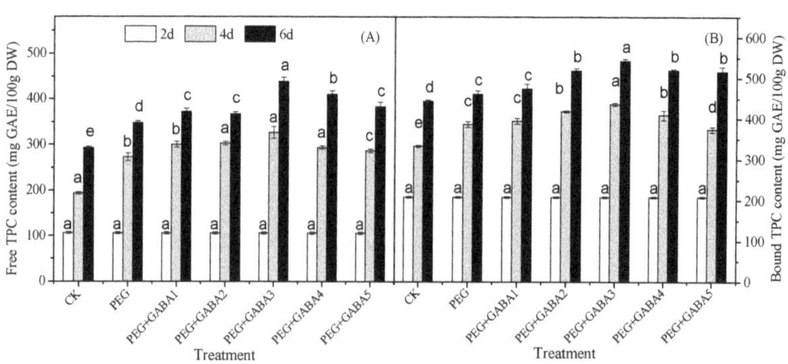

Figure 3. Effects of GABA on the free (**A**) and bound (**B**) total phenolic content of wheat seedlings under drought stress. Different letters following values indicate significant differences between treatments at the same germination time ($p < 0.05$). CK, PEG, PEG+GABA1, PEG+GABA2, PEG+GABA3, PEG+GABA4, and PEG+GABA5 indicate distilled water treatment, 100 g/L PEG 6000 treatment, 100 g/L PEG 6000 + 0.25 mM GABA treatment, 100 g/L PEG 6000 + 0.5 mM GABA treatment, 100 g/L PEG 6000 + 1.0 mM GABA treatment, 100 g/L PEG 6000 + 2.0 mM GABA treatment, and 100 g/L PEG 6000 + 4.0 mM GABA treatment, respectively.

2.4. Effects of GABA the on Phenolic Acids Content of Wheat Seedlings under Drought Stress

As can be seen from Figure 4, phenolic acids are mainly in the bonded form in germinated wheat. Six phenolic acids were found in combined form, including *p*-hydroxybenzoic acid, vanillic acid, syringic acid, *p*-coumaric acid, ferulic acid, and sinapic acid. However, only syringic acid and *p*-coumaric acid were found in free form.

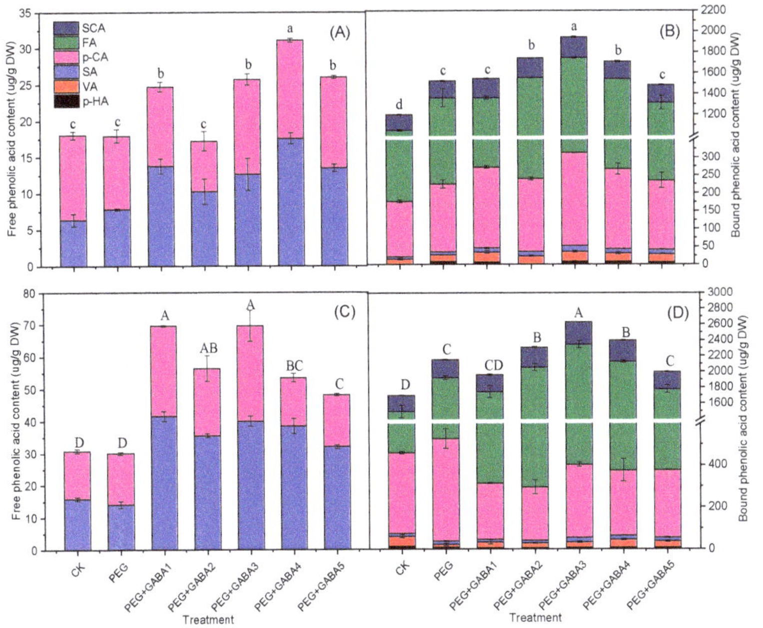

Figure 4. Effects of GABA on phenolic acids in wheat seedlings under drought stress: (**A**) free phenolic acid content of wheat seedlings on day 4, (**B**) bound phenolic acid content of wheat seedlings on day 4, (**C**) free phenolic acid content of wheat seedlings on day 6, (**D**) bound phenolic acid content of wheat seedlings on day 6. Values with different letters are significantly different at $p < 0.05$. Lowercase letters represent significant differences among treatment factors on day 4, and capital letters represent significant differences among treatment factors on day 6 of germination. CK, PEG, PEG+GABA1, PEG+GABA2, PEG+GABA3, PEG+GABA4, and PEG+GABA5 indicate distilled water treatment, 100 g/L PEG 6000 treatment, 100 g/L PEG 6000 + 0.25 mM GABA treatment, 100 g/L PEG 6000 + 0.5 mM GABA treatment, 100 g/L PEG 6000 + 1.0 mM GABA treatment, 100 g/L PEG 6000 + 2.0 mM GABA treatment, and 100 g/L PEG 6000 + 4.0 mM GABA treatment, respectively.

Ferulic acid was the main phenolic acid in wheat seedlings. Drought stress induced a significant increase of 30% and 35% of bound ferulic acid on day 4 and day 6 of germination, respectively, compared with that of the control. A comparison of the 100 g/L PEG 6000 plus 1.0 mM GABA treatment with the drought stress on wheat seedlings showed that the bound ferulic acid content significantly increased by 26% and 39%, compared with drought stress on day 4 and day 6 of germination, respectively. Ferulic acid content also increased by 26% and 25% on day 6 of germination under 0.5 mM and 2.0 mM GABA treatment, but there was no significant difference under 0.25 mM and 5.0 mM GABA treatment compared with that of drought stress.

p-Coumaric acid existed in both free and bound forms in wheat seedlings, but the bound form is the main state of existence. Under drought stress, the content of free *p*-coumaric acid did not increase significantly, but the content of bound *p*-coumaric acid was 1.23 and 1.27 times higher than the control on day 4 and day 6 of germination. The content of *p*-coumaric acid varied with different concentrations of GABA at different germination stages. The exogenous application of 1.0 mM GABA induced a significant increase of 38% of total *p*-coumaric acid on day 4 of germination, but led to a decrease of 26% on day 6 of germination, compared with that under drought stress.

Sinapic acid was only detected in bound form. Exogenous application of 1.0 mM GABA under drought stress significantly increased the sinapic acid content, which were 1.23 and 1.24 times higher than that under only drought stress, respectively, on day 4 and day 6 of germination. Other GABA concentrations also had different effects on sinapic acid content, which increased significantly at concentrations of 0.5 mM, 1.0 mM, and 2.0 mM in the late stage of germination, but there was no significant difference under 0.25 mM and 5.0 mM GABA treatment, compared with that of drought stress.

Other phenolic acids in wheat seedlings had a lower content. *p*-Hydroxybenzoic acid and vanillic acid were only detected in bound form, and syringic acid was found both in free and bound form. *p*-Hydroxybenzoic acid was not found in the control treatment at the early stage of germination until day 6. However, *p*-Hydroxybenzoic acid was detected both under drought stress and PEG plus GABA treatment, and its content decreased significantly compared with the control. The content of vanillic acid was significantly increased at different concentrations of GABA treatment in the initial stage of germination, but decreased at the later stage compared with that of drought stress. Syringic acid increased significantly when treated with exogenous GABA. The content of the bound form was significantly increased at the early stage of germination, and the free form was significantly increased at the late stage of germination.

2.5. Effects of GABA on the PAL, C4H, and 4CL Activity of Wheat Seedlings under Drought Stress

As shown in Figure 5, the activities of PAL, C4H, and 4CL enzymes increased significantly at the early stage of germination. However, the activities of PAL and 4CL decreased at the late stage of germination (Figure 5). The activities of PAL, C4H, and 4CL in wheat seedlings under drought stress were 1.25, 1.30, and 1.23 times, and 1.34, 1.29, and 1.35 times higher than those of the control, respectively, on day 4 and day 6 of germination. The addition of GABA significantly enhanced the activities of PAL, which increased by 35%, 51%, 63%, 27%, and 11% compared with that of drought stress, respectively, under 0.25 mM, 0.5 mM, 1.0 mM, 2.0 mM, and 4.0 mM GABA at the 6 d of germination. GABA treatment also significantly increased the activities of C4H and 4CL, which were 1.40 and 1.94 times higher than those of the control, respectively, on day 6 of germination.

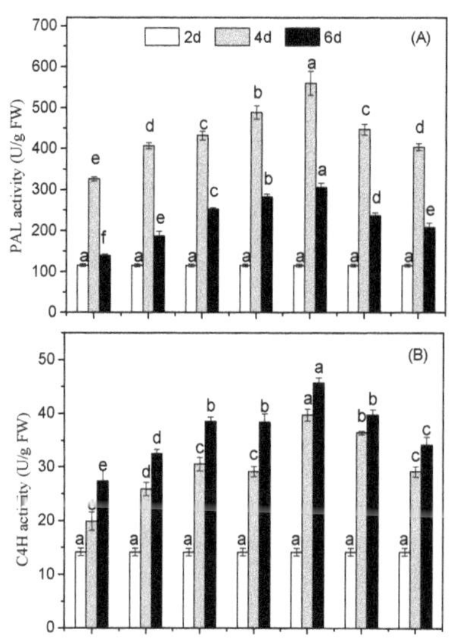

Figure 5. *Cont.*

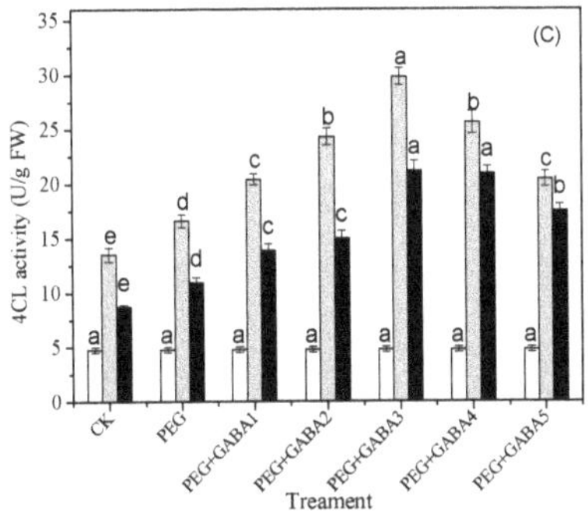

Figure 5. Effects of GABA on the PAL (**A**), C4H (**B**) and 4CL (**C**) activity of wheat seedlings under drought stress. Different letters following values indicate significant differences between treatments at the same germination time ($p < 0.05$). CK, PEG, PEG+GABA1, PEG+GABA2, PEG+GABA3, PEG+GABA4, and PEG+GABA5 indicate distilled water treatment, 100 g/L PEG 6000 treatment, 100 g/L PEG 6000 + 0.25 mM GABA treatment, 100 g/L PEG 6000 + 0.5 mM GABA treatment, 100 g/L PEG 6000 + 1.0 mM GABA treatment, 100 g/L PEG 6000 + 2.0 mM GABA treatment, and 100 g/L PEG 6000 + 4.0 mM GABA treatment, respectively.

2.6. *Effects of GABA on the POD, CAT and APX Activity of Wheat Seedlings under Drought Stress*

As shown in Figure 6, the activity of the antioxidant enzymes of POD, CAT, and APX was determined. Higher levels of antioxidant enzyme activity were observed under drought stress. Compared with drought stress, exogenous GABA increased the activities of POD and CAT during the whole experiment. Under NaCl + 0.1 mM GABA treatment, APX activity increased significantly on day 4, but there was no significant difference on day 6 when compared with that of drought stress.

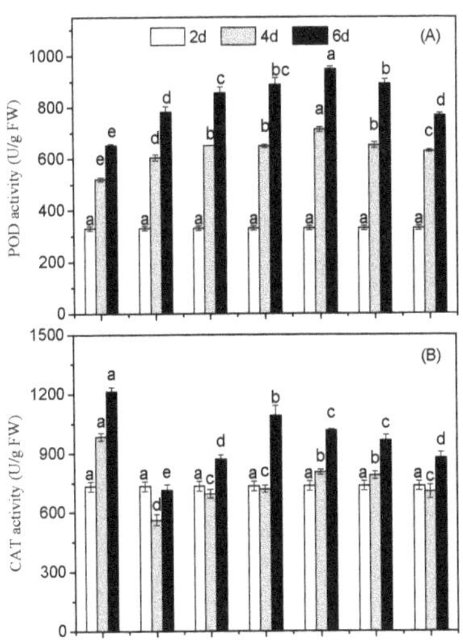

Figure 6. *Cont.*

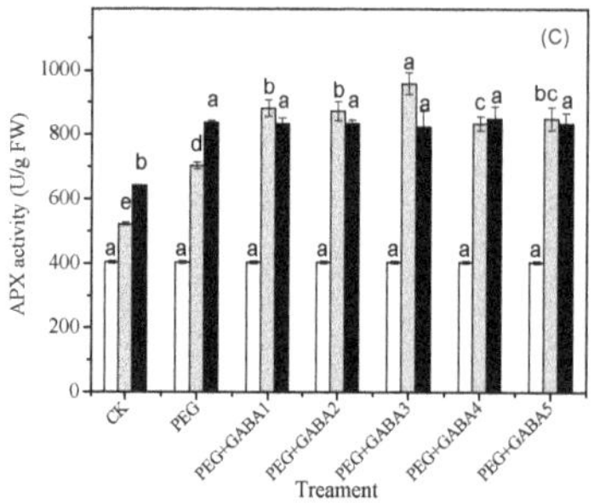

Figure 6. Effects of GABA on the POD (**A**), CAT (**B**) and APX (**C**) activity of wheat seedlings under drought stress. Different letters following values indicate significant differences between treatments at the same germination time ($p < 0.05$). CK, PEG, PEG+GABA1, PEG+GABA2, PEG+GABA3, PEG+GABA4, and PEG+GABA5 indicate distilled water treatment, 100 g/L PEG 6000 treatment, 100 g/L PEG 6000 + 0.25 mM GABA treatment, 100 g/L PEG 6000 + 0.5 mM GABA treatment, 100 g/L PEG 6000 + 1.0 mM GABA treatment, 100 g/L PEG 6000 + 2.0 mM GABA treatment, and 100 g/L PEG 6000 + 4.0 mM GABA treatment, respectively.

2.7. *Effects of GABA on the Antioxidant Capacity of Wheat Seedlings under Drought Stress*

As shown in Figure 7, the antioxidant capacity of wheat seedlings under different treatments was expressed by ABTS and DPPH radical scavenging capacity. The radical scavenging ability of wheat seedlings was enhanced under drought stress, which was similar to the change in total phenolic content. There was a concentration gradient effect of GABA on free radical scavenging activity. Exogenous GABA enhanced the radical scavenging activity of wheat seedlings, with 1.0 mM being the most effective concentration. The 0.1 mM GABA treatment increased the ABTS and DPPH values by 24% and 28% on day 4, and 30% and 34% on day 6 of germination, compared with those of the wheat seedlings under drought stress alone.

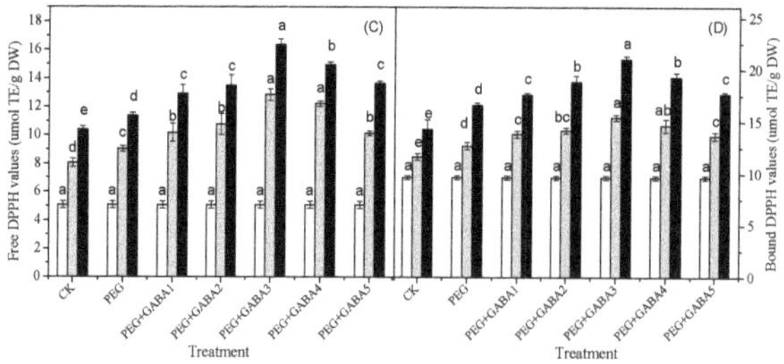

Figure 7. Effects of GABA on the ABTS values (**A**,**B**) and DPPH values (**C**,**D**) of wheat seedlings under drought stress. Different letters following values indicate significant differences between treatments at the same germination time ($p < 0.05$). CK, PEG, PEG+GABA1, PEG+GABA2, PEG+GABA3, PEG+GABA4, and PEG+GABA5 indicate distilled water treatment, 100 g/L PEG 6000 treatment, 100 g/L PEG 6000 + 0.25 mM GABA treatment, 100 g/L PEG 6000 + 0.5 mM GABA treatment, 100 g/L PEG 6000 + 1.0 mM GABA treatment, 100 g/L PEG 6000 + 2.0 mM GABA treatment, and 100 g/L PEG 6000 + 4.0 mM GABA treatment, respectively.

3. Discussion

Drought stress is one of the most serious problems for plants all over the world. It has a negative effect on plants because the available water in the root zone is lower than the water required to maintain optimal growth and development [31]. In the present study, drought stress caused a noticeable increase in MDA levels and electrolyte leakage (Figure 2), indicating that drought stress caused osmotic stress. The increased accumulation of EL and MDA was considered to be an indicator of membrane damage and lipid peroxidation caused by excessive ROS production during stress [23]. As a result, plant dehydration and water uptake were inhibited, and the concentration of cellular components increased to levels that led to membrane damage and even cell death [12,31], which resulted in significant reductions in seedling length and fresh weight throughout growth (Figure 1). The application of exogenous GABA alleviated the growth stress caused by drought stress effectively (Figure 1). This was consistent with the results of Abd El-Gawad et al., in which exogenous GABA increased the shoot weight and leaf area of *Phaseolus vulgaris* L [32]. The positive effects of the exogenous application of GABA on mitigating drought stress has been demonstrated in a variety of plant species from different perspectives. Among them, improving osmotic regulation and maintaining membrane stability were recognized as protective effects [19,20,32]. Numerous studies have explored the role of GABA in wheat seedlings under abiotic stress, and there are many similarities between salt stress and drought stress. For example, GABA could resist the damage caused by salt or drought stress by promoting the production of proline, soluble sugar, and other osmotic regulating substances in wheat seedlings [20,27]. The difference is that GABA also reduced the activity loss or degeneration of some enzymes and proteins under drought stress, so as to maintain the balance of cell metabolism [28]. Under salt stress, GABA not only made the osmotic regulation system and antioxidant system respond, but also changed the signal transduction pathway [27]. Studies showed that GABA alleviated drought stress with a concentration gradient effect, in which the electrolyte leakage and MDA content of wheat seedlings increased more slowly under 1.0 mM of GABA treatment. The results showed that GABA could protect wheat seedlings from oxidative damage and thus improve drought resistance.

It is well known that drought stress triggers enzymatic and non-enzymatic antioxidant defenses leading to redox homeostasis. The proven antioxidant activity of phenolics enables them to act as active oxygen scavengers. Therefore, their synthesis is usually triggered by biotic/abiotic stresses, especially under salt and drought stress [33]. In the present study, our results depicted that drought stress resulted in a marked increase in total phenolic (Figure 3) and phenolic acid content (Figure 4). Interestingly, our study showed that the increase in the variety and content of bound phenolic acids was a major contributor to the increase in phenolic acids. Phenolic acids mainly existed in their bonded form in germinated wheat. Six phenolic acids were found in combined form, but only syringic acid and *p*-coumaric acid were found in free form. This may have something to do with the type of wheat we chose. It was generally believed that the composition of phenolic acids of wheat seedlings varied dynamically under different varieties and culture conditions, but the main phenolic acids such as ferulic acid and *p*-coumaric acid tended to increase with increasing germination time [8–10]. By the same token, we achieved the same result. The accumulation of phenolic acids in response to drought stress was mainly caused by the increase in two major compounds, ferulic acid and *p*-coumaric acid. Drought stress can damage the photosynthetic organs of plants through the interaction with ultraviolet or visible light, while ferulic acid can transform the light falling on the leaves into blue fluorescence to make it no longer destructive and provide protection for photosynthesis [34]. In addition, a new phenolic acid, *p*-hydroxybenzoic acid, emerged under drought stress, which was not detected in the early stage of wheat seed germination. Similar results were found for other plant species, where drought also increased the content and diversity of phenolic compounds [12,35,36].

The addition of GABA significantly improved the activity of PAL, C4H, and 4CL and displayed a higher total phenolic (Figure 5) and phenolic acids content (Figure 4) than that of drought stress, which was consistent with previous studies [29,30]. The increase in total phenolic and phenolic acid content was attributed to the enhanced activity of key enzymes involved in phenolic synthesis. PAL was involved in the first step of the phenylpropionate pathway and has been suggested to act as a rate-limiting enzyme regulating the overall flux into phenylpropionate metabolism [37]. The change in PAL activity was consistent with the content of total phenolic at the initial stage of germination. However, on the 6th day of germination, the activity decreased, but the total phenolic content was still increasing, which may be related to the cumulative effect of phenolic compounds. The cytochrome P450 monooxygenase, cinnamate 4-hydroxylase (C4H), is the second major gene in the phenylpropanoid pathway branch [37]. This enzyme catalyzes the hydroxylation of trans-cinnamic acid to *p*-coumaric acid, which provides the feedstock for eventual conversion to various phenolic acids, including ferulic acid and vanillic acid [36,37]. In fact, our study showed a high correlation between C4H and the total phenolic and total phenolic acid ($r = 0.927$ or 0.923, $p < 0.01$) (Table S1), and the activity of C4H was also highly positively correlated with the accumulation of major phenolic acids such as ferulic acid, *p*-coumaric acid, and sinapic acid ($r = 0.938$, 0.592, and 0.802, respectively, $p < 0.01$ or $p < 0.05$) (Table S1). These results help to confirm that C4H activity is closely related to the accumulation of phenolic acids.

Plants respond to drought stress by increasing their antioxidant defenses, which involve several compounds, such as anthocyanins, flavonoids, and phenolic acids. In addition, antioxidant enzymes are involved, such as peroxidase (POD), catalase (CAT), ascorbate peroxidase (APX), etc. Ascorbate is a ubiquitous non-enzymatic antioxidant that not only has significant potential to scavenge reactive oxygen species, but can also affect some key processes in plants under both stressed and non-stressed conditions [37,38]. In this study, drought stress caused a significant increase in the POD and APX enzyme activities (Figure 6), suggesting that they played an integrative role in regulating drought resistance in wheat seedlings. However, drought stress resulted in a marked decrease in the CAT enzyme activity, which was possibly due to the inactivation of the enzyme by overproduction of reactive oxygen species or due to interference with the synthesis of the enzyme or its subunits [39]. Similar findings were observed in sunflowers (*Helianthus annuus* L.) subjected to drought and heat stress [40]. Exogenous application of GABA improved the activities of POD, CAT, and APX enzymes when subjected to drought stress. Previous reports have confirmed the role of GABA in the regulation of antioxidant enzyme defense in other plants [38,40], suggesting that GABA plays a key role as a mitigating agent against drought-induced oxidative stress and the up-regulation of defense responses.

The antioxidant capacity of germinated wheat was determined in vitro using two different antioxidant assays (ABTS and DPPH) (Figure 7). Both the ABTS and DPPH radical scavenging activity of wheat seedlings increased in response to drought stress, which was consistent with the change in the antioxidant capacity of Amaranthus leafy vegetable [35]. Exogenous GABA enhanced the radical scavenging activity of wheat seedlings, with 1.0 mM being the most effective concentration. It has been proved that the increase in antioxidant capacity is due to the enrichment of phenolic compounds [41]. Pearson's correlation coefficient revealed that antioxidant capacity measured by ABTS and DPPH was strongly correlated with syringic acid, *p*-coumaric acid, ferulic acid, and sinapic acid, which means that these phenolic acids were the highest contributors to the antioxidant activity determined by ABTS and DPPH (Table S2). The results showed that GABA was involved in the enhancement of antioxidant systems in wheat seedlings.

4. Materials and Methods

4.1. Material and Experimental Design

The wheat variety Qiule 168 was used as the research object. "Qiule 168" is a wheat variety independently cultivated by Henan Qiule Seed Industry Science and Technology

Co., Ltd. using the genealogy method, and approved by Henan Provincial Clothing Crop Variety Examination Committee in 2019 (Yumai 20190008). "Qiule 168" is a hybrid bred from "Yumai 34" as female parent and "Zhoumai 13" as male parent.

Wheat seeds were soaked in 0.5% sodium hypochlorite for 15 min and then washed. The seeds were then immersed in deionized water at 25 °C for 6 h at a ratio of 1:5 (w/v). All seeds were then evenly scattered on an automatic spray system tray and placed in a growing chamber at 25 °C. The seeds were treated with deionized water for 2 days first. Then, different treatments were carried out from the 3rd day as follows: (1) CK: the wheat was sprayed with distilled water. (2) PEG: the wheat was sprayed with 100 g/L PEG 6000. (3) PEG + GABA1: 100 g/L PEG 6000 and 0.25 mM GABA were sprayed on the wheat. (4) PEG + GABA2: 100 g/L PEG 6000 and 0.5 mM GABA were sprayed on the wheat. (5) PEG + GABA3: 100 g/L PEG 6000 and 1.0 mM GABA were sprayed on the wheat. (6) PEG + GABA4: 100 g/L PEG 6000 and 2.0 mM GABA were sprayed on the wheat. (7) PEG + GABA5: 100 g/L PEG 6000 and 4.0 mM GABA were sprayed on the wheat. The wheat seeds germinated for a total of 6 days.

4.2. Measurement of Length and Weight

The wheat seedlings' lengths were measured with vernier calipers, and twenty wheat seedlings were selected and measured. The result was expressed in mm/seedlings.

The weight was determined by randomly selecting 100 wheat seedlings as the sampling group, a total of 3 groups were selected to take the average value, and the results were expressed as g/100 seedlings. For the weight of fresh wheat seedlings, 100 fresh wheat seedlings were weighed in an analytical balance. For the weight of dry wheat seedlings, 100 fresh wheat seedlings were dried in an electrically heated air-drying oven and then cooled in a dryer.

4.3. Determination of Malondialdehyde (MDA) Content and Electrolyte Leakage

MDA content was determined by the method of Hodges [42]. Briefly, fresh wheat seedlings (1 g) were mashed and transferred to a solution of 10% trichloroacetic acid (5 mL). After centrifugation at 10,000× g for 20 min at 4 °C, the supernatant was mixed with 0.67% thiobarbituric acid, and the mixture was immersed in boiling water for 30 min. After cooling, the supernatant was centrifuged again. The absorbance of the supernatant was measured at 450, 532, and 600 nm. MDA content was calculated using the following equation:

$$\text{MDA concentration } (\mu\text{mol/L}) = 6.45 \times (\text{A532} - \text{A600}) - 0.56 \times \text{A450}$$
$$\text{MDA concentration } (\mu\text{mol/g FW}) = \frac{c \times 5}{1}$$

Electrolyte leakage was measured using a conductivity meter. Fresh wheat seedlings were cut to 3 mm, placed in a tube and mixed with water. After oscillating for 1 h at 25 °C, the measured conductivity denoted EC1. The tube was then placed in a boiling water bath. After cooling, the measured boiling conductivity denoted EC2. The electrolyte leakage (EL) was calculated using the equation:

$$\text{EL } (\%) = \text{EC1/EC2} \times 100$$

4.4. Extraction of Phenolic Compounds

Free and bound phenolic compounds were extracted based on the methods reported by Ma et al. [29]. The dried extracts were redissolved with 50% methanol to a volume of 10 mL and used as crude free phenolic extracts. The extracts were stored at −20 °C for subsequent use.

4.5. Determination of Total Phenolic Content

The total phenolic content was determined using the Folin–Ciocalteu colorimetric method [29]. Methanol was used as the blank, and gallic acid (GA) was used as the standard.

The total phenolic content was expressed as gallic acid equivalents (mg GAE/100 g DW, dry weight).

4.6. Determination of Free and Bound Phenolic Acid Content

Free and bound phenolic compounds were filtered through a 0.45 μm membrane filter for a Shimadzu LC-2030 high-performance liquid chromatography (HPLC) system with a Zorbax SB-C18 column (5 μm particle size, 250 mm × 4.6 mm, Agilent Technologies Inc., Palo Alto, CA, USA) and a Variable Wavelength Detector (VWD). The mobile phase A was 1% acetic acid in water and the mobile phase B was methanol. The allowed time of separation of phenolic acids was 45 min at a constant temperature of 30 °C with 0.8 mL/min flow rate. A 45 min gradient was programmed as follows: 0–22 min, 18–28% B; 22–40 min, 28–40% B; 40–45 min, 40–18% B.

4.7. Determination of Phenylpropanoid Metabolism-Related Enzymes Activity

The PAL activity was determined as described by Assis et al. [43]. One unit of PAL activity was equal to a change of 0.01 at 290 nm per min, and expressed as U/g FW, dry weight.

Cinnamate-4-hydroxylase (C4H) activity was assayed following the method of Lamb and Rubery [44] with some modification. The 0.5 g of fresh tissue was blended with 5 mL of ice-cold 100 mM phosphate buffer (pH 7.6, containing 2 mM β-mercaptoethanol, 0.5 mM EDTA, 0.25 M sucrose) and ground at 4 °C. After centrifugation (12,000× g, 4 °C, 30 min), the supernatant was used for the enzyme assay. The reaction mixture contained 0.2 mL of 50 mM trans-cinnamic acid, 2 mL of 0.5 g/L $NADPNa_2$, 3 mL of 100 mM phosphate buffer, and 0.1 mL crude enzyme. One unit of C4H activity was equal to a change of 0.01 at 340 nm per min, and expressed as U/g FW.

The activity of 4-coumarate coenzyme A ligase (4CL) was assayed based on Han et al. [45]. The absorbance was measured at 333 nm. One unit of 4CL activity was equal to a change of 0.01 in absorbance per min, and expressed as U/g FW.

4.8. Determination of Antioxidant Enzyme Activity

The activities of peroxidase (POD), catalase (CAT), and ascorbate peroxidase (APX) were measured according to the method reported previously [9]. One unit (U) of POD activity was defined as a change of 1 at 470 nm per minute and the results were expressed as U/g FW. One unit of CAT activity was defined as a change of 0.01 at 240 nm per minute and the results were expressed as U/g FW, and one unit of APX activity was defined as the change of 0.01 at 290 nm per minute and the results were expressed as U/g FW.

4.9. Determination of Antioxidant Capacity

The ABTS and DPPH radical scavenging activity assay was determined following the method of Chen et al. [9]. A plot of Trolox concentration with ABTS and DPPH radical scavenging rate was used as the standard curve. They were expressed as micromoles of Trolox equivalent (TE) per gram of sample (μmol TE/g DW).

4.10. Statistical Analysis

The results were presented as mean ± standard deviation (SD) of duplicated determination. The data were analyzed by one-way ANOVA using SPSS 18.0 (SPSS Inc., Chicago, IL, USA). The significance level was defined at $p < 0.05$ between treatments at the same germination time.

5. Conclusions

GABA induced the accumulation of phenolic compounds, especially ferulic acid, *p*-coumaric acid, and sinapic acid, by regulating the activities of PAL, C4H, and 4CL involved in phenylpropanoid metabolism, with 1.0 mM being the most effective concentration. A total of 1.0 mM of GABA enhanced the content of total phenolic acids by 28% and 22%, respectively, compared with that of 100 g/L PEG 6000 treatment, on day 4 and day 6 of germination. Additionally, the activities of POD, CAT, and APX were induced in wheat seedlings. Pearson's correlation coefficient revealed that antioxidant capacity was strongly correlated with syringic acid, *p*-coumaric acid, ferulic acid, and sinapic acid. The results showed that GABA was involved in the drought resistance of wheat seedlings through enhancing antioxidant capacity, which was achieved by inducing the accumulation of phenolic compounds and increasing the level of antioxidant enzyme activity.

Supplementary Materials: The following supporting information can be downloaded at: https://www.mdpi.com/article/10.3390/plants12132495/s1. Table S1 Pearson correlations for variables of phenolic content and enzymes of phenolic biosynthesis in wheat seedlings with GABA treatment; Table S2 Pearson correlations for variables of phenolic content and antioxidant capacity in wheat seedlings with GABA treatment.

Author Contributions: X.H. and L.S. designed the research. Q.Z. and Y.M. conducted the experiments. Q.Z. and Y.M. analyzed the data and wrote the manuscript. N.L. and M.Q. revised the manuscript. T.L. and D.H. did data curation. Y.C. improved method. All authors have read and agreed to the published version of the manuscript.

Funding: Financial support was provided by the Natural Science Foundation of Henan Province (Grant Number: 222300420179), the National Natural Science Foundation of China (Grant Number: 32201942), the Science and Technology Innovation Fund project of Henan Agricultural University (Grant Number: KJCX2021C04), the Postgraduate Joint Training Base of Henan Province (Grant Number: YJS2022JD16), and the Innovative Research Team (in Science and Technology) in the University of Henan Province (Grant Number: 23IRTSTHN023).

Data Availability Statement: Not applicable.

Conflicts of Interest: The authors declare no conflict of interest.

References

1. Márcia, C.; Irene, G.; Isaura, C.; Manuela, M.; Eduardo, R.; Carnide, V.; Ana Barros, A. Drought stress effect on polyphenolic content and antioxidant capacity of cowpea pods and seeds. *J. Agron. Crop Sci.* **2021**, *207*, 197–207.
2. AL-Quraan, N.A.; Al-Ajlouni, Z.I.; Qawasma, N.F. Physiological and Biochemical Characterization of the GABA Shunt Pathway in Pea (*Pisum sativum* L.) Seedlings under Drought Stress. *Horticulturae* **2021**, *7*, 125. [CrossRef]
3. Hajlaoui, H.; Denden, M.; El, A.N. Differential responses of two maize (*Zea mays* L.) varieties to salt stress: Changes on polyphenols composition of foliage and oxidative damages. *Ind. Crops Prod.* **2009**, *30*, 144–151.
4. Ghumman, A.; Singh, N.; Kaur, A. Chemical, nutritional and phenolic composition of wheatgrass and pulse shoots. *Int. J. Food Sci. Technol.* **2017**, *52*, 2191–2200. [CrossRef]
5. Singh, A.; Sharma, S. Bioactive components and functional properties of biologically activated cereal grains: A bibliographic review. *Crit. Rev. Food Sci. Nutr.* **2015**, *57*, 3051–3071. [CrossRef]
6. Mishra, N.P.; Fatma, T.; Singhal, G.S. Development of antioxidative defense system of wheat seedlings in response to high light. *Physiol. Plant.* **2006**, *95*, 77–82. [CrossRef]
7. Charoenthaikij, P.; Jangchud, K.; Jangchud, A.; Prinyawiwatkul, W.; Tungtrakul, P. Germination Conditions Affect Selected Quality of Composite Wheat-Germinated Brown Rice Flour and Bread formulations. *J. Food Sci.* **2010**, *75*, 312–318. [CrossRef]
8. Gawlik-Dziki, U.; Dziki, D.; Nowak, R.; Świeca, M.; Olech, M.; Pietrzak, W. Influence of sprouting and elicitation on phenolic acids profile and antioxidant activity of wheat seedlings. *J. Cereal Sci.* **2016**, *70*, 221–228. [CrossRef]
9. Chen, Z.; Wang, P.; Weng, Y.; Ma, Y.; Gu, Z.; Yang, R. Comparison of phenolic profiles, antioxidant capacity and relevant enzyme activity of different Chinese wheat varieties during germination. *Food Biosci.* **2017**, *20*, 159–167. [CrossRef]
10. Ceccaroni, D.; Alfeo, V.; Bravi, E.; Sileoni, V.; Perretti, G.; Marconi, O. Effect of the time and temperature of germination on the phenolic compounds of *Triticum aestivum*, L. and *Panicum miliaceum*, L. *LWT-Food Sci. Technol.* **2020**, *127*, 109396. [CrossRef]
11. Crozier, A.; Jaganath, I.B.; Clifford, M.N. *Phenols, Polyphenols and Tannins: An Overview*; Blackwell Publishing Ltd.: Hoboken, NJ, USA, 2007.
12. Espadas, J.L.; Castaño, E.; Marina, M.L.; Rodríguez, L.C.; Plaza, M. Phenolic compounds increase their concentration in *Carica papaya* leaves under drought stress. *Acta Physiol. Plant.* **2019**, *41*, 180. [CrossRef]
13. Zobayed, S.; Afreen, F.; Kozai, T. Phytochemical and physiological changes in the leaves of St. John's wort plants under a water stress condition. *Environ. Exp. Bot.* **2007**, *59*, 109–116. [CrossRef]
14. Swigonska, S.; Amarowicz, R.; Król, A.; Mostek, A.; Badowiec, A.; Weidner, S. Influence of abiotic stress during soybean germination followed by recovery on the phenolic compounds of radicles and their antioxidant capacity. *Acta Soc. Bot. Pol.* **2014**, *83*, 209–218. [CrossRef]

15. Laddomada, B.; Blanco, A.; Mita, G.; D'Amico, L.; Singh, R.P.; Ammar, K.; Crossa, J.; Guzmán, C. Drought and heat stress impacts on phenolic acids accumulation in durum wheat cultivars. *Foods* **2021**, *10*, 2142. [CrossRef]
16. Wang, W.; Liu, S.; Yan, M. Synthesis of γ-Aminobutyric acid-modified chitooligosaccharide derivative and enhancing salt resistance of wheat seedlings. *Molecules* **2022**, *27*, 3068. [CrossRef]
17. Kinnersley, A.M.; Turano, F.J. Gamma aminobutyric acid (GABA) and plant responses to Stress. *Crit. Rev. Plant Sci.* **2000**, *19*, 479–509. [CrossRef]
18. Roberts, M.R. Does GABA act as a signal in plants?: Hints from molecular studies. *Plant Signal. Behav.* **2007**, *2*, 408–409. [CrossRef]
19. Elbar, O.H.A.; Elkelish, A.; Niedbała, G.; Farag, R.; Wojciechowski, T.; Mukherjee, S.; Abou-Hadid, A.F.; El-Hennawy, H.M.; El-Yazied, A.A.; El-Gawad, H.G.A.; et al. Protective effect of γ-aminobutyric acid against chilling stress during reproductive stage in tomato plants through modulation of sugar metabolism, chloroplast integrity, and antioxidative defense systems. *Front. Plant Sci.* **2021**, *12*, 663750. [CrossRef]
20. Vijayakumari, K.; Puthur, J.T. γ-Aminobutyric acid (GABA) priming enhances the osmotic stress tolerance in *Piper nigrum* linn. plants subjected to PEG-induced stress. *Plant Growth Regul.* **2016**, *78*, 57–67. [CrossRef]
21. Sheteiwy, M.S.; Shao, H.; Qi, W.; Hamoud, Y.A.; Shaghaleh, H.; Khan, N.U.; Yang, R.; Tang, B. GABA-alleviated oxidative injury induced by salinity, osmotic stress and their combination by regulating cellular and molecular signals in rice. *Int. J. Mol. Sci.* **2019**, *20*, 5709. [CrossRef]
22. Li, Z.; Yu, J.; Peng, Y.; Huang, B. Metabolic pathways regulated by γ -aminobutyric acid (GABA) contributing to heat tolerance in creeping bentgrass (*Agrostis stolonifera*). *Sci. Rep.* **2016**, *6*, 30338. [CrossRef]
23. Yong, B.; Xie, H.; Li, Z.; Li, Y.; Zhang, Y.; Nie, G.; Zhang, X.; Ma, X.; Huang, L.; Yan, Y.; et al. Exogenous application of GABA improves PEG-induced drought tolerance positively associated with GABA-shunt, polyamines, and proline metabolism in white clover. *Front. Physiol.* **2017**, *8*, 1107. [CrossRef]
24. Mekontso, F.N.; Duan, W.; Cisse, E.H.M.; Chen, T.; Xu, X. Alleviation of postharvest chilling injury of carambola fruit by γ-aminobutyric acid: Physiological, biochemical, and structural characterization. *Front. Nutr.* **2021**, *8*, 752583. [CrossRef]
25. Kumari, S.; Nazir, F.; Jain, K.; Khan, M.I.R. GABA and Potassium Modulates Defence Systems, Assimilation of Nitrogen and Carbon, and Yield Traits Under Salt Stress in Wheat. *J. Plant Growth Regul.* **2023**. [CrossRef]
26. Li, M.; Guo, S.; Yang, X.; Meng, Q.; Wei, X. Exogenous gamma-aminobutyric acid increases salt tolerance of wheat by improving photosynthesis and enhancing activities of antioxidant enzymes. *Biol. Plant.* **2016**, *60*, 123–131. [CrossRef]
27. Khanna, R.R.; Jahan, B.; Iqbal, N.; Khan, N.A.; AlAjmi, M.F.; Rehman, M.T.; Khan, M.I.R. GABA reverses salt-inhibited photosynthetic and growth responses through its influence on NO-mediated nitrogen-sulfur assimilation and antioxidant system in wheat. *J. Biotechnol.* **2021**, *325*, 73–82. [CrossRef]
28. Qiu, Y.; Yu, D. Over-expression of the stress-induced OsWRKY45 enhances disease resistance and drought tolerance in Arabidopsis. *Environ. Exp. Bot.* **2009**, *65*, 35–47. [CrossRef]
29. Ma, Y.; Wang, P.; Wang, M.; Sun, M.; Gu, Z.; Yang, R. GABA mediates phenolic compounds accumulation and the antioxidant system enhancement in germinated hulless barley under NaCl stress. *Food Chem.* **2019**, *270*, 593–601. [CrossRef]
30. Zhao, Y.; Xie, C.; Wang, P.; Gu, Z.; Yang, R. GABA Regulates Phenolics Accumulation in Soybean Sprouts under NaCl Stress. *Antioxidants* **2021**, *10*, 990. [CrossRef]
31. Deikman, J.; Petracek, M.; Heard, J.E. Drought tolerance through biotechnology: Improving translation from the laboratory to farmers' fields. *Curr. Opin. Biotechnol.* **2012**, *23*, 243–250. [CrossRef]
32. Abd El-Gawad, H.G.; Mukherjee, S.; Farag, R.; Abd Elbar, O.H.; Hikal, M.; Abou El-Yazied, A.; Abd Elhady, S.A.; Helal, N.; ElKelish, A.; El Nahhas, N.; et al. Exogenous γ -aminobutyric acid (GABA)-induced signaling events and field performance associated with mitigation of drought stress in *Phaseolus vulgaris* L. *Plant Signal. Behav.* **2020**, *16*, 1853384. [CrossRef]
33. Caliskan, O.; Radusiene, J.; Temizel, K.E.; Staunis, Z.; Cirak, C.; Kurt, D.; Odabas, M.S. The effects of salt and drought stress on phenolic accumulation in greenhouse-grown *Hypericum pruinatum*. *Ital. J. Agron.* **2017**, *12*, 271–275. [CrossRef]
34. Hura, T.; Grzesiak, S.; Hura, K.; Thiemt, E.; Tokarz, K.; Wedzony, M. Physiological and biochemical tools useful in drought-tolerance detection in genotypes of winter triticale: Accumulation of ferulic acid correlates with drought tolerance. *Ann. Bot.* **2007**, *100*, 767–775. [CrossRef]
35. Sarker, U.; Oba, S. Drought stress enhances nutritional and bioactive compounds, phenolic acids and antioxidant capacity of *Amaranthus* leafy vegetable. *BMC Plant Biol.* **2018**, *18*, 258. [CrossRef]
36. Wang, A.; Zhu, M.; Luo, Y.; Liu, Y.; Li, R.; Kou, M.; Wang, X.; Zhang, Y.; Meng, X.; Zheng, Y.; et al. A sweet potato cinnamate 4-hydroxylase gene, *IbC4H*, increases phenolics content and enhances drought tolerance in tobacco. *Acta Physiol. Plant.* **2017**, *39*, 276. [CrossRef]
37. Singh, K.; Kumar, S.; Rani, A.; Gulati, A.; Ahuja, P.S. Phenylalanine ammonia-lyase (PAL) and cinnamate 4-hydroxylase (C4H) and catechins (flavan-3-ols) accumulation in tea. *Funct. Integr. Genom.* **2009**, *9*, 125–134. [CrossRef]
38. Hasan, M.M.; Alabdallah, N.M.; Alharbi, B.M.; Waseem, M.; Yao, G.; Liu, X.; El-Gawad, H.G.A.; El-Yazied, A.A.; Ibrahim, M.F.M.; Jahan, M.S.; et al. GABA: A Key Player in Drought Stress Resistance in Plants. *Int. J. Mol. Sci.* **2021**, *22*, 10136. [CrossRef]
39. Pirzadah, T.B.; Malik, B.; Tahir, I.; Rehman, R.U.; Hakeem, K.R.; Alharby, H.F. Aluminium stress modulates the osmolytes and enzyme defense system in *Fagopyrum* species. *Plant Physiol. Biochem.* **2019**, *144*, 178–186. [CrossRef]

40. Razik, E.S.A.; Alharbi, B.M.; Pirzadah, T.B.; Alnusairi, G.S.H.; Soliman, M.H.; Hakeem, K.R. γ-Aminobutyricacid (GABA) mitigates drought and heat stress in sunflower (*Helianthus annuus* L.) by regulating its physiological, biochemical and molecular pathways. *Physiol. Plant.* **2021**, *172*, 505–527. [CrossRef]
41. Hung, P.V. Phenolic Compounds of Cereals and Their Antioxidant Capacity. *Crit. Rev. Food Sci.* **2016**, *56*, 25–35. [CrossRef]
42. Hodges, D.M.; DeLong, J.M.; Forney, C.F.; Prange, R.K. Improving the thiobarbituric acid-reactive-substances assay for estimating lipid peroxidation in plant tissues containing anthocyanin and other interfering compounds. *Planta* **1999**, *207*, 604–611. [CrossRef]
43. Assis, J.S.; Maldonado, R.; Muñoz, T.; Escribano, M.A.I.; Merodio, C. Effect of high carbon dioxide concentration on PAL activity and phenolic contents in ripening cherimoya fruit. *Postharvest Biol. Technol.* **2001**, *23*, 33–39. [CrossRef]
44. Lamb, C.J.; Rubery, P.H. A spectrophotometric assay for trans-cinnamic acid 4-hydroxylase activity. *Anal. Biochem.* **1975**, *68*, 554–561. [CrossRef]
45. Han, C.; Li, J.; Jin, P.; Li, X.; Wang, L.; Zheng, Y. The effect of temperature on phenolic content in wounded carrots. *Food Chem.* **2017**, *215*, 116–123. [CrossRef]

plants

MDPI

Article

Comparative Physiological and Transcriptome Analyses Reveal Mechanisms of Salicylic-Acid-Reduced Postharvest Ripening in 'Hosui' Pears (*Pyrus pyrifolia* Nakai)

Jing Zhang, Mengmeng Wen, Rong Dai, Xiao Liu and Chunlei Wang *

College of Horticulture and Landscape Architecture, International Research Laboratory of Agriculture and Agri-Product Safety, Key Laboratory of Plant Functional Genomics of the Ministry of Education, Yangzhou University, 48 Wenhui East Road, Yangzhou 225009, China; zhangj45@yzu.edu.cn (J.Z.); wenmm1014@163.com (M.W.); dairong1100@163.com (R.D.); liuxiao@yzu.edu.cn (X.L.)

* Correspondence: wangcl@yzu.edu.cn; Tel.: +86-514-87979344

Abstract: Postharvest ripening of sand pear fruit leads to quality deterioration, including changes in texture, flavor, and fruit color. Salicylic acid (SA), an important defense-related hormone, delays fruit ripening and maintains fruit quality, but the underling mechanism remains unclear. Herein, we evaluated the efficacy of SA in delaying the ripening process of *Pyrus pyrifolia* cv. 'Hosui' pear fruit, as evidenced by the reduction in fruit weight loss, inhibition of firmness loss, cell wall degradation and soluble sugars, and retention of total phenols. Based on comparative transcriptomic data, a total of 3837 and 1387 differentially expressed genes (DEGs) were identified during room-temperature storage of control fruit and between SA-treated and control fruit, respectively. Further KEGG analysis revealed that the DEGs were mainly implicated in plant hormone signal transduction, starch and sugar metabolism, and cell wall modification. Moreover, exogenous SA treatment also altered the expression of many transcription factor (TF) families, including those in the ethylene-responsive factor (ERF), NAM, ATAF, CUC (NAC), basic helix-loop-helix (bHLH), basic leucine zipper (bZIP), and v-myb avian myeloblastosis viral oncogene homolog (MYB) families. Together, the results offer important insights into the role of SA-responsive genes in controlling fruit ripening in sand pears.

Keywords: sand pear; salicylic acid; fruit ripening; transcriptome

1. Introduction

Pears (*Pyrus* spp.) are an economically important fruit crop with a juicy taste and rich nutrition, and they are widely planted around the world [1], with an annual production of 25.66 million tons in 2021 (FAOSTAT). Presently, pear cultivars are divided into five species, including *P. communis*, *P. pyrifolia*, *P. bretschneideri*, *P. sinkiangensis*, and *P. ussuriensis* [2]. Of these, *P. pyrifolia*—the Japanese pear or sand pear—is a major species of cultivated pear that is mainly distributed in China, Japan, and Korea. To ensure a year-round supply and extend the shelf life of sand pear fruits, they are typically not consumed immediately after harvest. Instead, most of the harvested pear fruits undergo a more or less prolonged postharvest period where they are stored and then supplied to the market at the appropriate time. However, fruit storage potential is closely related to the maximum level of ethylene production [3]. Ethylene production in different sand pear cultivars varies during fruit ripening, suggesting that there are both climacteric and non-climacteric types [4–6]. In typical climacteric sand pear cultivars, rapid postharvest ripening progression that is regulated by ethylene accelerates ripening-related characteristics such as water loss, flesh softening, browning, and decay, resulting in fruit quality deterioration and reduced shelf life [7,8]. Furthermore, sand pear fruits predominantly ripen during the mid-summer season, and continuous exposure to high temperatures and undesirable storage conditions can hasten tissue disintegration and cell death [9,10]. Hence, developing efficient storage techniques to maintain postharvest quality properties are of great significance in sand pear fruit industrialization. In past decades, several treatment methods, including low temperature, 1-methylcyclopropene (1-MCP), modified atmosphere packaging, phytohormones, and edible coatings, have been utilized on pears (including several sand pear cultivars) to extend their postharvest storability and minimize their deterioration [9–14].

Fruit ripening is the final stage of fruit development. Fleshy fruit undergoes a genetically programmed and developmentally regulated postharvest ripening process that involves various physiological and biochemical changes, such as discoloration, texture changes, flavor loss, and nutritional value reduction, which significantly reduce fruit quality and consumer acceptance [15–17]. In pears, softening changes like reductions in firmness, cell wall degradation, and decreases in cell turgor pressure have been described as key physiological processes during postharvest ripening [4,6]. Fruit softening is closely related to the degradation and remodeling of primary cell walls that are composed of cellulose, hemicellulose, and pectin [18,19]. Pectin is the main component of the glue layer in the primary cell wall, and different forms of pectin, such as water-soluble pectin (WSP), ionic soluble pectin (ISP), and covalent binding pectin (CSP), can be extracted in different ways [20]. In the process of pear fruit's softening, the contents of cellulose, hemicellulose, and CSP usually decrease, while the abundance of WSP increases [14,21,22]. As for fruit flavor, the composition and amounts of soluble sugars (including sucrose, glucose, fructose, and sorbitol) play key roles in determining the sweetness of pear fruit [23]. During postharvest pear storage, significant changes were observed in individual soluble sugars [23–25]. To date, several ripening-related genes have been identified in pears, including those correlated with cell wall metabolism (e.g., *polygalacturonase 1/2* (*PG1/2*) from *P. pyrifolia* cv. 'Nijisseiki' and 'Starkrimson'; *pectin methylesterase 2/3* (*PME2/3*), *pectin lyase* (*PL*), *β-galactoside 1/2/4* (*β-gal1/2/4*), and *α-arabinofuranase 1* (*ARF1*) from *P. communis* cv. 'Docteur Jules Guyot') [8,14,21,22]; sweetness (*acid invertase 1* (*Ac-Inv1*), *invertase inhibitor 5* (*II5*) and *sucrose phosphate synthase 1* (*SPS1*) from *P. pyrifolia* cv. 'Hosui'; and *neutral invertase* (*NI*), *sucrose synthase* (*SUS*), and *SPS* from *Pyrus ussuriensis* cv. 'Nanguo') [23–25].

SA is a natural hormone that is widely distributed in plants and plays important roles in plants' growth, development, and responses to biotic and abiotic stresses [26]. In addition, several studies have shown that treatments with exogenous SA or its derivatives (acetylsalicylate (ASA) and methyl salicylate (MeSA)) can maintain fruit quality and extend shelf life by decreasing ethylene production and respiration rates, activating the antioxidant system and regulating the metabolism of cell wall components, sugars, acids, and aromatic volatiles of the fruit [27–30]. For instance, SA delayed tomato and kiwifruit ripening by suppressing the expression levels of 1-aminocyclopropane-1-carboxylic acid oxidase (ACO) and 1-aminocyclopropane-1-carboxylic acid synthase (ACS) and preventing the conversion of 1-aminocyclopropane-1-carboxylic acid (ACC) to ethylene, which is the most important plant hormone responsible for fruit ripening [31,32]. Exogenous SA enhanced the expression levels of *PpLOX*, *PpHPL*, *PpADH*, and *PpAAT* in peaches (*Prunus persica* L. Batsch., cv. 'Hujingmilu') during cold storage and increased the amounts of fruity-note volatile esters and lactones [33]. To date, several studies on pear fruit have revealed that SA treatment with optimal concentrations could also be effective in alleviating chilling injury, maintaining postharvest quality, and reducing fruit decay and tissue browning [34–37]. However, the molecular mechanism underlying SA-alleviated pear fruit ripening is less known.

'Hosui' (*Pyrus pyrifolia* Nakai) is an important sand pear cultivar with high yield, high pest-resistance capacity, and good consumption quality, and is widely planted in China [12]. As a typical climacteric fruit, 'Hosui' pears exhibit rapid increases in respiratory rate and ethylene production during postharvest storage, accompanied by severe fruit quality deterioration [38], and they were selected for this study. In the present paper, the influences of SA on maintaining fruit quality and delaying ripening of 'Hosui' pear fruit under room-temperature storage were analyzed, and RNA sequencing (RNA-Seq) was conducted to reveal the molecular mechanisms underlying SA-mediated preservation of sand pear fruit. These results offer important insights into the role of SA-responsive genes in controlling fruit ripening.

2. Results

2.1. Effect of SA Treatment on the Postharvest Ripening Process of Pear Fruit

In this study, 'Hosui' pear fruits' weight and firmness changes during postharvest ripening are shown in Figure 1. The fruit weight of 'Hosui' pears continuously decreased throughout the whole storage period, regardless of treatment, but SA treatment alleviated the reduction in fruit weight (Figure 1A). Consistent with the pear fruit weight trend, fruit firmness declined throughout the storage period, and the softening trajectory was retarded by SA treatment (Figure 1B).

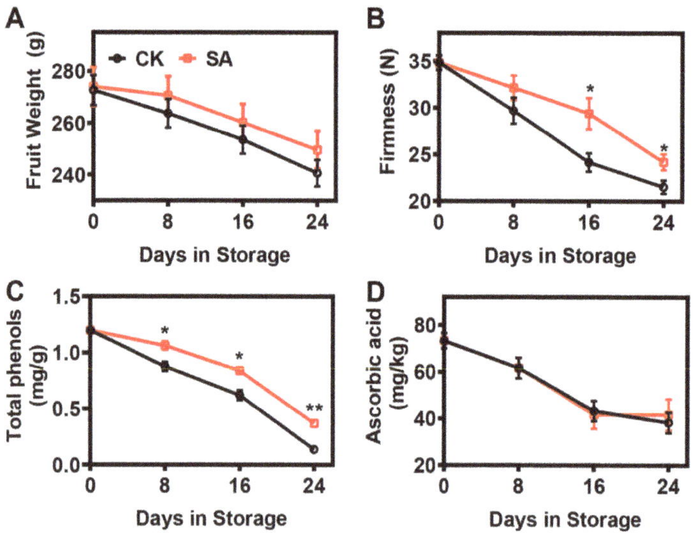

Figure 1. Physiological changes in pear fruit treated with sterile water (CK) and SA during room-temperature storage: (**A**) Fruit weight. (**B**) Firmness. (**C**) Total phenols. (**D**) Ascorbic acid. Error bars indicate SEs from 15 (for fruit weight), 9 (for firmness), and 3 (for total phenols and ascorbic acid) replicates. The statistical analysis was performed using two-tailed Student's *t*-tests. The asterisks indicate significant differences: * $p < 0.1$; ** $p < 0.01$.

Total phenols and ascorbic acid are important antioxidant components that are tightly associated with tissue damage and fruit aging [39]. In this study, the total phenols and ascorbic acid in 'Hosui' pear fruit presented decreasing tendencies during postharvest storage (Figure 1C,D). The trend of total phenols was significantly mitigated by SA treatment (Figure 1C). However, no significant difference in ascorbic acid was observed between the control and SA-treated pear fruit (Figure 1D).

2.2. Effect of SA Treatment on the Cell Wall Composition of Pear Fruit during Room-Temperature Storage

In this research, statistical analysis demonstrated that the cellulose, hemicellulose, and CSP contents continuously decreased during the postharvest storage period, while the rate of decrease was significantly inhibited in SA-treated pear fruit (Figure 2). At the 24th day of storage, the cellulose contents fell to 47% and 77%, the hemicellulose contents dropped to 59% and 73%, and the CSP contents decreased to 73% and 86% of the levels measured on day 0 for the control and SA-treated pear fruit, respectively (Figure 2). In contrast, the contents of WSP increased remarkably during postharvest storage, and the level was lower in SA-treated pear fruit (Figure 2D). Furthermore, the ISP contents in the two groups fluctuated during postharvest storage, and the ISP content in the control pear fruit was significantly higher at the 24th day of storage compared with the SA-treated pear fruit (Figure 2E).

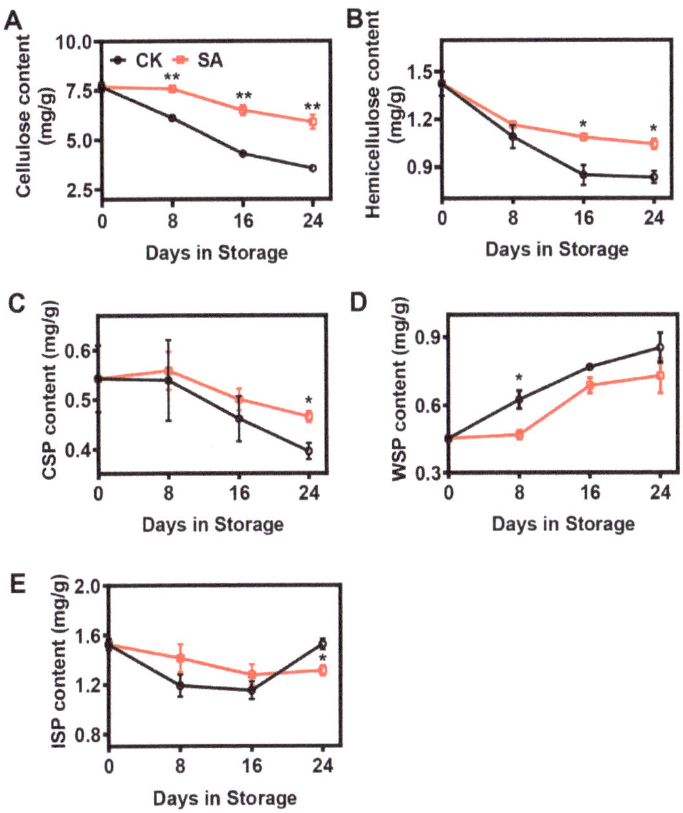

Figure 2. Effects of SA treatment on the cell wall material (CWM) composition of pear fruit during room-temperature storage: (**A**) Cellulose content. (**B**) Hemicellulose content. (**C**) CSP content. (**D**) WSP content. (**E**) ISP content. Error bars indicate SEs from 3 replicates. The statistical analysis was performed using two-tailed Student's *t*-tests. The asterisks indicate significant differences: * $p < 0.1$; ** $p < 0.01$.

2.3. *Effect of SA Treatment on the Soluble Sugar Composition of Pear Fruit during Room-Temperature Storage*

Here, the contents of soluble sugars were analyzed throughout postharvest storage. In control fruit, the contents of sucrose, fructose, and sorbitol tended to increase during the first 8 d and then remained steady, while the glucose content first increased and then decreased progressively (Figure 3). In SA-treated fruit, the changes in sugar contents during storage were strongly affected relative to those in the controls. The level of glucose in SA-treated fruit gradually declined after the 8th day of storage, while a reverse phenomenon was observed for fructose and sorbitol; on the other hand, the sucrose content remained relatively constant throughout storage (Figure 3). Overall, SA treatment strongly diminished the soluble sugar contents compared with the controls (Figure 3).

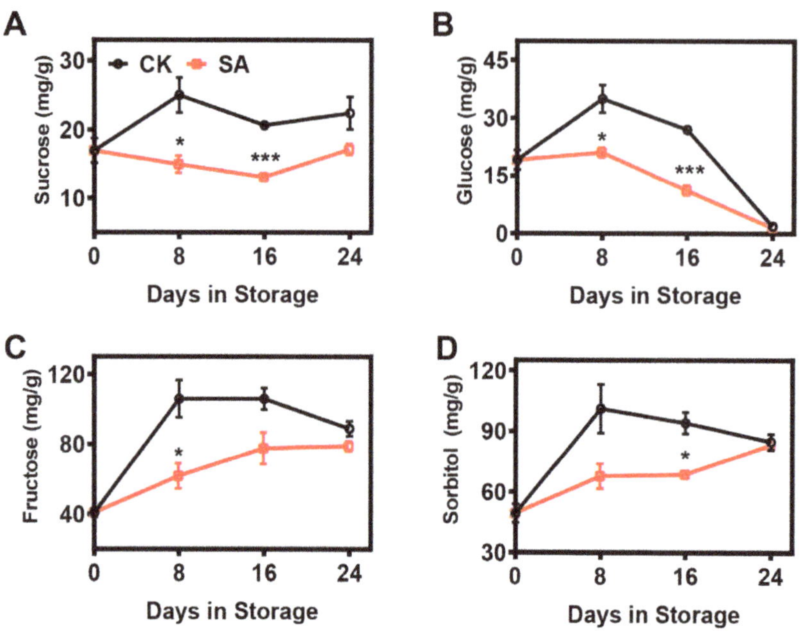

Figure 3. Effects of SA treatment on the soluble sugar composition of pear fruit during room-temperature storage: (**A**) Sucrose content. (**B**) Glucose content. (**C**) Fructose content. (**D**) Sorbitol content. Error bars indicate SEs from 3 replicates. The statistical analysis was performed using two-tailed Student's *t*-tests. The asterisks indicate significant differences: * $p < 0.1$; *** $p < 0.001$.

2.4. Characterization of Transcriptional Changes of Pear Fruit in Response to Salicylic Acid by RNA-Seq

To identify key genes contributing to the observed physiological traits, and to clarify the potential metabolism in response to exogenous SA treatment, we performed a comparative RNA-Seq analysis. As pear fruit exhibited the most significant physiological changes during 0–16 days of storage, pear fruit samples, including CK (0, 8, 16) and SA (8, 16), were selected for RNA extraction and RNA-Seq analysis, with three biological replicates. The details regarding the sequencing data for each sample are listed in Table S1. An average of 6.95 billion clean reads were obtained per library, of which approximately 77.65% were mapped to the pear genome (Table S1). The RNA-Seq results indicated that the three biological replicates were highly correlated ($R^2 \geq 0.93$), and the samples at different timepoints could be clearly distinguished ($0.97 \geq R^2 \geq 0.81$) (Figure 4A). Using the criteria listed in the Materials and Methods section, a total of 4308 DEGs were identified among the following four comparison groups: CK8, CK16 vs. CK0, SA8 vs. CK8, and SA16 vs. CK16 (Figure 4B, Table S2). Of these, 3175 DEGs (including 795 upregulated and 2380 downregulated) and 2722 DEGs (including 756 upregulated and 1966 downregulated) appeared at the 8th day and the 16th day compared to day 0 after treatment for the control group, respectively. Meanwhile, SA treatment induced dramatic changes in the transcript abundances of pear fruit, including 1276 and 153 DEGs identified in fruit treated with SA at the 8th day and the 16th day, respectively, compared to non-treated pear fruits (Figure 4B).

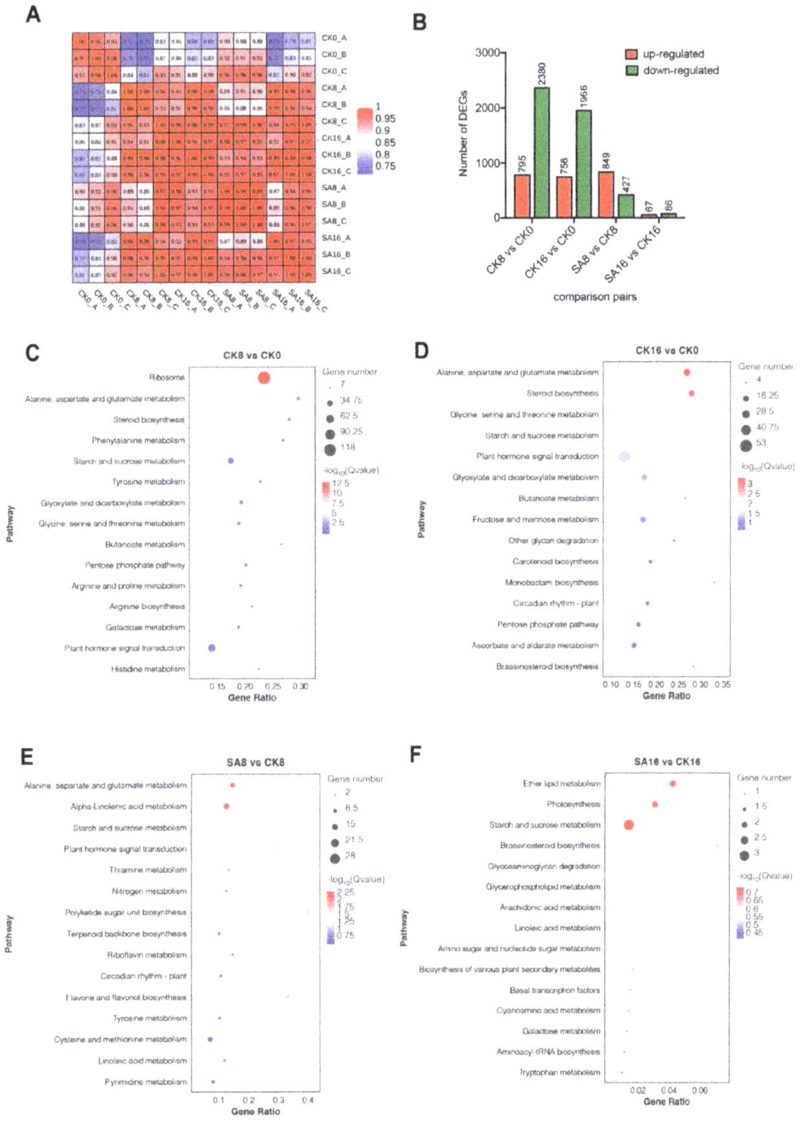

Figure 4. Identification and expression analysis of the DEGs involved in pear fruit ripening and SA responses: (**A**) Correlation analysis between the fifteen samples (CK 0, 8, 16 d and SA 8, 16 d, with three biological replicates). (**B**) Comparison of the DEGs among different comparison groups and the numbers of up− and downregulated DEGs in the individual comparison groups. (**C**,**D**) KEGG pathway analysis of DEGs during pear fruit ripening at 8 d (**C**) and 16 d (**D**) compared to 0 d after treatment for the control group. (**E**,**F**) KEGG pathway analysis of DEGs between pairwise comparisons of SA treatment and the control group at 8 d (**E**) and 16 d (**F**). Gene ratio represents the ratio of the number of significantly expressed genes in a particular pathway to the total number of genes in the pathway.

A KEGG enrichment analysis showed that 'Ribosome', 'Plant hormone signal transduction', and 'Starch and sucrose metabolism' were enriched in the CK8 vs. CK0 group (Figure 4C). In addition, 'Alanine, aspartate and glutamate metabolism', 'steroid biosynthesis', 'Starch and sucrose metabolism', and 'Plant hormone signal transduction' were significantly enriched in the CK16 vs. CK0 group (Figure 4D). For exogenous SA treatment, 'Alanine, aspartate and glutamate metabolism', 'Alpha-Linolenic acid metabolism', 'Starch and sucrose metabolism', and 'Plant hormone signal transduction' were the most enriched pathways on the 8th day compared with the control group (Figure 4E). 'Ether lipid metabolism', 'Photosynthesis', and 'Starch and sucrose metabolism' were the most enriched pathways on the 16th day (Figure 4F). It can be seen from the KEGG analysis that 'plant hormone signal transduction' and 'starch and sucrose metabolism' were significantly enriched in both pear fruit ripening and exogenous SA treatment (Figure 4C–F), indicating that plant hormone signal transduction and sugar transformation may play important roles in SA-reduced pear fruit ripening.

2.5. DEGs Related to Plant Hormone Signal Transduction

To obtain a deeper understanding of the effect of SA treatment on plant hormone signal transduction during pear fruit ripening, the expression patterns of 78 DEGs were annotated to different plant hormone signal transduction pathways. As shown in Figure 5, the auxin signal transduction pathway was remarkably affected during SA-reduced pear fruit ripening. A total of 42 genes mapped to the auxin signal transduction pathway, including 2 *AUX1* (auxin influx carrier) genes, 13 *AUX/IAA* (auxin responsive protein) genes, 3 *ARF* (auxin response factor) genes, 22 *SAUR* (SAUR family protein) genes, and 2 *GH3* (Gretchen Hagen 3) genes, of which 8 DEGs were upregulated and 31 DEGs were downregulated in the CK8 vs. CK0 and CK16 vs. CK0 comparison groups, while 14 DEGs were upregulated and 9 DEGs were downregulated in the SA8 vs. CK8 and SA16 vs. CK16 comparison groups (Figure 5 and Table S3). The brassinosteroid signal transduction pathway contained eight genes, including *BSK* (BR-signaling kinase), *BIN2* (brassinosteroid-insensitive 2), *BZR1/BES1* (brassinazole-resistant 1/BRI1-ethylmethylsulfone-suppressor 1), and *CYCD3* (cyclin D3, plant), of which one DEG was upregulated and seven DEGs were downregulated in the CK8 vs. CK0 and CK16 vs. CK0 comparison groups, while only one upregulated DEG was identified in the SA8 vs. CK8 comparison group (Figure 5, Table S3). Six genes were identified in the ethylene signal transduction pathway, including two genes annotated as ETR (ethylene receptor), one gene annotated as EIN3 (ethylene insensitive 3), two genes annotated as EBF2 (EIN3-binding F-box protein), and one gene annotated as ERF (TF), of which five DEGs were upregulated in the CK8 vs. CK0 and CK16 vs. CK0 comparison groups, whereas one DEG was upregulated and two DEGs were downregulated in the SA8 vs. CK8 and SA16 vs. CK16 comparison groups (Figure 5, Table S3). Additionally, most of the DEGs associated with particular signal transduction pathways, including the cytokinin, abscisic acid, gibberellin, and jasmonic acid signal transduction pathways, were significantly downregulated in pear fruit during postharvest ripening, and several of the identified DEGs, such as one *A-ARR* (two-component response regulator ARR-A family), one *B-ARR* (two-component response regulator ARR-B family), one *PP2C* (protein phosphatase 2C), two *GIDs* (gibberellin-insensitive dwarf), and two *MYC2s* (TF), were upregulated in SA-treated pear fruit (Figure 5, Table S3). The above results suggest that SA treatment has vital effects on plant hormone signal transduction during pear fruit ripening.

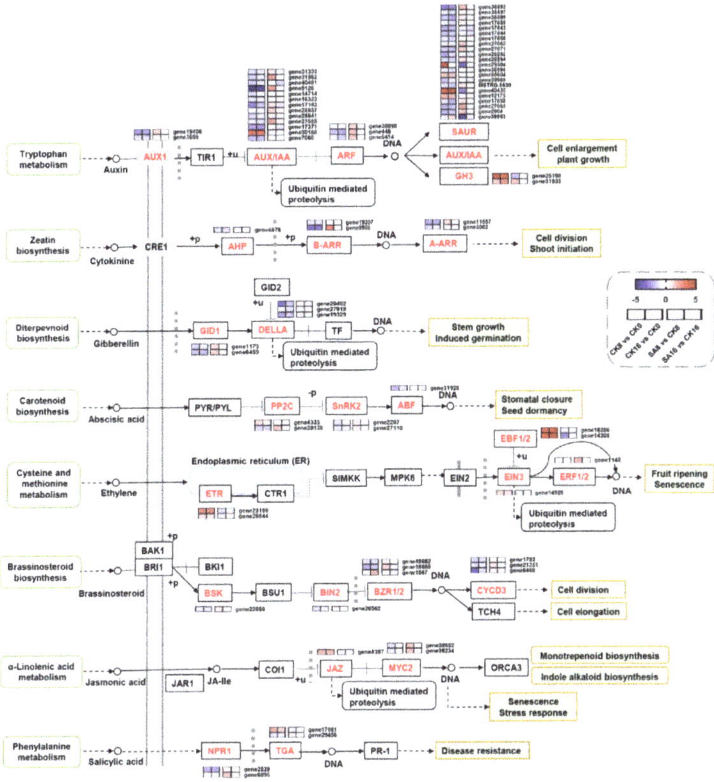

Figure 5. Expression profiles of DEGs associated with different plant hormones' signal transduction. Gradient colors represent the $\log_2$ fold change in gene expression between different times and treatments. The red and blue colors indicate the up− and downregulated genes, respectively.

2.6. DEGs Related to Sugar Metabolism

SA treatment dramatically reduced the accumulation of soluble sugars during postharvest ripening (Figure 3). Since DEGs were highly enriched in the 'Starch and sucrose metabolism' pathway (Figure 4), we further investigated the regulation of starch and sucrose metabolism pathways and identified 43 genes that were significantly differentially expressed (Figure 6 and Table S4). The expression levels of DEGs encoding invertase (INV), SPS, and fructokinase (FRK) were significantly upregulated in pear fruits undergoing postharvest ripening and were highly correlated with the changes in soluble sugars in control 'Hosui' fruit (Figure 6 and Table S4). Among them, two *INVs* (*gene22918* and *gene32825*) exhibited the most significant fold changes between the control and SA-treated pear fruit, reaching fivefold (Figure 6, Table S4). In contrast, some other key genes involved in soluble sugar biosynthesis showed the opposite expression pattern, including sucrose synthase (SUS), sucrose-phosphatase (SPP), sucrose transport protein (SUT), and hexokinase (HXK) (Figure 6). Additionally, the DEGs associated with starch synthesis and degradation, including ADP-glucose pyrophosphorylase (AGPase), starch synthase (SS), granule-bound starch synthase (GBSS), starch-branching enzyme (SBE), α-amylase (AMY), β-amylase (BAM), glucan water dikinase (GWD), isoamylase (ISA), and α-glucan phosphorylase (α-GP), exhibited differentiated expression patterns (Figure 6). These results suggest that changes in genes related to sugar synthesis and metabolism contribute to pear fruit ripening and exogenous SA treatment.

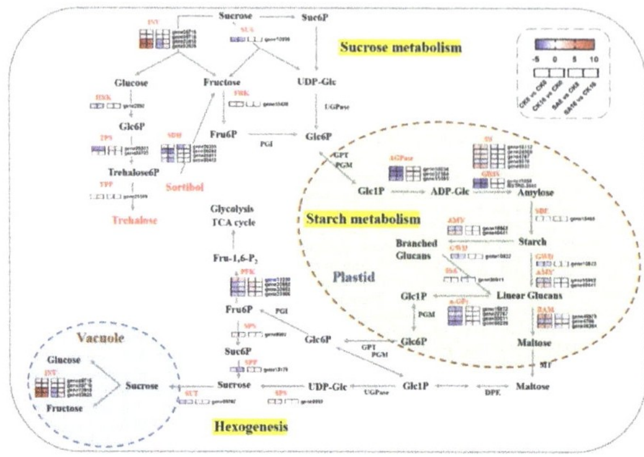

Figure 6. Expression profiles of DEGs associated with sugar metabolism. Gradient colors represent the $\log_2$ fold change in gene expression between different times and treatments. The red and blue colors indicate the up− and downregulated genes, respectively.

2.7. DEGs Related to Fruit Softening

As is known, pear fruit's postharvest softening is closely related to cell wall degradation (Figures 1B and 2). To investigate the mechanism of exogenous SA regulating pear fruit's postharvest softening, we further analyzed the DEGs related to cell wall degradation. A total of 30 target genes were identified from RNA-Seq, including xyloglucan endotransglucosylase/hydrolase (XTH), endo-1,4-β-glucanases (eGase), PG, β-gal, PL, PME, and expansin (EXP) (Figure 7 and Table S5). Here, the expression levels of two *XTH* genes (*gene22067* and *gene37501*), three *PG* genes (*gene18181*, *gene8667*, and *gene8668*), two *β-gal* genes (*gene12403* and *gene5253*), and one *EXP* gene (*gene20093*) were apparently upregulated during the storage of the control fruit (Figure 7, Table S5). In comparison to the control pear fruit, exogenous SA treatment obviously inhibited the expression of *gene37501* (encoding XTH), *gene12403* (encoding β-gal), *gene12246* (encoding PME), and *gene20093* and *gene33653* (encoding EXP) (Figure 7 and Table S5).

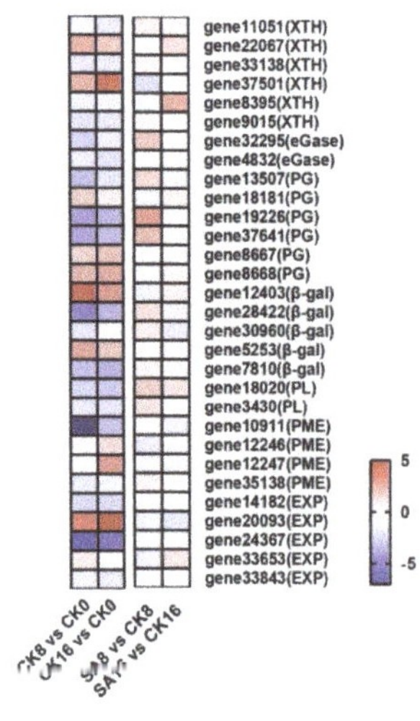

Figure 7. Expression profiles of DEGs associated with cell wall degradation. Gradient colors represent the $\log_2$ fold change in gene expression between different times and treatments. The red and blue colors indicate the up− and downregulated genes, respectively.

2.8. Differentially Expressed TFs

To investigate the regulatory networks influencing pear fruit's ripening and SA response, we detected 264 differentially expressed TFs from 41 TF families, consisting of 235 differentially expressed TFs in the CK8 vs. CK0 and CK16 vs. CK0 comparison groups and 120 differentially expressed TFs in the SA8 vs. CK8 and SA16 vs. CK16 comparison groups (Table S6). The TF families with the most identified DEGs were ERF (19, 7.2%), NAC (19, 7.2%), bHLH (18, 6.8%), bZIP (16, 6%), and MYB (15, 5.7%) (Figure 8A). Meanwhile, eighty-nine TFs were differentially expressed in both the control and SA-treated pear fruits, and we subsequently analyzed their gene expression levels and clustered the genes into two groups: group 1 contained most of the TFs that were upregulated in the CK8 vs. CK0 and CK16 vs. CK0 comparison groups but downregulated in the SA8 vs. CK8 and SA16 vs. CK16 comparison groups; group 2 contained most of the TFs that exhibited the opposite expression pattern compared with group 1 (Figure 8B). The results confirm that a series of differentially expressed TFs are involved in the regulation of pear fruit's ripening and SA response.

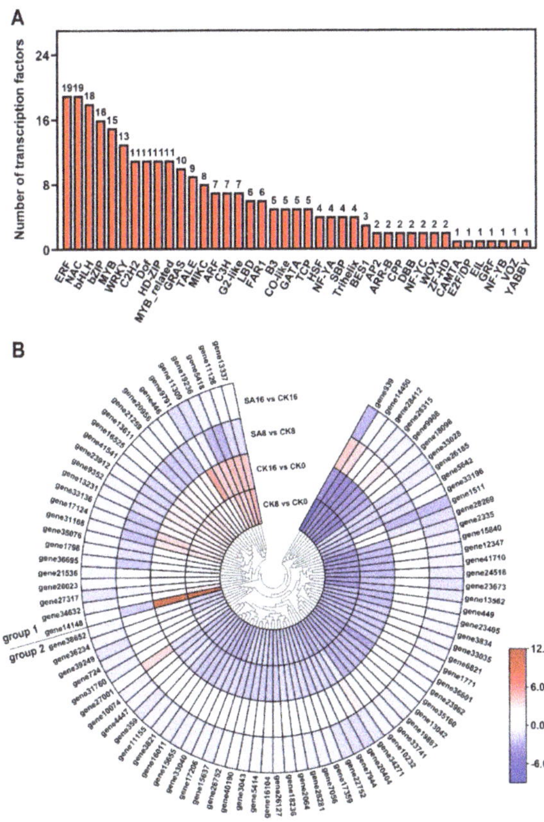

Figure 8. TFs that were differentially expressed during pear fruit ripening and in response to SA treatment: (**A**) The numbers of differentially expressed TFs in different families. (**B**) Expression heatmaps of the DEGs encoding the TFs. Group 1 and group 2 indicate opposite expression patterns of TFs in response to SA treatment. Gradient colors represent the $\log_2$ fold change in gene expression between different times and treatments. The red and blue colors indicate the up– and downregulated genes, respectively.

2.9. Validation of RNA-Seq by qRT-PCR Analysis

To confirm the reliability of the DEGs identified by RNA-Seq, twelve significantly expressed DEGs that were putatively involved in plant hormone signal transduction, starch and sucrose metabolism, and cell wall degradation, along with three differentially expressed TFs, were selected for qRT-PCR analysis. As shown in Figure 9, most of the DEGs showed significantly different expression profiles between the SA treatment and controls during pear fruit storage. Except for *gene8895* (*non-expresser pathogenesis-related gene* (*NPR*)), *gene28138* (*hypersensitive to ABA* (*HAB*)), and *gene36397* (*MIKC*), the expression of other genes was repressed by SA, especially on the 8th day of room-temperature storage. Moreover, the relative expression levels of *gene37501* (*XTH*), *gene20093* (*EXP*), and *gene5418* (*NAC*) gradually increased in control pear fruit and reached maximal expression after 16 days of room-temperature storage while they remained relatively lower in SA-treated pear fruit (Figure 9). In addition, the relative expression levels of the selected DEGs were consistent with the overall trend results of the transcriptomic datasets, suggesting the high accuracy of the RNA-Seq data (Figure 9).

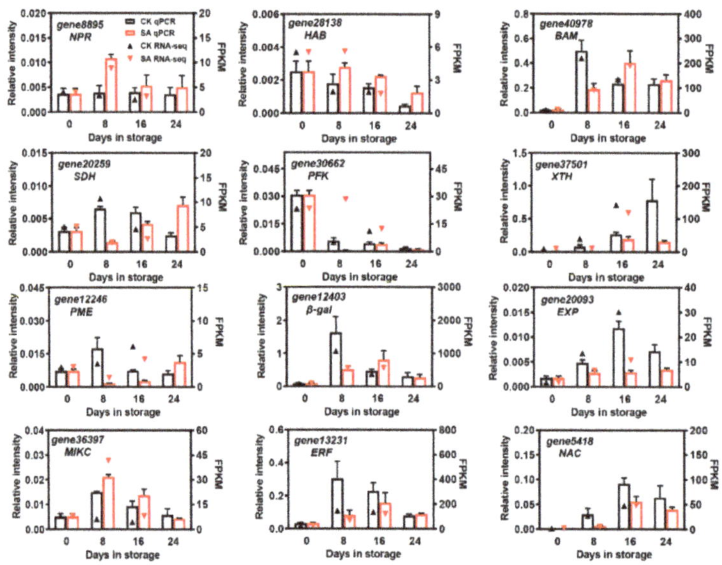

Figure 9. The qRT-PCR validation of DEGs. Relative expression levels from qRT-PCR were calculated using actin as a standard. The columns represent the data obtained from qRT-PCR, while the arrows represent the data obtained from RNA-Seq.

3. Discussion

Rapid postharvest ripening progression in pears strongly shortens their shelf life and restricts their industrial development [1,6,33]. To date, attempts to extend shelf life and reduce quality loss have been widely explored in pears, and several effective techniques, including low temperature, 1-MCP, controlled atmosphere/modified atmosphere packaging, phytohormones, and edible coatings, have been utilized to extend postharvest storability and maintain fruit quality [9–14,40]. For example, postharvest fumigation in fruits of several pear cultivars effectively reduced the fruits' respiration and substantially suppressed their quality deterioration, including acid metabolism, weight loss, flesh firmness, and browning during the post-storage ripening period [4,12,41,42]. Modified atmosphere packaging reduced ethylene production, ascorbic acid degradation, and cell membrane peroxidation and increased the storage life of 'Doyenne du Comice' (*P. communis*) pears by up to 2 months [43]. In 'Kosui' (*P. pyrifolia*) pear fruits, chitosan/alginate-based layer-by-layer coatings minimized the fruits' respiration and ethylene production rates, inhibited flesh firmness loss, prevented peel color changes, and prolonged shelf life [44]. As a critical defense-related hormone, SA and its derivatives also show marked effects in delaying the progression of postharvest ripening and extending pear fruit's shelf life, compared with other storage techniques such as editable coatings [34–37]. However, the precise mechanism is less known. In the current study, exogenous application of SA to harvested 'Hosui' pears significantly delayed their ripening progression, including reducing weight loss, alleviating firmness reduction and cell wall degradation, and retaining higher total phenols simultaneously (Figures 1 and 2), which is consistent with previous studies on postharvest fruits [45,46]. Meanwhile, exogenous application of SA diminished the soluble sugar contents during storage compared with the controls (Figure 3). The effects of SA on soluble sugars vary between different species and cultivars. In pomegranate [47] and peach [48] fruits, postharvest treatment with SA reduced the total soluble solids (TSS) and soluble sugars during storage, while Davarynejad et al. [49] and Wang et al. [32] reported that exogenous SA significantly increased the fruit TSS in plums and kiwifruit.

Subsequent transcriptome analysis showed that there were 3837 and 1387 DEGs identified during room-temperature storage in control fruit and between SA-treated and control fruit, respectively, and many of the selected DEGs (78 DEGs) were significantly enriched in the 'plant hormone signal transduction' pathway (Figures 4 and 5). Among them, genes involved in ethylene signal transduction cannot be neglected, as ethylene plays a critical role in climacteric fruit ripening [50]. Here, five ethylene-signal-transduction-related genes (two *ETR*, one *EIN3*, and two *EBF2*) were identified as upregulated during pear fruit ripening, while three of them (one *ETR* and two *EBF2*) were negatively regulated by SA treatment (Figure 5 and Table S3), suggesting their potential vital roles in pear fruit's ripening and SA response. Moreover, 42 genes (53.8%) involved in the auxin signal transduction pathway were identified as differentially expressed during pear fruit ripening and in response to exogenous SA treatment (Figure 5 and Table S3). Auxin plays an inhibitory role in fruit ripening, and reduced auxin signaling activity could increase the sensitivity of fruit to ethylene [51,52]. In the present study, most of the auxin-signal-transduction-pathway-related genes were downregulated during pear fruit ripening and upregulated in SA-treated pear fruits, and one *AUX/IAA* gene (*gene5126*) was most dramatically changed, implying its potential role in SA-reduced pear fruit ripening (Figure 5 and Table S3). In papaya, the auxin-responsive genes *CpIAA9/17/27* may be candidate genes in fruit ripening [53]. Next to auxin, eight genes involved in the brassinosteroid signal transduction pathway and five genes involved in the cytokinin, abscisic acid, and gibberellin signal transduction pathway were identified as DEGs (Figure 5). These data indicate that the mechanism of SA application delaying 'Hosui' pear fruit's ripening may be through various plant hormones.

Starch and soluble sugars are the primary flavor components of fleshy fruits, which determine their postharvest shelf life and quality [54]. Previous studies have recognized several sugar biosynthesis- and metabolism-related genes that were involved in the postharvest storage of pear fruit, e.g., *PpAc-Inv1*, *PpII5*, *PuAI*, *PuNI*, *PuSS*, and *PuSPS* [23–25]. Here, a series of DEGs were also highly enriched in the 'Starch and sucrose metabolism' pathway, and the expression levels two *INVs* (*gene22918* and *gene32825*) were highly correlated with the changes in soluble sugars in postharvest 'Hosui' fruit (Figures 3, 4 and 6). INV irreversibly hydrolyzes sucrose into glucose and fructose [55,56]. It can be inferred from the results that exogenous SA may inhibit the expression of DEGs encoding INV, which can reduce the soluble sugar contents.

The modification and breakdown of plant cell wall polysaccharides, including cellulose, hemicellulose, and pectin, are responsible for fruit's softening and textural changes during fruit ripening, and extensive studies have focused on cell-wall-degrading and -modifying enzymes and corresponding genes, which synergistically contribute to the fruit softening phenotype [12,57,58]. In this study, transcriptome data showed that several DEGs associated with cell wall metabolism, including *XTH* (*gene37501*), *β-gal* (*gene12403*), *PME* (*gene20093*), and *EXP* (*gene33653*), were upregulated during the storage of control fruit and downregulated by exogenous SA, indicating their potential roles in the postharvest softening of 'Hosui' pear fruit (Figure 7). XTH is involved in hemicellulose metabolism, while β-gal and PME are responsible for pectin modification [58]. In tomato, overexpression or loss-of-function of *SlXTH1* and *SlPMEU1* altered the fruit ripening and softening process [59,60]. The antisense of strawberry *FaβGal4* significantly reduced fruit softening [61]. EXP is involved in cell wall relaxation, and modification of *SlEXP1* in tomato influenced cell wall polymer metabolism during fruit ripening [62].

Fruit-quality-related TFs have been reported in diverse fruit species, and their detailed regulatory effects have also been defined, including for tomato *SlNOR* (NAC TF) in fruit ripening and softening [63], banana *MaDREB2* (ERF TF) and *MabHLH6* in aroma production and starch degradation [64,65], and strawberry *FvMYB10* in anthocyanin accumulation [66]. Here, we identified many TFs that were differentially expressed in the control and SA-treated 'Hosui' pear fruits, e.g., *ERF*, *NAC*, *bHLH*, *bZIP*, and *MYB* genes (Figure 8A). These results are consistent with those of previous studies showing that exogenous SA could enhance the antioxidant systems and maintain the storage quality of winter jujube fruit during cold storage by activating many TFs, including the *MYB*, *ERF*, *C2H2 (Cys-2/His-2)*, and *WRKY* genes [67]. Furthermore, different TFs exhibited diverse expression patterns in the control and SA-treated pear fruits, indicating their potential function in 'Hosui' pear fruits' ripening and SA response (Figure 8B).

The present study showed that exogenous SA treatment effectively maintained fruit quality and delayed postharvest 'Hosui' pear fruit ripening. Moreover, transcriptome analysis showed that salicylic acid altered the transcripts of the genes related to hormone signaling, sugar metabolism, cell wall modification, and TFs, providing important information for identifying the genes involved in 'Hosui' pear fruit ripening. Further functional analyses of these candidate genes will be required to reveal the SA-mediated molecular mechanisms underlying 'Hosui' pear fruit ripening. These findings will guide the application of SA in sand pear fruit and could be a helpful reference for fruit preservation technology in food engineering.

4. Materials and Methods

4.1. Plant Material and Treatments

Commercial mature 'Hosui' pear fruits (approximately 150 d after full bloom) were harvested from homogeneous trees in a commercial orchard in Yangzhou (Jiangsu Province, China) in 2022. After transportation to the laboratory, two hundred uniform and defect-free fruits were selected and randomly divided into two groups for the control and SA treatments. SA treatment was applied by immersing the pear fruits for 20 min in a solution of 2 mmol/L SA (analytical reagent, 99.5%, Macklin Biochemical Technology, Shanghai, China), while pear fruits in the control group were immersed in sterile water, and then the fruits were dried in air and held at 20 °C for room-temperature storage. Fruits were sampled at 0, 8, 16, and 24 days, and at each sampling point, fruit flesh from three replicate samples, each of which consisted of four fruits, were sampled for each treatment, cut into small pieces, frozen in liquid nitrogen, and stored at −80 °C for further experiments.

4.2. Weight and Firmness

Fifteen fruits for each treatment were weighed at each sampling point. Fruit weight was recorded using a BSA124S-CW balance (Sartorius, Gottingen, Germany).

Fruit firmness was measured with a GY-4 penetrometer (Zhejiang Top Instrument, Hangzhou, China) fitted with an 8 mm probe to puncture the fruit at a 10 mm sample depth and determine the maximum penetration force of the 'Hosui' pear fruit after the removal of a small piece of peel. The firmness of each fruit was averaged from two measurements 90° apart at the fruit equator. Fruit firmness was expressed in newtons (N), and 9 individual fruit replicates were used.

4.3. Determination of Total Phenols and Ascorbic Acid

Total phenols were determined using the Folin–Ciocâlteu (FC) method [68]. Briefly, frozen fruit flesh samples were thawed overnight at room temperature before drying. The fruit samples were dried using a hot-air dryer set at 65 °C until constant weight and then ground and screened through an 80-mesh sieve. Dried fruit flesh (0.1 g per sample) was ground into a powder, and 2 mL of ethanol (60%) was added to extract the total phenols. The levels of total phenols were calculated from the standard curve drawn using pure gallic acid and were expressed as gallic acid equivalents (GAE)/g dry weight basis. The total phenols of each sample were replicated 3 times ($n = 3$).

The levels of ascorbic acid were determined following the procedure described previously by Hughes [69]. Briefly, 10 g of frozen fruit flesh was homogenized with an equal volume of 2% oxalic acid solution. The filtrate was titrated with 2,6-dichloroindophenol until the color was pale pink and did not fade within 15–30 s. There were 3 replicates for each treatment.

4.4. Determination of Cell Wall Components

Cell wall components were extracted and measured as described previously [70,71]. Approximately 1 g of frozen fruit flesh per replicate was washed and centrifuged sequentially with 80% (*v:v*) ethanol and acetone. After air-drying the residue at 20 °C for 2 d, the CWM was obtained. Approximately 15 mg of dried CWM was fractionated by extraction with distilled water, 50 mM CDTA and sodium acetate (pH 6.5), and 50 mM Na_2CO_3 (including 2 mM CDTA) in sequence to obtain WSP, ISP, and CSP, respectively. Subsequently, the residue was extracted with 4 M KOH at 20 °C for 5 h of continuous shaking, and the supernatant was designated as the hemicellulose content. Finally, the insoluble residue was washed with 0.3 M acetic acid and 80% (*v*/*v*) ethanol and then centrifuged. The pellet was dried at 40 °C and weighed as α-cellulose.

Pectin was determined by the carbazole method [72]. First, 3 mL of H_2SO_4 was added to 0.5 mL of extracted solution and boiled for 20 min. After the mixture was cooled down to room temperature, 0.2 mL of carbazole-anhydrous ethanol (1.5 g/L) was added. After 30 min of standing, the absorbance was measured at 530 nm. Hemicellulose and cellulose were extracted and measured according to the anthrone–sulfuric acid method [73]. The extracted cellulose and hemicellulose solutions were each mixed with 5 mL of anthrone reagent (2 g of anthrone dissolved in 80% [*v*/*v*] H_2SO_4 and diluted with 80% H_2SO_4 to 1000 mL). The mixtures were then heated in a 100 °C water bath for 10 min. After cooling to room temperature, absorbance measurements were carried out at 625 nm. The results were expressed in mg/g. Each component was replicated 3 times ($n = 3$).

4.5. Determination of Sugar Components

Soluble sugars, including sucrose, glucose, fructose, and sorbitol, were measured as described by Miao et al. [74], with slight modifications. Briefly, 1 g fruit flesh samples were ground into homogenate with 1 mL of deionized water and centrifuged at 12,000 g for 10 min at 4 °C. Then, the supernatant was passed through a 0.22 mm aqueous membrane filter for further analyses. The levels of individual sugars were determined using high-performance liquid chromatography (HPLC, Agilent 1200-6460 QQQ, Santa Clara, CA, USA) equipped with a carbohydrate analysis column (5 μm, 300 × 6.5 mm, Waters-SugarPak1, Milford, MA, USA). The flow rate of the mobile phase (deionized water) was 0.5 mL/min. Each component was replicated 3 times ($n = 3$).

4.6. RNA Extraction and RNA-Seq

The total RNA of 'Hosui' pear fruit obtained from each sampling point was extracted by the hexadecyl trimethyl ammonium bromide (CTAB) method as described previously [75]. In order to evaluate the transcriptomic dynamics underlying different treatments, RNA-Seq and bioinformatics analyses were performed using samples (0, 8, and 16 d, including control and SA treatment) with three biological replicates. The experimental parameters were similar to those used in our previous reports [76]. Briefly, RNA-Seq libraries were generated with the NEB Next® Ultra™ RNA Library Prep Kit for Illumina (NEB, Ipswich, MA, USA) and were then sequenced using the Illumina Novaseq6000 platform by Gene Denovo Biotechnology (Guangzhou, China). The clean reads were obtained by removing poor-quality reads (unknown nucleotides > 10% or low Q values $\leq$ 20%) and mapped to the *Pyrus bretschneideri* 'DangshanSuli' genome (version 1.1, https://www.rosaceae.org/species/pyrus_bretschneideri/genome_v1.1, accessed on 15 December 2022). Genes with average FPKM $\geq$ 1, $|\log2$ (fold change)$| \geq 1$, and p-values < 0.05 were assigned as the DEGs.

4.7. cDNA Synthesis and qRT-PCR Analysis

Approximately 1 µg of total RNA was reverse-transcribed into first-strand cDNA using a PrimeScript™ RT reagent kit in conjunction with gDNA Eraser (Takara, Tokyo, Japan). Twelve significantly expressed DEGs were selected for qRT-PCR analysis, and gene-specific primers were designed using the online software Primer3 (the primer sequences are listed in Table S7). Then, qRT-PCR was carried out as described in our previous reports, with the same system and program [76]. The relative expression of the examined genes was normalized using the Ct value corresponding to the pear actin gene (GU830959.1), with three biological replicates.

4.8. Statistical Analysis

Data analysis was performed with Microsoft Excel. The significant differences between treatments were identified using two-tailed Student's t-tests. The asterisks indicate significant differences: * $p < 0.1$; ** $p < 0.01$; *** $p < 0.001$. Bar graphs and heatmaps were constructed using GraphPad Prism 7.0 scientific software (San Diego, CA, USA).

Supplementary Materials: The following supporting information can be downloaded at: https://www.mdpi.com/article/10.3390/plants12193429/s1, Table S1: Statistics on the quality and output of the RNA-Seq libraries; Table S2: Differentially expressed genes identified through transcriptome analysis; Table S3: Differentially expressed genes involved in pathways related to plant hormone signaling transduction; Table S4: Differentially expressed genes involved in starch and sucrose metabolism pathways; Table S5: Differentially expressed genes involved in cell wall metabolism; Table S6: Differentially expressed transcription factors between the control and SA-treated pear fruits during postharvest storage; Table S7: Primers used for qRT-PCR analysis in this study.

Author Contributions: J.Z., M.W. and C.W. designed the experiments and wrote the paper. J.Z., M.W. and R.D. performed the experiments. J.Z. and M.W. analyzed the data. X.L. revised the paper. All authors have read and agreed to the published version of the manuscript.

Funding: This work was supported by the National Natural Science Foundation of China (grant number 32202445), Key Research and Development Program (Modern Agriculture) of Jiangsu Province (grant number BE2022381), and Natural Science Foundation of Jiangsu Province (grant number BK20190896). J.Z. was funded by the National Natural Science Foundation of China (grant number 32202445) and Natural Science Foundation of Jiangsu Province (grant number BK20190896). C.-L.W. was funded by Key Research and Development Program (Modern Agriculture) of Jiangsu Province (grant number BE2022381).

Data Availability Statement: The RNA-Seq datasets generated during the current study have been deposited at NCBI with the project ID PRJNA1002124, and other data supporting the results are included in this published article and its Supplementary Materials.

Conflicts of Interest: The authors declare no conflict of interest.

References

1. Li, J.M.; Zhang, M.Y.; Li, X.L.; Khan, A.; Kumar, S.; Allan, A.C.; Lin-Wang, K.; Espley, R.V.; Wang, C.H.; Wang, R.Z.; et al. Pear genetics: Recent advances, new prospects, and a roadmap for the future. *Hortic. Res.* **2022**, *9*, uhab040. [CrossRef]
2. Wu, J.; Wang, Y.T.; Xu, J.B.; Korbanet, S.S.; Feial, Z.J.; Tao, S.T.; Ming, R.; Tai, S.S.; Khan, A.M.; Postman, J.D.; et al. Diversification and independent domestication of Asian and European pears. *Genome Biol.* **2018**, *19*, 77. [CrossRef] [PubMed]

3. Itai, A.; Kawata, T.; Tanabe, K.; Tamura, F.; Uchiyama, M.; Tomomitsu, M.; Shiraiwa, N. Identification of 1-aminocyclopropane-1-carboxylic acid synthase genes controlling the ethylene level of ripening fruit in Japanese pear (*Pyrus pyrifolia* Nakai). *Mol. Genom. Genet.* **1999**, *262*, 42–49. [CrossRef] [PubMed]
4. Cheng, Y.; Dong, Y.; Yan, H.; Ge, W.; Shen, C.; Guan, J.; Liu, L.; Zhang, Y. Effects of 1-MCP on chlorophyll degradation pathway-associated genes expression and chloroplast ultrastructure during the peel yellowing of Chinese pear fruits in storage. *Food Chem.* **2012**, *135*, 415–422. [CrossRef]
5. Xu, H.Y.; Chen, Y.Y.; Wang, L.B.; Xie, Z.H.; Gu, C.; Zhang, S.L. Transcriptome analysis reveals a regulation of ethylene-induced post-harvest senescence in pear fruit. *Sci. Hortic.* **2018**, *240*, 585–591. [CrossRef]
6. Itai, A.; Kotaki, T.; Tanabe, K.; Tamura, F.; Kawaguchi, D.; Fukuda, M. Rapid identification of 1-aminocyclopropane-1-carboxylate (ACC) synthase genotypes in cultivars of Japanese pear (*Pyrus pyrifolia* Nakai) using CAPS markers. *Theor. Appl. Genet.* **2003**, *106*, 1266–1272. [CrossRef] [PubMed]
7. Liu, R.; Lai, T.; Xu, Y.; Tian, S. Changes in physiology and quality of Laiyang pear in long time storage. *Sci. Hortic.* **2013**, *150*, 31–36. [CrossRef]
8. Zhang, S.L.; Ma, M.; Zhang, H.P.; Zhang, S.L.; Qian, M.; Zhang, Z.; Luo, W.Q.; Fan, J.B.; Liu, L.Q.; Wang, L.B. Genome-wide analysis of polygalacturonase gene family from pear genome and identification of the member involved in pear softening. *BMC Plant Biol.* **2019**, *19*, 587. [CrossRef] [PubMed]
9. Nath, A.; Bidyut, C.D.; Singh, A.; Patel, R.K.; Paul, D.; Misra, L.K.; Ojha, H. Extension of shelf life of pear fruits using different packaging materials. *J. Food Sci. Technol.* **2012**, *49*, 556–563. [CrossRef]
10. Singh, V.; Jawandha, S.K.; Gill, P.P.S.; Gill, M.S. Suppression of fruit softening and extension of shelf life of pear by putrescine application. *Sci. Hortic.* **2019**, *256*, 108623. [CrossRef]
11. Golias, J.; Letal, J.; Balik, J.; Koziskova, J. Effect of controlled atmosphere storage on production of volatiles and ethylene from cv. Zaosuli pears. *Hortic. Sci.* **2016**, *43*, 117–125. [CrossRef]
12. Wang, L.B.; Ma, M.; Zhang, Y.R.; Wu, Z.F.; Guo, L.; Luo, W.Q.; Wang, L.; Zhang, Z.; Zhang, S.L. Characterization of the genes involved in malic acid metabolism from pear fruit and their expression profile after postharvest 1-MCP/Ethrel treatment. *J. Agric. Food Chem.* **2018**, *66*, 8772–8782. [CrossRef] [PubMed]
13. Adhikary, T.; Gill, P.P.S.; Jawandha, S.K.; Kaur, N. Postharvest quality response of pears with beeswax coatings during long term cold storage. *J. Hortic. Sci. Biotechnol.* **2022**, *97*, 785–798. [CrossRef]
14. Li, J.Z.; Dai, X.N.; Li, Q.Y.; Jiang, F.D.; Xu, X.F.; Guo, T.T.; Zhang, H.X. Low temperatures inhibit the pectin degradation of 'Docteur Jules Guyot' pear (*Pyrus communis* L.). *Int. J. Biol. Macromol.* **2023**, *242*, 124719. [CrossRef] [PubMed]
15. Iqbal, N.; Khan, N.A.; Ferrante, A.; Trivellini, A.; Francini, A.; Khan, M.I.R. Ethylene role in plant growth, development and senescence: Interaction with other phytohormones. *Front. Plant Sci.* **2017**, *8*, 475. [CrossRef] [PubMed]
16. Kou, X.H.; Feng, Y.; Yuan, S.; Zhao, X.Y.; Wu, C.E.; Wang, C.; Xue, Z.H. Different regulatory mechanisms of plant hormones in the ripening of climacteric and non-climacteric fruits: A review. *Plant Mol. Biol.* **2021**, *107*, 477–497. [CrossRef]
17. Li, X.M.; Wang, X.M.; Zhang, Y.; Zhang, A.H.; You, C.X. Regulation of fleshy fruit ripening: From transcription factors to epigenetic modifications. *Hortic. Res.* **2022**, *9*, uhac013. [CrossRef]
18. Song, B.B.; Li, X.L.; Cao, B.B.; Zhang, M.Y.; Korban, S.S.; Yu, L.A.; Yang, W.X.; Zhao, K.J.; Li, J.M.; Wu, J. An identical-by-descent segment harbors a 12-bp insertion determining fruit softening during domestication and speciation in Pyrus. *BMC Biol.* **2022**, *20*, 215. [CrossRef]
19. Tucker, G.; Yin, X.R.; Zhang, A.D.; Wang, M.M.; Zhu, Q.G.; Liu, X.F.; Xie, X.L.; Chen, K.S.; Grierson, D. Ethylene and fruit softening. *Food Qual. Saf.* **2017**, *1*, 253–267. [CrossRef]
20. Zhang, A.D.; Wang, W.Q.; Tong, Y.; Li, M.J.; Grierson, D.; Ferguson, I.; Chen, K.S.; Yin, X.R. Transcriptome analysis identifies a zinc finger protein regulating starch degradation in kiwifruit. *Plant Physiol.* **2018**, *178*, 850–863. [CrossRef]
21. Hiwasa, K.; Nakano, R.; Hashimoto, A.; Matsuzaki, M.; Murayama, H.; Inaba, A.; Kubo, Y. European, Chinese and Japanese pear fruits exhibit differential softening characteristics during ripening. *J. Exp. Bot.* **2004**, *55*, 2281–2290. [CrossRef] [PubMed]
22. Song, L.; Wang, Z.; Wang, Z.; Meng, G.; Zhai, R.; Cai, M.; Ma, F.; Xu, L. Screening of cell wall-related genes that are expressed differentially during ripening of pears with different softening characteristics. *Postharvest Biol. Technol.* **2016**, *115*, 1–8. [CrossRef]
23. Itai, A.; Tanahashi, T. Inhibition of sucrose loss during cold storage in Japanese pear (*Pyrus pyrifolia* Nakai) by 1-MCP. *Postharvest Biol. Technol.* **2008**, *48*, 355–363. [CrossRef]
24. Ma, M.; Wang, L.B.; Zhang, S.L.; Guo, L.; Zhang, Z.; Li, J.L.; Sun, Q.; Zhang, S.L. Acid vacuolar invertase 1 (PbrAc-Inv1) and invertase inhibitor 5 (PbrII5) were involved in sucrose hydrolysis during postharvest pear storage. *Food Chem.* **2020**, *320*, 126635. [CrossRef]
25. Wang, J.W.; Dong, S.Z.; Jiang, Y.G.; He, H.S.; Liu, T.; Lv, M.; Ji, S.J. Influence of long-term cold storage on phenylpropanoid and soluble sugar metabolisms accompanied with peel browning of 'Nanguo' pears during subsequent shelf life. *Sci. Hortic.* **2020**, *260*, 108888. [CrossRef]
26. Rivas-San Vicente, M.; Plasencia, J. Salicylic acid beyond defence: Its role in plant growth and development. *J. Exp. Bot.* **2011**, *62*, 3321–3338. [CrossRef]

27. Asghari, M.; Aghdam, M.S. Impact of salicylic acid on postharvest physiology of horticultural crops. *Trends Food Sci. Technol.* **2010**, *21*, 502–509. [CrossRef]
28. Zhou, Y.; Ma, J.; Xie, J.; Deng, L.; Yao, S.; Zeng, K. Transcriptomic and biochemical analysis of highlighted induction of phenylpropanoid pathway metabolism of citrus fruit in response to salicylic acid, Pichia membranaefaciens and oligochitosan. *Postharvest Biol. Technol.* **2018**, *142*, 81–92. [CrossRef]
29. Chen, C.J.; Sun, C.C.; Wang, Y.H.; Gong, H.S.; Zhang, A.D.; Yang, Y.Q.; Guo, F.J.; Cui, K.B.; Fan, X.G.; Li, X.L. The preharvest and postharvest application of salicylic acid and its derivatives on storage of fruit and vegetables: A review. *Sci. Hortic.* **2023**, *312*, 111858. [CrossRef]
30. Trejo, E.J.O.; Brizzolara, S.; Cardillo, V.; Ruperti, B.; Bonghi, C.; Tonutti, P. The impact of PGRs applied in the field on the postharvest behavior of fruit crops. *Sci. Hortic.* **2023**, *318*, 112103. [CrossRef]
31. Kumar, N.; Tokas, J.; Raghavendra, M.; Singal, H.R. Impact of exogenous salicylic acid treatment on the cell wall metabolism and ripening process in postharvest tomato fruit stored at ambient temperature. *Int. J. Food Sci. Technol.* **2021**, *56*, 2961–2972. [CrossRef]
32. Wang, J.; Liu, X.; Zhang, H.; Allan, A.C.; Wang, W.; Yin, X. Transcriptional and post-transcriptional regulation of ethylene biosynthesis by exogenous acetylsalicylic acid in kiwifruit. *Hortic. Res.* **2022**, *9*, uhac116. [CrossRef] [PubMed]
33. Yang, C.; Duan, W.; Xie, K.; Ren, C.; Zhu, C.; Chen, K.; Zhang, B. Effect of salicylic acid treatment on sensory quality, flavor-related chemicals and gene expression in peach fruit after cold storage. *Postharvest Biol. Technol.* **2020**, *161*, 111089. [CrossRef]
34. Hassan, I.; Zhang, Y.; Du, G.; Wang, G.; Zhang, J. Effect of salicylic acid (SA) on delaying fruit senescence of Huang Kum pear. *Front. Agric. China* **2007**, *1*, 456–459.
35. Adhikary, T.; Gill, P.S.; Jawandha, S.K.; Bhardwaj, R.D.; Anurag, R.K. Browning and quality management of pear fruit by salicylic acid treatment during low temperature storage. *J. Sci. Food Agric.* **2020**, *101*, 853–862. [CrossRef]
36. Sinha, A.; Gill, P.P.S.; Jawandha, S.K.; Grewal, S.K. Composite coating of chitosan with salicylic acid retards pear fruit softening under cold and supermarket storage. *Food Res. Int.* **2022**, *160*, 11724. [CrossRef]
37. Luo, M.L.; Ge, W.Y.; Sun, H.J.; Yang, Q.X.; Sun, Y.Y.; Zhou, X.; Zhou, Q.; Ji, S.J. Salicylic acid treatment alleviates diminished ester production in cold-stored Nanguo' pear by promoting the transcription of PuAAT. *Postharvest Biol. Technol.* **2022**, *187*, 111849. [CrossRef]
38. Liu, D. Effect of 1-MCP on postharvest physiology and quality of 'Hosui' pear produced from three gorges reservoir areas. *Storage Process* **2016**, *16*, 27–30. (In Chinese)
39. Nie, Z.; Wan, C.; Chen, C.; Chen, J. Comprehensive evaluation of the postharvest antioxidant capacity of majiayou pomelo harvested at different maturities based on PCA. *Antioxidants* **2019**, *8*, 136. [CrossRef]
40. Zhang, H.; Zhao, L.; Fan, C.; Wang, P.; Cui, M.; Liu, L.; Yang, H.; Wang, J. Impact of methyl salicylate on storage quality, ethylene action, and protein profiling of 'Zaosu' pear (*Pyrus bretschneideri*). *Sci. Hortic.* **2020**, *264*, 109196. [CrossRef]
41. Mahajan, B.V.C.; Singh, K.; Dhillon, W.S. Effect of 1-methylcyclopropene (1-MCP) on storage life and quality of pear fruits. *J. Food Sci. Technol.* **2010**, *47*, 351–354. [CrossRef] [PubMed]
42. Lwin, H.P.; Rudell, R.D.; Lee, J. Metabolism and cold chain performance of 'Chuhwangbae' Asian pears as impacted by 1-MCP treatment. *Sci. Hortic.* **2021**, *288*, 110357. [CrossRef]
43. Wang, Y.; Sugar, D. Ripening behavior and quality of modified atmosphere packed 'Doyenne du Comice' pears during cold storage and simulated transit. *Postharvest Biol. Technol.* **2013**, *81*, 51–59. [CrossRef]
44. Hira, N.; Mitalo, O.W.; Okada, R.; Sangawa, M.; Masuda, K.; Fujita, N.; Ushijima, K.; Akagi, T.; Kubo, Y. The effect of layer-by-layer edible coating on the shelf life and transcriptome of 'Kosui' Japanese pear fruit. *Postharvest Biol. Technol.* **2022**, *185*, 111787. [CrossRef]
45. Jiang, B.; Liu, R.L.; Fang, X.J.; Tong, C.; Chen, H.J.; Gao, H.Y. Effects of salicylic acid treatment on fruit quality and wax composition of blueberry (*Vaccinium virgatum* Ait). *Food Chem.* **2023**, *368*, 130757. [CrossRef]
46. Chen, C.Y.; Huang, Q.; Peng, X.; Wan, C.P.; Zeng, J.K.; Zhang, Y.J.; Chen, J.Y. Alleviatory effects of salicylic acid on postharvest softening and cell wall degradation of 'Jinshayou' pummelo (*Citrus maxima* Merr.): A comparative physiological and transcriptomic analysis. *Food Chem.* **2023**, *424*, 136428. [CrossRef] [PubMed]
47. Koyuncu, M.A.; Erbas, D.; Onursal, C.E.; Secmen, T.; Guneyli, A.; Uzumcu, S.S. Postharvest treatments of salicylic acid, oxalic acid and putrescine influences bioactive compounds and quality of pomegranate during controlled atmosphere storage. *J. Food Sci. Technol.* **2019**, *56*, 350–359. [CrossRef]
48. Zhao, Y.Y.; Song, C.C.; Brimmel, D.A.; Qi, S.N.; Lin, Q.; Bi, J.F.; Duan, Y.Q. Salicylic acid treatment mitigates chilling injury in peach fruit by regulation of sucrose metabolism and soluble sugar content. *Food Chem.* **2021**, *358*, 129867. [CrossRef]
49. Davarynejad, G.H.; Zarei, M.; Nasrabadi, M.E.; Ardakani, E. Effects of salicylic acid and putrescine on storability, quality attributes and antioxidant activity of plum cv. 'Santa Rosa'. *J. Food Sci. Technol.* **2015**, *52*, 2053–2062. [CrossRef]
50. Li, S.; Chen, K.; Grierson, D. Molecular and hormonal mechanisms regulating fleshy fruit ripening. *Cells* **2021**, *10*, 1136. [CrossRef]
51. Hao, Y.; Hu, G.; Breitel, D.; Liu, M.; Mila, I.; Frasse, P.; Fu, Y.; Aharoni, A.; Bouzayen, M.; Zouine, M. Auxin Response Factor SlARF2 Is an Essential Component of the Regulatory Mechanism Controlling Fruit Ripening in Tomato. *PLoS Genet.* **2015**, *11*, e1005649. [CrossRef] [PubMed]

52. Shin, J.H.; Mila, I.; Liu, M.; Rodrigues, M.A.; Vernoux, T.; Pirrello, J.; Bouzayen, M. The RIN-regulated Small Auxin-Up RNA SAUR 69 is involved in the unripe-to-ripe phase transition of tomato fruit via enhancement of the sensitivity to ethylene. *New Phytol.* **2018**, *222*, 820–836. [CrossRef]
53. Liu, K.; Yuan, C.; Feng, S.; Zhong, S.; Li, H.; Zhong, J.; Shen, C.; Liu, J. Genome-wide analysis and characterization of Aux/IAA family genes related to fruit ripening in papaya (*Carica papaya* L.). *BMC Genom.* **2017**, *18*, 351. [CrossRef] [PubMed]
54. Yu, J.; Tseng, Y.; Pham, K.; Liu, M.; Beckles, D.M. Starch and sugars as determinants of postharvest shelf life and quality: Some new and surprising roles. *Curr. Opin. Biotechnol.* **2022**, *78*, 102844. [CrossRef] [PubMed]
55. Sasaki, H.; Ichimura, K.; Imada, S.; Yamaki, S. Sucrose synthase and sucrose phosphate synthase, but not acid invertase, are regulated by cold acclimation and deacclimation in cabbage seedlings. *J. Plant Physiol.* **2001**, *158*, 847–852. [CrossRef]
56. Kanayama, Y. Sugar metabolism and fruit development in the tomato. *Hortic. J* **2017**, *86*, 417–425. [CrossRef]
57. Payasi, A.; Mishra, N.N.; Chaves, A.L.; Singh, R. Biochemistry of fruit softening: An overview. *Physiol. Mol. Biol. Plants* **2009**, *15*, 103–113. [CrossRef] [PubMed]
58. Shi, Y.; Li, B.J.; Su, G.; Zhang, M.X.; Grierson, D.; Chen, K.S. Transcriptional regulation of fleshy fruit texture. *J. Integr. Plant Biol.* **2022**, *64*, 1649–1672. [CrossRef]
59. Miedes, E.; Herbers, K.; Sonnewald, U.; Lorences, E.P. Overexpression of a cell wall enzyme reduces xyloglucan depolymerization and softening of transgenic tomato fruits. *J. Agric. Food Chem.* **2010**, *58*, 5708–5713. [CrossRef]
60. Wang, D.D.; Samsulrizal, N.H.; Yan, C.; Allcock, N.S.; Craigon, J.; Blanco Ulate, B.; Ortega-Salazar, I.; Marcus, S.E.; Bagheri, H.M.; Perez Fons, L.; et al. Characterization of CRISPR mutants targeting genes modulating pectin degradation in ripening tomato. *Plant Physiol.* **2019**, *179*, 544–557.
61. Paniagua, C.; Blanco Portales, R.; Barcelo-Muñoz, M.; Garcia-Gago, J.A.; Waldron, K.W.; Quesada, M.A.; Muñoz-Blanco, J.; Mercado, J.A. Antisense down regulation of the strawberry β-galactosidase gene FaβGal4 increases cell wall galactose levels and reduces fruit softening. *J. Exp. Bot.* **2016**, *67*, 619–631. [CrossRef] [PubMed]
62. Brummell, D.A.; Harpster, M.H.; Civello, P.M.; Palys, J.M.; Bennett, A.B.; Dunsmuir, P. Modification of expansin protein abundance in tomato fruit alters softening and cell wall polymer metabolism during ripening. *Plant Cell* **1999**, *11*, 2203–2216. [CrossRef] [PubMed]
63. Gao, Y.; Wei, W.; Fan, Z.Q.; Zhao, X.D.; Zhang, Y.P.; Jing, Y.; Zhu, B.Z.; Zhu, H.L.; Shan, W.; Chen, J.Y.; et al. Re-evaluation of the nor mutation and the role of the NAC-NOR transcription factor in tomato fruit ripening. *J. Exp. Bot.* **2020**, *71*, 3560–3574. [CrossRef] [PubMed]
64. Kuang, J.F.; Chen, J.Y.; Liu, X.C.; Han, Y.C.; Xiao, Y.Y.; Shan, W.; Tang, Y.; Wu, K.Q.; He, J.X.; Lu, W.J. The transcriptional regulatory network mediated by banana (*Musa acuminata*) dehydration-responsive element binding (MaDREB) transcription factors in fruit ripening. *New Phytol.* **2017**, *214*, 762–781. [CrossRef] [PubMed]
65. Xiao, Y.Y.; Kuang, J.F.; Qi, X.N.; Ye, Y.J.; Wu, Z.X.; Chen, J.Y.; Lu, W.J. A comprehensive investigation of starch degradation process and identification of a transcriptional activator MabHLH6 during banana fruit ripening. *Plant Biotechnol. J.* **2018**, *16*, 151–164. [CrossRef]
66. Mao, W.W.; Han, Y.; Chen, Y.T.; Sun, M.Z.; Feng, Q.Q.; Li, L.; Liu, L.P.; Zhang, K.K.; Wei, L.Z.; Han, Z.H.; et al. Low temperature inhibits anthocyanin accumulation in strawberry fruit by activating FvMAPK3-induced phosphorylation of FvMYB10 and degradation of chalcone synthase 1. *Plant Cell* **2022**, *34*, 1226–1249. [CrossRef]
67. Sang, Y.Y.; Liu, Y.X.; Tang, Y.S.; Yang, W.T.; Guo, M.R.; Chen, G.G. Transcriptome sequencing reveals mechanism of improved antioxidant capacity and maintained postharvest quality of winter jujube during cold storage after salicylic acid treatment. *Postharvest Biol. Technol.* **2022**, *189*, 111929. [CrossRef]
68. Sekhon-Loodu, S.; Warnakulasuriya, S.N.; Rupasinghe, H.V.; Shahidi, F. Antioxidant ability of fractionated apple peel phenolics to inhibit fish oil oxidation. *Food Chem.* **2013**, *140*, 189–196. [CrossRef]
69. Hughes, D.E. Titrimetric determination of ascorbic acid with 2,6-dichlorophenol indophenol in commercial liquid diets. *J. Pharm. Sci.* **1983**, *72*, 126–129. [CrossRef]
70. Lin, Y.F.; Lin, Y.Z.; Lin, Y.X.; Lin, M.S.; Chen, Y.H.; Wang, H.; Lin, H.T. A novel chitosan alleviates pulp breakdown of harvested longan fruit by suppressing disassembly of cell wall polysaccharides. *Carbohydr. Polym.* **2019**, *217*, 126–134. [CrossRef]
71. Yang, H.Y.; Li, J.Z.; Li, X.H.; Wu, R.; Zhang, X.L.; Fan, X.G.; Li, G.T.; Gong, H.S.; Yin, X.R.; Zhang, A.D. The mechanism of gibberellins treatment suppressing kiwifruit postharvest ripening processes by transcriptome analysis. *Postharvest Biol. Technol.* **2023**, *198*, 112223. [CrossRef]
72. Kintner, P.K., III; Van Buren, J.P. Carbohydrate interference and its correction in pectin analysis using the m-hydroxydiphenyl method. *J. Food Sci.* **1982**, *47*, 756–759. [CrossRef]
73. Yemm, E.W.; Willis, A.J. The estimation of carbohydrates in plant extracts by anthrone. *Biochem. J.* **1954**, *57*, 508–514. [CrossRef] [PubMed]
74. Miao, M.M.; Xu, X.F.; Chen, X.H.; Xue, L.B.; Cao, B.S. Cucumber carbohydrate metabolism and translocation under chilling night temperature. *J. Plant Physiol.* **2007**, *164*, 621–628. [CrossRef] [PubMed]

75. Chang, S.; Puryear, J.; Cairney, J. A simple and efficient method for isolating RNA from pine trees. *Plant Mol. Biol. Rep.* **1993**, *11*, 113–116. [CrossRef]
76. Zhang, J.; Zhang, Y.F.; Zhang, P.F.; Bian, Y.H.; Liu, Z.Y.; Zhang, C.; Liu, X.; Wang, C.L. An integrated metabolic and transcriptomic analysis reveals the mechanism through which fruit bagging alleviates exocarp semi-russeting in pear fruit. *Tree Physiol.* **2021**, *41*, 1306–1318. [CrossRef] [PubMed]

plants

Article

Investigation into the Relationship between Spermidine Degradation and Phenolic Compounds Biosynthesis in Barley Seedlings under Ultraviolet B Stress

Chong Xie [1], Yahui Zhu [3], Chaoqun Leng [1], Qiaoe Wang [2], Pei Wang [1] and Runqiang Yang [1,*]

1 College of Food Science and Technology, Whole Grain Food Engineering Research Center, Nanjing Agricultural University, Nanjing 210095, China; xiechong@njau.edu.cn (C.X.); wangpei@njau.edu.cn (P.W.)

2 College of Food Science and Technology, Tibet Agriculture and Animal Husbandry University, Linzhi 860000, China; wangqe@th.btbu.edu.cn

3 Beijing Key Lab of Plant Resource Research and Development, Beijing Technology and Business University, Beijing 100048, China; zhuyahui@xza.edu.cn

* Correspondence: yangrq@njau.edu.cn; Tel./Fax: +86-025-84396293

Abstract: Barley germination under ultraviolet B (UV-B) illumination stress induces effective accumulation of phenolic compounds in the barley. Spermidine can enhance the biosynthesis of phenolic compounds and alleviate the oxidative damage caused by UV-B. To better understand the function of spermidine, inhibitors of enzymes that are involved in the degradation of spermidine and the synthesis of gamma-aminobutyric acid (GABA), the product of spermidine degradation, were applied to barley germinated under UV-B treatment. The results showed a more severe oxidative damage, and a decrease in phenolic acid contents were observed when spermidine degradation was inhibited. However, GABA application did attenuate an increase in electrolyte permeability and MDA content caused by UV-B induced oxidative damage and improved the respiration rate. Meanwhile, GABA application can elevate the accumulation of phenolic compounds by ca. 20%, by elevating the activities of some key enzymes. Furthermore, the application of GABA, together with the inhibitor of spermidine degradation, can alleviate its suppression of the synthesis of phenolic acids, and resistance to UV-B stress. In conclusion, spermidine alleviated oxidative damage and enhanced the accumulation of phenolic compounds using its degradation product.

Keywords: barley seedling; spermidine; phenolic compounds; gamma-aminobutyric acid

1. Introduction

Phenolic compounds play important roles in plant growth, development and defense against environmental stress [1,2]. Meanwhile, these substances have various physiological activities in the human body and epidemiological studies have shown that long-term consumption of foods rich in phenolic compounds may provide protection against many chronic diseases [3,4]. Therefore, food products that are high in phenolic compounds have gained increased popularity.

Barley is a nutritious cereal that offers several health benefits, but, when compared to wheat and rice, it does have palatability and processing issues, which limit their uses to the beer industry. Germination is an effective way to increase the nutrient content and alter the flavor profiles of cereals and some abiotic stresses can be used to enhance the accumulation of phenolic compounds during germination [5–7]. UV-B (ultraviolet B light) is part of the electromagnetic spectrum that has a wavelength ranging from 280 to 320 nm; it can induce oxidative damage in grains, which leads to a higher accumulation of phenolic compounds but a significantly lower length and fresh weight.

Polyamines, including putrescine, spermidine and spermine, have diverse roles in cellular processes in plants, such as apoptosis, cell proliferation, DNA synthesis, gene expression and signal transduction [8]. In our previous study, exogenous spermidine not only enhanced the accumulation of phenolic compounds, but also alleviated the oxidative damage caused by UV-B radiation [9]. In plants, polyamine degradation is catalyzed by polyamine oxidase and renders aminobutyraldehyde, which is subsequently converted into gamma-aminobutyric acid (GABA) by the action of aminoaldehyde dehydrogenase (AMADH) [10]. GABA, a four-carbon molecule, plays various roles in plants, including resistance to both abiotic and biotic stress [11]. GABA has been shown to influence the biosynthesis of phenolic compounds and antioxidant systems in soybean sprouts germinated under NaCl stress [7].

To better understand the role of spermidine in phenolic compound biosynthesis and the establishment of antioxidant adaptive responses of barley during germination, whether the effect of spermidine is directly from the action of itself or by its degradation product, i.e., GABA, should be clarified. Therefore, in the present study, barley seedlings germinated under UV-B stress were treated with GABA (G seedling), aminoguanidine hydrochloride (AG), the inhibitor of polyamine oxidase (A seedling), 1-ethyl-(3-dimethylaminopropyl), carbon diimide hydrochloride (EDC), the inhibitor of AMADH (E seedling) and EDC together with GABA (EG seedling), respectively. The growth, content of phenolic compounds and antioxidant capacity of seedlings with different treatments were investigated.

2. Results

2.1. Growth of Barley Seedlings

As shown in Figure 1A,B, treatments with both AG and EDC resulted in significantly ($p < 0.05$) shorter lengths of seedling and root. Application of GABA led to a 20% increase in seedling length. The barley seedling treated with GABA and EDC together exhibited a significantly ($p < 0.05$) longer seedling length (ca. 47 mm) than E seedling. The CK seedling had ca. 17 g/100 seedlings of fresh weight and ca. 3.1 g/100 seedlings of dry weight (Figure 1C,D). The highest fresh weight of seedlings was found in the G seedling, which was approximately 19 g/100 fresh seedlings, and the lowest fresh weight was observed in the A seedling (13 g/100 fresh seedlings). However, the A seedling had the highest level of dry weight (3.62 g/100 fresh seedling) and the lowest level of dry weight (3.14 g/100 fresh seedling) was found in the CK seedling.

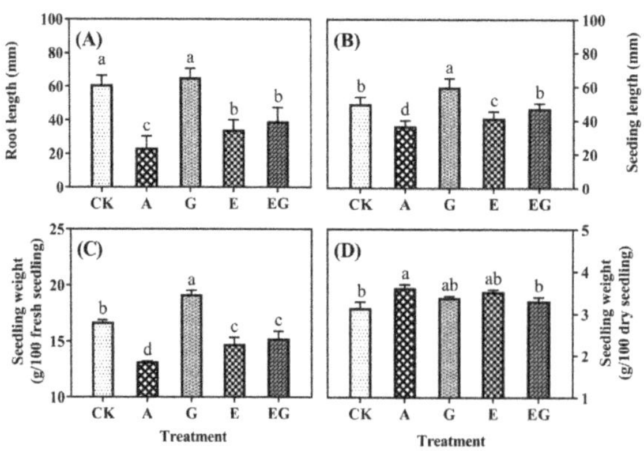

Figure 1. Root length (**A**), seedling length (**B**), fresh weight (**C**) and dry weight (**D**) of barley seedlings. CK are seedlings treated with distilled water. A, G, E and EG are seedlings treated with aminoguanidine hydrochloride, gamma-aminobutyric acid (GABA), 1-ethyl-(3-dimethylaminopropyl) carbon diimide hydrochloride (EDC) and GABA together with EDC, respectively. Different lowercase letters (a–d) represent significant difference among treatments (n = 3; $p < 0.05$).

2.2. Content of Malondialdehyde, Electrolyte Leakage, Respiratory Rate and Spermidine Content

As shown in Figure 2A–C, the application of AG resulted in a significantly ($p < 0.05$) higher level of MDA content (5.76 nmol/g FW) and electrolyte leakage (38%) than the CK seedling. Both GABA and EDC can significantly ($p < 0.05$) decrease the MDA content and electrolyte leakage. The highest and lowest respiratory rates were found in the G and A seedlings, respectively, and EG seedling had a ca. 40% higher respiratory rate than E seedling.

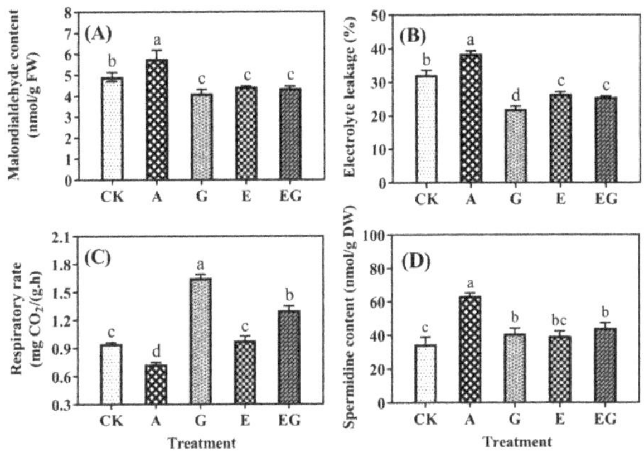

Figure 2. Content of malondialdehyde (**A**), electrolyte leakage (**B**), respiratory rate (**C**) and contents of spermidine (**D**) in barley seedlings (n = 3). Different lowercase letters (a–d) represent significant difference among treatments (n = 3; $p < 0.05$).

The content of spermidine in the CK seedling was ca. 35 nmol/g DW (Figure 2D). Application of AG and GABA significantly ($p < 0.05$) increased the content of spermidine by 82.79% and 18.38%, respectively. EDC had no significant ($p > 0.05$) effect on spermidine content but a significantly ($p < 0.05$) higher spermidine content was observed when it was applied in combination with GABA.

2.3. Content of Phenolic Compounds and Phenolic Acids

As shown in Figure 3, the A seedling had a significantly ($p < 0.05$) higher level of free phenolic compounds (1140 mg GAE/g DW) and a lower level of bound phenolic compounds (ca. 960 mg GAE/g DW) than the CK seedling. The addition of GABA resulted in a significant ($p < 0.05$) increase in the content of both free phenolic compounds (increased by 21.18%) and total phenolic compounds (increased by 13.42%) when compared to the control. EDC treatment also led to a significant ($p < 0.05$) increase in free phenolic compounds and decrease in bound phenolic compounds. However, the content of total phenolic compounds in E seedling (1958 mg GAE/g DW) was significantly ($p < 0.05$) lower than the control seedling (2043 mg GAE/g DW). The addition of GABA to EDC resulted in the lowest level of free (778 mg GAE/g DW) and total phenolic compounds (1849 mg GAE/g DW).

As shown in Table 1, ferulic acid exhibited the highest content among all the detected phenolic acids in all the seedlings, followed by *p*-coumaric acid, sinapic acid and caffeic acid. Among them, ferulic acid and *p*-coumaric acid were mainly present in the bound forms but caffeic acid was mainly present in the free form. The contents of vanillic acid, *p*-hydroxybenzoic acid, protocatechuic acid and syringic acid were present in relatively low levels. After germination, the content of total ferulic acid decreased by 19.49% under AG treatment but increased by 29.80% when GABA was added under UV-B stress. A 20.16% decrease in total ferulic acid content was observed by the addition of EDC and this inhibitory effect of EDC was alleviated by exogenous GABA. A similar trend was observed in other types of phenolic acids, such as *p*-hydroxybenzoic acid, vanillic acid and sinapic acid. Free protocatechuic acid was not detected under EDC treatment, and free syringic acid was only detected under UV-B stress with the addition of GABA.

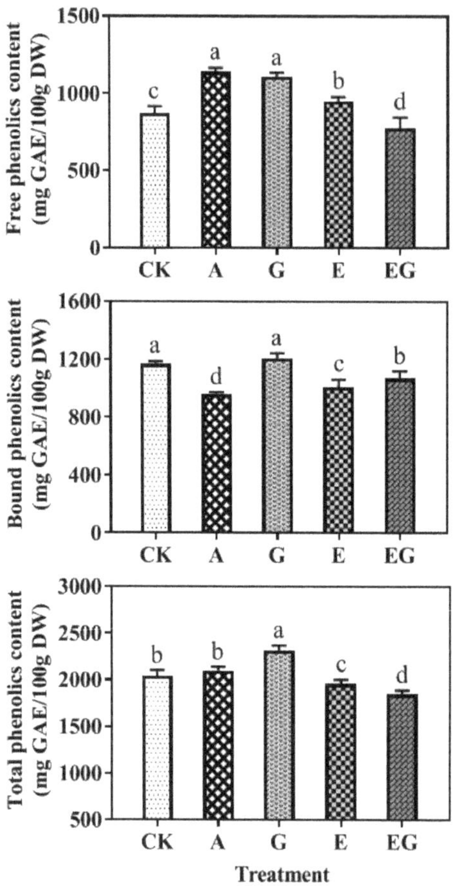

Figure 3. Content of free, bound and total phenolic compounds in barley seedlings (n = 3). Different lowercase letters (a–d) represent significant difference among different treatments ($p < 0.05$).

Table 1. Content of phenolic acids (μg/g DW) in barley seedlings.

Phenolic Acid	Treatments	Phenolic Acids Content (μg/g DW)		
		Free	Bound	Total
Ferulic acid	CK	32.36 ± 1.40 ^{b}	1817.78 ± 15.75 ^{b}	1850.14 ± 14.35 ^{b}
	A	24.76 ± 1.18 ^{d}	1464.88 ± 3.73 ^{c}	1489.64 ± 2.55 ^{c}
	G	37.60 ± 1.50 ^{a}	2363.78 ± 52.77 ^{a}	2401.39 ± 54.27 ^{a}
	E	23.63 ± 0.13 ^{d}	1453.55 ± 47.84 ^{c}	1477.18 ± 47.71 ^{c}
	EG	28.00 ± 0.22 ^{c}	1785.29 ± 26.96 ^{b}	1813.29 ± 38.45 ^{b}
p-Coumaric acid	CK	6.01 ± 0.01 ^{a}	228.01 ± 2.74 ^{b}	234.02 ± 2.73 ^{b}
	A	3.93 ± 0.12 ^{b}	146.96 ± 2.78 ^{d}	150.89 ± 2.90 ^{d}
	G	5.89 ± 0.08 ^{a}	256.27 ± 3.01 ^{a}	262.16 ± 2.93 ^{a}
	E	3.75 ± 0.52 ^{b}	172.69 ± 7.36 ^{c}	176.44 ± 7.88 ^{c}
	EG	4.21 ± 0.36 ^{b}	190.64 ± 9.99 ^{c}	194.84 ± 14.63 ^{c}

Table 1. *Cont.*

Phenolic Acid	Treatments	Phenolic Acids Content (μg/g DW)		
		Free	Bound	Total
Sinapic acid	CK	78.06 ± 1.71 b	63.67 ± 0. 32 c	141.73 ± 2.03 b
	A	66.65 ± 0.26 d	63.93 ± 0.11 c	130.58 ± 0.15 d
	G	86.03 ± 1.35 a	69.40 ± 0.11 a	155.43 ± 1.46 a
	E	66.17 ± 0.07 d	62.78 ± 0.49 d	128.95 ± 0.43 d
	EG	73.21 ± 0.10 c	65.91 ± 0.38 b	139.12 ± 0.69 c
Caffeic acid	CK	38.25 ± 0.82 a	2.24 ± 0.03 c	40.49 ± 0.78 a
	A	18.55 ± 0.24 d	2.33 ± 0.01 bc	20.88 ± 0.25 d
	G	29.07 ± 2.44 b	2.28 ± 0.08 b c	31.34 ± 2.36 b
	E	23.79 ± 0.13 c	2.56 ± 0.23 ab	26.34 ± 0.36 c
	EG	29.03 ± 2.14 b	2.73 ± 0.08 a	31.76 ± 2.92 b
Vanillic acid	CK	8.00 ± 0.24 b	6.50 ± 0.12 a	14.51 ± 0.50 b
	A	4.49 ± 0.09 d	5.24 ± 0.04 b	9.74 ± 0.07 d
	G	9.62 ± 0.15 a	6.44 ± 0.06 a	16.06 ± 0.13 a
	E	4.50 ± 0.09 d	4.80 ± 0.17 c	9.31 ± 0.37 d
	EG	6.27 ± 0.07 c	4.99 ± 0.04 c	11.25 ± 0.04 c
p-Hydroxybenzoic acid	CK	4.43 ± 0.97 b	6.76 ± 0.07 a	11.19 ± 0.90 b
	A	4.72 ± 0.20 b	3.49 ± 0.11 b	8.20 ± 0.31 c
	G	7.87 ± 0.46 a	7.74 ± 0.77 a	15.61 ± 1.23 a
	E	3.84 ± 0.06 b	4.16 ± 0.22 b	7.99 ± 0.16 c
	EG	4.28 ± 0.10 b	6.68 ± 0.36 a	10.97 ± 0.38 b
Protocatechuic acid	CK	4.23 ± 0.14 b	1.35 ± 0.01 b	5.58 ± 0.22 b
	A	4.16 ± 0.37 b	0.28 ± 0.02 c	4.44 ± 0.55 b
	G	12.37 ± 0.90 a	3.43 ± 0.56 a	15.80 ± 1.83 a
	E	ND	1.00 ± 0.53 b	1.00 ± 0.53 c
	EG	5.36 ± 0.56 b	1.30 ± 0.17 b	6.66 ± 1.04 b
Syringic acid	CK	2.50 ± 0.08 a	4.38 ± 0.06 b	6.87 ± 0.14 b
	A	ND	3.82 ± 0.09 c	3.82 ± 0.09 d
	G	2.40 ± 0.01 a	4.77 ± 0.01 a	7.17 ± 0.01 a
	E	ND	4.28 ± 0.01 b	4.28 ± 0.01 c
	EG	ND	4.01 ± 0.19 c	4.01 ± 0.19 cd

Different lowercase letters (a–d) represent significant difference of each phenolic acid content among different treatments (n = 3; $p < 0.05$).

2.4. Activities of Key Enzymes Related to Synthesis of Phenolic Acid

Under UV-B stress, AG treatment significantly ($p < 0.05$) reduced the activities of phenylalanine ammonia lyase (PAL) and caffeic acid-o-methyltransferase (COMT) but increased the activities of 4-coumarate coenzyme A ligase (4CL), coumarate 3-hydroxylase (C3H), cinnamic acid 4-hydroxylase (C4H) and ferulate 5-hydroxylase (F5H), as shown in Figure 4. The addition of GABA significantly ($p < 0.05$) increased the activities of C3H and COMT by 52.17% and 111.17%, respectively, but had no significant ($p > 0.05$) influence on the activities of other enzymes. Meanwhile, the highest activities of PAL and COMT were found in the G seedling. EDC treatment decreased the activities of PAL and C4H by 20.99% and 19.14%, respectively, and the lowest activities of C4H and F5H were found in the E seedling. The EG seedling had significantly ($p < 0.05$) higher activities of PAL, C3H, COMT and F5H than the E seedling, and the highest activities of C3H and F5H were observed in the EG seedling.

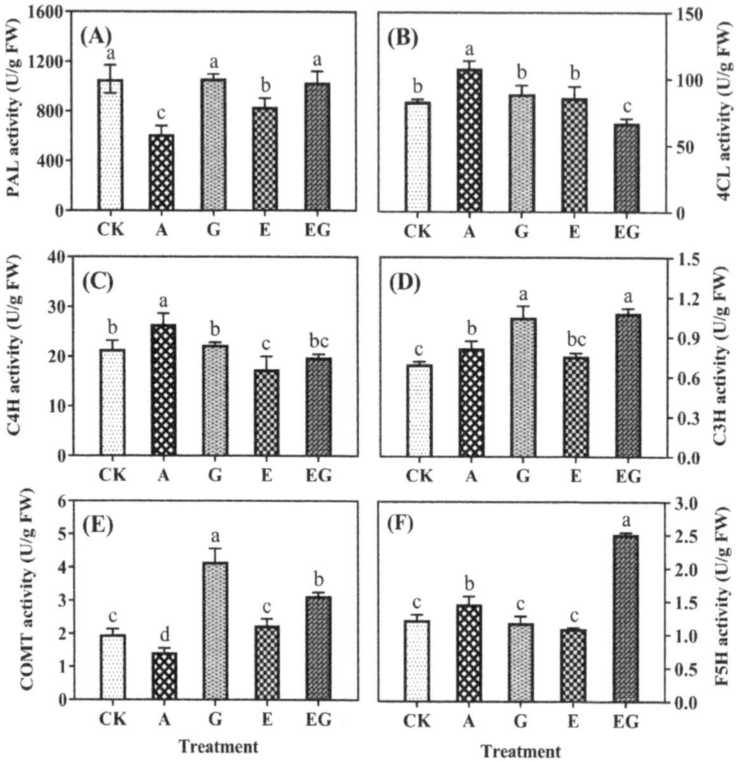

Figure 4. Activities of PAL (**A**), 4CL (**B**), C4H (**C**), C3H (**D**), COMT (**E**) and F5H (**F**) in barley seedlings (n = 3). Different lowercase letters (a–d) represent significant difference among treatments ($p < 0.05$).

2.5. Antioxidant Capacity

As shown in Table 2, total ABTS radical scavenging in all seedlings ranged from 77.06 μmol TE/g DW (EG seedling) to 83.80 μmol TE/g DW (A seedling). In the A and G seedlings, free phenolic compounds had a higher contribution than the bound forms. Total DPPH radical scavenging in all seedlings ranged from 75.79 μmol TE/g DW (A seedling) to 43.53 μmol TE/g DW (G seedling) and bound phenolic compounds had a higher contribution than the free forms in all seedlings.

Table 2. Free radical scavenging ability of ABTS and DPPH in barley seedlings.

Treatments	ABTS Values (μmol TE/g DW)			DPPH Values (μmol TE/g DW)		
	Free	Bound	Total	Free	Bound	Total
CK	39.80 ± 3.02 ^{bc}	43.57 ± 2.96 ^{a}	83.37 ± 3.25 ^{a}	14.89 ± 2.09 ^{bc}	25.01 ± 0.39 ^{b}	39.90 ± 2.29 ^{b}
A	48.20 ± 3.03 ^{a}	35.61 ± 2.58 ^{b}	83.80 ± 2.85 ^{a}	14.52 ± 0.54 ^{c}	21.27 ± 0.49 ^{d}	35.79 ± 0.87 ^{d}
G	41.17 ± 1.27 ^{b}	36.37 ± 3.10 ^{b}	77.54 ± 3.08 ^{b}	17.87 ± 0.66 ^{a}	25.66 ± 0.47 ^{a}	43.53 ± 0.89 ^{a}
E	37.04 ± 1.62 ^{c}	44.13 ± 2.84 ^{a}	81.18 ± 3.64 ^{ab}	16.37 ± 0.47 ^{ab}	23.55 ± 0.31 ^{c}	39.91 ± 0.49 ^{b}
EG	33.51 ± 3.18 ^{d}	43.55 ± 1.49 ^{a}	77.06 ± 4.06 ^{b}	13.92 ± 1.16 ^{c}	24.05 ± 0.24 ^{c}	37.97 ± 1.14 ^{c}

Different lowercase letters (a–d) represent significant difference of each phenolic acid content among different treatments (n = 3; $p < 0.05$).

3. Discussion

Polyamines are involved in various biological processes in plants, such as growth, development and adaption to environmental stresses [12]. The previous study showed the application of exogenous spermidine increased the length and fresh weight of barley seedling under UV-B stress [9]. In the present study, the addition of either AG or EDC significantly ($p < 0.05$) elevated the content of endogenous spermidine through the inhibition on polyamine oxidase or AMADH (Figure 2D). However, significant ($p < 0.05$) decreases in the length and fresh weight of barley seedlings under UV-B stress after treatments of AG or EDC suggested the degradation product of spermidine, i.e., GABA, may have a role in the adaption of the germinating barley seed to abiotic stress, as this has been shown to occur in other plants [13–15].

GABA is involved in the regulation of various physiological plant responses, such as stress tolerance, growth regulation and signal transduction [16]. The results of the present study showed that GABA enhanced the growth of barley seedlings under UV-B stress by mitigating the increase in electrolyte permeability and MDA content caused by oxidative damage (Figure 2). Additionally, GABA promoted the respiration rate, which has been shown to positively correlate with seedling growth [17]. Meanwhile, exogenous GABA also increased the content of spermidine, which can be attributed to its suppression on polyamine oxidase as reported in the literature [18].

Phenolic compounds are important secondary metabolites ubiquitously present in plants, and their biosynthesis primarily begins with the conversion of the amino acid phenylalanine to cinnamic acid by the action of PAL [19]. Then, cinnamic acid undergoes hydroxylation through the catalyzation of C4H to form *p*-coumaric acid, which serves as a precursor for other phenolic acids generated via enzymatic reactions facilitated by 4CL, C3H, COMT and F5H [20]. Our previous study showed exogenous spermidine enhanced the biosynthesis of phenolic acids in barley during germination through elevating the activities of these enzymes and an opposite trend was observed when the inhibitor of endogenous spermidine synthesis was applied [9].

In the present study, although the content of endogenous spermidine was significantly ($p < 0.05$) increased when AG or EDC was applied, the content of phenolic acids, such as ferulic acid and *p*-coumaric acid were notably reduced (Table 1) even though total phenolic content did not significantly ($p > 0.05$) differ from that of the control seedling (Figure 1). The reduction in phenolic acid biosynthesis can be explained by the inhibition of the activity of PAL, the pivotal enzyme in phenolics biosynthesis [21]. Furthermore, it was shown that the application of exogeneous GABA can alleviate the suppression of EDC in the synthesis of phenolic acids. Taken together, these results suggest that the effect of spermidine on biosynthesis of phenolic acid in barley germinated under UV-B stress may mainly result from its degradation product (see Figure 5 below).

Phenolic acids in cereals exist in both free and bound forms, and free phenolic acids are generally located in the outer layer of the grains, while bound phenolic acids are covalently bound with structural molecules, such as celluloses, hemicelluloses and proteins [22]. In barley grains, the cell wall polysaccharides, such as arabinoxylans, can interact with phenolic acids, resulting in their retarded digestibility in the upper part of gastrointestinal tract [23]. Interestingly, treatments of AG and EDC resulted in the conversion of bound phenolic compounds and certain phenolic acids to the free forms even though their contents were not increased. In the seedlings treated with GABA, the contents of phenolic compounds or phenolic acids were mainly increased in the bound forms.

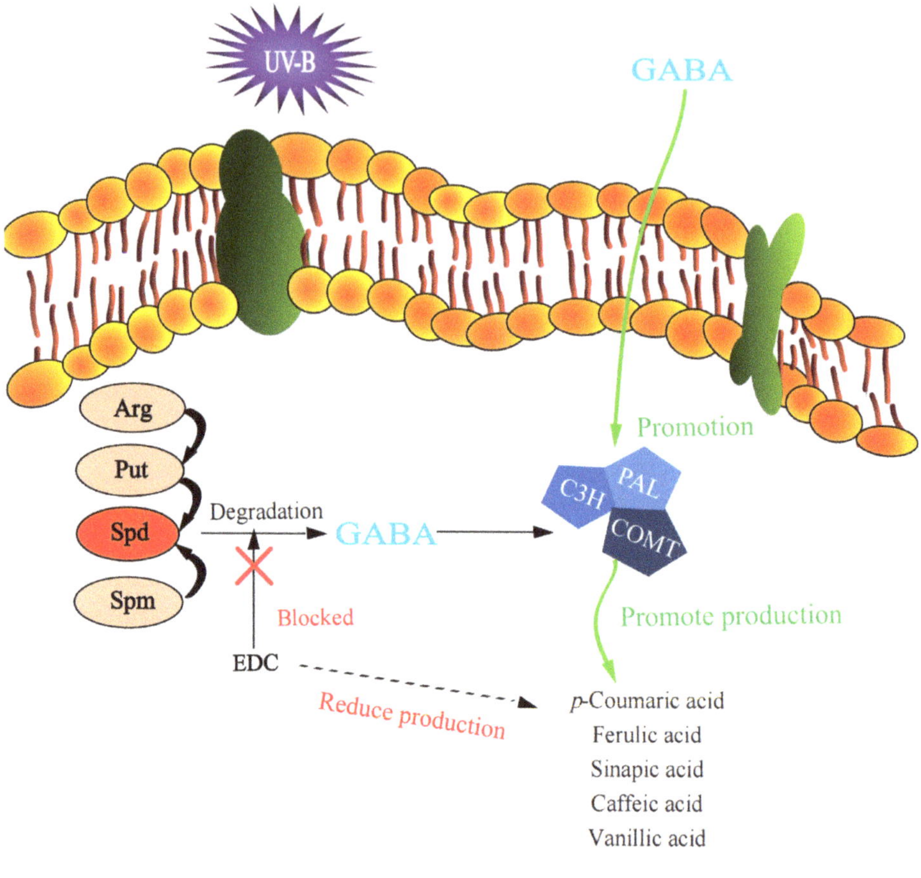

Figure 5. The speculative schema of the phenolic compound synthesis. Arg, Put, Spd and Spm were arginine, putrescine, spermidine and spermine, respectively.

4. Materials and Methods

4.1. Materials and Reagents

Barley grains (cultivar: Su Hullless 2) provided by the Jiangsu Academy of Agricultural Sciences were harvested in Jiangsu Province, China, in 2020 and were stored at −20 °C before use. All the chemicals were purchased from Sigma-Aldrich (St. Louis, MO, USA).

4.2. Experimental Design

The barley grains were immersed in a sodium hypochlorite solution (0.5% *v*/*v*, 15 min) after rinsing with deionized water. Then, the grains were rinsed with distilled water until they attained a neutral pH. After soaking for 6 h in deionized water at 25 °C, barley grains were evenly distributed in a dark incubator maintained at 25 °C to initiate the germination process. Following a one-day germination without light, grains underwent exposure to UV-B (intensity was 38 μW/cm^2) for a duration of 6 h every 24 h. Throughout the duration of the experiment, the culture solutions were changed at intervals of 8 h. Five distinct types of solution were utilized, which included distilled water as the control (CK), a solution of 2.5 mM AG (A), a solution of 5 mM GABA (G), a solution of 0.5 mM EDC (E) and a combination of 0.5 mM EDC and 5 mM GABA (EG). On the fourth day of germination, a portion of the barley seedlings were randomly selected for measurement as a fresh sample and the remaining seedlings were stored at −80 °C after being rinsed with deionized water.

4.3. Determination of Average Seedling Length and Weight, MDA Content, Electrolyte Leakage and Respiratory Rate

The lengths and weights of seedlings were measured by vernier caliper and balance, respectively, as reported in the previous study [9].

The MDA content, electrolyte leakage and respiratory rate were determined as described in the previous study [15]. For determination of MDA content, 10 fresh seedlings were mixed with 5 mL 10% trichloroacetic acid solution after grinding and centrifuged (10,000× *g*, 20 min). Then, 2 mL of the supernatant was combined with 2 mL of 0.67% thiobarbituric acid. After boiling for 30 min, the mixtures were centrifuged and the supernatants (4000× *g*, 15 min) were used for measurement of absorbances (450 nm, 532 nm and 600 nm). The MDA content was calculated as $c = 6.45 \times (A532 - A600) - 0.56 \times A450$.

For determination of electrolyte leakage, approximately 1 g of fresh seedlings that had been cut into lengths of 3 mm were immersed in 30 mL distilled deionized water. The electrolyte leakage was calculated based on the change in conductivity before and after the boiling water bath for 10 min. Respiratory rate was determined by the CO_2 production from 5 g of barley seedlings in a jar during 30 min, and expressed as mg $CO_2/(g \cdot h)$.

4.4. Determination of Spermidine Content

To determine the content of spermidine, approximately 1 g of barley seedlings were ground in an ice bath after being added to 4 mL of pre-cooled 5% HClO4 (*v*/*v*) and placed in the ice bath for 1 h. Subsequently, the mixture was centrifuged (15,000× *g*) for 20 min at 4 °C and supernatant was utilized for the determination of spermidine by the HPLC method according to Hu et al. [24]. Briefly, 1 mL supernatant was combined with 2 mL 2 M NaOH and 10 µL of benzoyl chloride. The mixture was vortexed for 30 s and then incubated in a water bath (37 °C) for 30 min. Subsequently, 2 mL of saturated NaCl solution was added to the mixture, followed by the addition of 3 mL of cold diethyl ether for the purpose of extracting benzoyl polyamide. The resulting mixture was then subjected to centrifugation (10,000× *g*, 5 min). From the resulting centrifuged mixture, 1.5 mL of the ether phase was collected and dried using nitrogen gas. To prepare the samples for analysis, 1250 µL of 64% methanol was added to the dried ether phase, which was then passed through a 0.45 µm organic phase filter membrane. The determination of both the samples and the standard solutions was performed using an Agilent 1200 Series HPLC system, equipped with an XDB C18 column (4.6 × 250 mm, 5 µm) and a SPERMIDINE-6AV UV detector operating at a wavelength of 254 nm. The mobile phase consisted of a mixture of methanol and ultrapure water in a ratio of 64:36, and the column temperature was set to 30 °C. The flow rate was maintained at 0.6 mL/min, and an injection volume of 20 µL was used.

4.5. Determination of the Contents of Phenolic Compounds and Phenolic Acids

Extraction and quantification of PCs were conducted according to Xie et al. [25] with slight modification. The freeze-dried samples were finely ground and sieved using a 60-mesh sieve. Approximately 1 g of the resulting powders were extracted with methanol (80%, *v*/*v*) and shaken in the absence of light for one hour under a nitrogen atmosphere at a temperature of 25 °C and a stirring speed of 200 rpm for 3 cycles. Following the shaking process, the mixture underwent centrifugation at 10,000× *g* (4 °C, 15 min). The supernatants from 3 cycles were evaporated by a rotary evaporator and the dried extracts were dissolved in 10 mL of 50% methanol (*v*/*v*) as the free phenolic compound.

The residues remaining from the extraction of the free phenolics were hydrolyzed by reaction with 2 M NaOH. The hydrolysis process occurred in a shaker (25 °C, 200 rpm) for a duration of 4 h, in the dark. The resulting hydrolysates were adjusted to an acidic pH (1.5–2.0) by 6 M HCl. Subsequently, the mixtures were extracted with ethyl acetate for 15 min followed by centrifugation at 10,000× *g* (4 °C, 5 min). The ethyl acetate layers collected from three rounds of centrifugation were evaporated in a rotary evaporator and the dried extracts were redissolved in 10 mL of 50% methanol (v/v) and protected with nitrogen gas, constituting the extract containing the bound phenolic compounds.

The contents of bound and free phenolic compounds were determined by the Folin phenol method. The contents of free and bound phenolic acids were measured by a high-performance liquid chromatography (HPLC) method (Shimadzu LC-20A, column C18 110A, 5 μm particle size, 4.6 × 150 mm). The mobile phases were water (contained 0.1% acetic acid) and methanol (contained 0.1% acetic acid) and the flow rate was set at 0.9 mL/min with a column temperature set at 35 °C. The quantification of phenolic acids was conducted by measuring their absorbance at a wavelength of 280 nm and subsequently calculating the concentrations based on standard curves.

4.6. Determination of Enzyme Activities

To determine the activities of enzymes related to phenolic compounds biosynthesis, ground barley seedlings were mixed with 50 mM phosphate buffer (pH 7.4), which contained 5 mM dithiothreitol, 0.1 mM ethylenediamine tetraacetic acid and 1% polyvinyl pyrrolidone. After centrifugation (15,000× *g*, 15 min), The activities of C3H, COMT and F5H in supernatants were determined by the corresponding kits that purchased from Jiancheng Bioengineering Research Institute (Nanjing, China). The activities of 4CL, PAL and C4H were determined according to the previous study [25].

4.7. ABTS and DPPH Radical Scavenging Activity Assay

ABTS and DPPH's radical scavenging abilities were determined according to the literature [25] with minor modifications. Briefly, extracts containing free (50 μL) or bound (80 μL) phenolics were combined with ABTS (3.0 mL) or DPPH (3.8 mL) solution, respectively. After thorough vortexing (30 min), the mixtures were placed at room temperature, in the absence of light. The methanol solution (50%) was employed as the blank control and a standard curve was generated using Trolox solution. Absorbance values at 734 nm were determined and radical scavenging activities of ABTS and DPPH were calculated based on standard curve.

4.8. Data Statistics and Analysis

The results were reported as the mean ± standard deviation (SD) of three biological replicates. Statistical analysis was conducted by SPSS 24.0 (IBM Corporation, Armonk, NY, USA). To assess significant differences among the samples, a one-way analysis of variance (ANOVA) was employed, and the significance level was set at $p < 0.05$.

5. Conclusions

The present study showed that inhibiting spermidine degradation resulted in severe oxidative damage and a significant decrease in phenolic acid content in barley seedlings exposed to UV-B treatment. However, GABA application was found to alleviate the inhibition of phenolic acid biosynthesis by elevating the activities of key enzymes related to this process and thus ameliorated the adverse effects of UV-B stress by reducing increased electrolyte permeability and MDA content. In conclusion, spermidine mitigated oxidative damage and elevated accumulation of phenolic compounds in barley seedlings through degradation to GABA.

Author Contributions: Conceptualization, R.Y.; methodology, C.L. and C.X.; software, Y.Z.; validation, Q.W.; formal analysis, C.X.; investigation, P.W.; resources, R.Y. and Y.Z.; data curation, P.W.; writing—original draft preparation, C.X. and C.L.; writing—review and editing, P.W. and R.Y.; visualization, C.X.; supervision, R.Y.; project administration, R.Y.; funding acquisition, Y.Z., Q.W. and R.Y. All authors have read and agreed to the published version of the manuscript.

Funding: This work was funded by the joint project of Nanjing Agricultural University–Tibet Agriculture and Animal Husbandry University (KYYZ202004); the Open Research Fund Program of Beijing Key Lab of Plant Resource Research and Development, Beijing Technology and Business University (PRRD−2022-YB2); the National Natural Science Foundation of China (Grant No. 31871725); and a project funded by the Priority Academic Program Development of Jiangsu Higher Education Institutions (PAPD).

Data Availability Statement: The data presented in this study are available on request from the corresponding author.

Conflicts of Interest: The authors declare no conflict of interest.

References

1. Mandal, S.M.; Chakraborty, D.; Dey, S. Phenolic acids act as signaling molecules in plant-microbe symbioses. *Plant Signal. Behav.* **2010**, *5*, 359–368. [CrossRef]
2. Hassan, M.A.; Xiang, C.; Farooq, M.; Muhammad, N.; Yan, Z.; Hui, X.; Yuanyuan, K.; Bruno, A.K.; Lele, Z.; Jincai, L. Cold stress in wheat: Plant acclimation responses and management strategies. *Front. Plant Sci.* **2021**, *12*, 676884. [CrossRef]
3. Dai, J.; Mumper, R.J. Plant phenolics: Extraction, analysis and their antioxidant and anticancer properties. *Molecules* **2010**, *15*, 7313–7352. [CrossRef] [PubMed]
4. Saibabu, V.; Fatima, Z.; Khan, L.A.; Hameed, S. Therapeutic potential of dietary phenolic acids. *Adv. Pharmacol. Sci.* **2015**, *2015*, 823539. [CrossRef]
5. Benincasa, P.; Falcinelli, B.; Lutts, S.; Stagnari, F.; Galieni, A. Sprouted grains: A comprehensive review. *Nutrients* **2019**, *11*, 421. [CrossRef]
6. Devi, C.B.; Kushwaha, A.; Kumar, A. Sprouting characteristics and associated changes in nutritional composition of cowpea (*Vigna unguiculata*). *J. Food Sci. Technol.* **2015**, *52*, 6821–6827. [CrossRef]
7. Zhao, Y.; Xie, C.; Wang, P.; Gu, Z.; Yang, R. GABA regulates phenolics accumulation in soybean sprouts under NaCl stress. *Antioxidants* **2021**, *10*, 990. [CrossRef]
8. Handa, A.K.; Fatima, T.; Mattoo, A.K. Polyamines: Bio-molecules with diverse functions in plant and human health and disease. *Front. Chem.* **2018**, *6*, 10. [CrossRef]
9. Xie, C.; Wang, P.; Gu, Z.; Yang, R. Spermidine alleviates oxidative damage and enhances phenolic compounds accumulation in barley seedlings under UV-B stress. *J. Sci. Food Agric.* **2022**, *103*, 648–656. [CrossRef]
10. Cona, A.; Rea, G.; Angelini, R.; Federico, R.; Tavladoraki, P. Functions of amine oxidases in plant development and defence. *Trends Plant Sci.* **2006**, *11*, 80–88. [CrossRef]
11. Li, L.; Dou, N.; Zhang, H.; Wu, C. The versatile GABA in plants. *Plant Signal. Behav.* **2021**, *16*, 1862565. [CrossRef]
12. Yu, Z.; Jia, D.; Liu, T. Polyamine oxidases play various roles in plant development and abiotic stress tolerance. *Plants* **2019**, *8*, 184. [CrossRef]
13. Al-Quraan, N.A. GABA shunt deficiencies and accumulation of reactive oxygen species under UV treatments: Insight from Arabidopsis thaliana calmodulin mutants. *Acta Physiol. Plant.* **2015**, *37*, 86. [CrossRef]
14. Kumar, N.; Dubey, A.K.; Upadhyay, A.K.; Gautam, A.; Ranjan, R.; Srikishna, S.; Sahu, N.; Behera, S.K.; Mallick, S. GABA accretion reduces Lsi-1 and Lsi-2 gene expressions and modulates physiological responses in *Oryza sativa* to provide tolerance towards arsenic. *Sci. Rep.* **2017**, *7*, 8786. [CrossRef]
15. Xie, C.; Sun, M.; Wang, P.; Yang, R. Interaction of Gamma-aminobutyric acid and Ca(2+) on phenolic compounds bioaccumulation in soybean sprouts under NaCl stress. *Plants* **2022**, *11*, 3503. [CrossRef]
16. Suhel, M.; Husain, T.; Pandey, A.; Singh, S.; Dubey, N.K.; Prasad, S.M.; Singh, V.P. An appraisal of ancient molecule gaba in abiotic stress tolerance in plants, and its crosstalk with other signaling molecules. *J. Plant Growth Regul.* **2023**, *42*, 614–629. [CrossRef]
17. Woodstock, L.W.; Grabe, D.F. Relationships between seed respiration during imbibition and subsequent seedling growth in *Zea mays* L. *Plant Physiol.* **1967**, *42*, 1071–1076. [CrossRef]
18. Seifikalhor, M.; Aliniaeifard, S.; Bernard, F.; Seif, M.; Latifi, M.; Hassani, B.; Didaran, F.; Bosacchi, M.; Rezadoost, H.; Li, T. γ-Aminobutyric acid confers cadmium tolerance in maize plants by concerted regulation of polyamine metabolism and antioxidant defense systems. *Sci. Rep.* **2020**, *10*, 3356. [CrossRef]
19. Marchiosi, R.; dos Santos, W.D.; Constantin, R.P.; de Lima, R.B.; Soares, A.R.; Finger-Teixeira, A.; Mota, T.R.; Oliveira, D.M.; Foletto-Felipe, M.d.P.; Abrahão, J.; et al. Biosynthesis and metabolic actions of simple phenolic acids in plants. *Phytochem. Rev.* **2020**, *19*, 865–906. [CrossRef]
20. Srinivasulu, C.; Ramgopal, M.; Ramanjaneyulu, G.; Anuradha, C.M.; Suresh Kumar, C. Syringic acid (SA)—A review of its occurrence, biosynthesis, pharmacological and industrial importance. *Biomed. Pharmacother.* **2018**, *108*, 547–557. [CrossRef]
21. Sharma, A.; Shahzad, B.; Rehman, A.; Bhardwaj, R.; Landi, M.; Zheng, B. Response of phenylpropanoid pathway and the role of polyphenols in plants under abiotic stress. *Molecules* **2019**, *24*, 2452. [CrossRef]
22. Shahidi, F.; Hossain, A. Importance of insoluble-bound phenolics to the antioxidant potential is dictated by source material. *Antioxidants* **2023**, *12*, 203. [CrossRef]

23. Cirkovic Velickovic, T.D.; Stanic-Vucinic, D.J. The role of dietary phenolic compounds in protein digestion and processing technologies to improve their antinutritive properties. *Compr. Rev. Food Sci. Food Saf.* **2018**, *17*, 82–103. [CrossRef]
24. Hu, X.; Zhang, Y.; Shi, Y.; Zhang, Z.; Zou, Z.; Zhang, H.; Zhao, J. Effect of exogenous spermidine on polyamine content and metabolism in tomato exposed to salinity–alkalinity mixed stress. *Plant Physiol. Biochem.* **2012**, *57*, 200–209. [CrossRef]
25. Xie, C.; Wang, P.; Sun, M.; Gu, Z.; Yang, R. Nitric oxide mediates γ-aminobutyric acid signaling to regulate phenolic compounds biosynthesis in soybean sprouts under NaCl stress. *Food Biosci.* **2021**, *44*, 101356. [CrossRef]

plants MDPI

Article

Metabolic Mechanisms Underlying Heat and Drought Tolerance in Lentil Accessions: Implications for Stress Tolerance Breeding

Noureddine El Haddad [1,2,*], Youness En-nahli [1,3,4], Hasnae Choukri [1,2], Khawla Aloui [1,5], Rachid Mentag [6], Adil El-Baouchi [4], Kamal Hejjaoui [4], Karthika Rajendran [7], Abdelaziz Smouni [2], Fouad Maalouf [8] and Shiv Kumar [9]

[1] International Center for Agricultural Research in the Dry Areas (ICARDA), Rabat 10112, Morocco; younes.en-nahli@um6p.ma (Y.E.-n.); hasnae_choukri@um5.ac.ma (H.C.); khawlaaloui111@gmail.com (K.A.)
[2] Laboratoire de Biotechnologie et de Physiologie Végétales, Centre de Recherche BioBio, Faculté des Sciences, Mohammed V University Rabat, Rabat 10112, Morocco; abdelaziz.smouni@um5.ac.ma
[3] Materials Science Center, Ecole Normale Supérieure, LPCMIO, Mohammed V University of Rabat, Rabat 10100, Morocco
[4] AgroBioSciences Program (AgBS), College of Sustainable Agriculture and Environmental Science (CSAES), University Mohammed VI Polytechnic (UM6P), Ben Guerir 43150, Morocco; adil.elbaouchi@um6p.ma (A.E.-B.); kamal.hejjaoui@um6p.ma (K.H.)
[5] Laboratory of Ecology and Environment, Ben M'Sick Faculty of Sciences, University Hassan II, Casablanca 20800, Morocco
[6] Biotechnology Research Unit, Regional Center of Agricultural Research of Rabat, National Institute of Agricultural Research (INRA), Rabat 10090, Morocco; rachidmentag@yahoo.ca
[7] Vellore Institute of Technology (VIT), VIT School of Agricultural Innovations and Advanced Learning (VAIAL), Vellore 632014, India; karthika.rajendran@vit.ac.in
[8] International Center for Agricultural Research in the Dry Areas (ICARDA), Beirut 1108 2010, Lebanon; f.maalouf@cgiar.org
[9] International Center for Agricultural Research in the Dry Areas (ICARDA), New Delhi 110012, India; sk.agrawal@cgiar.org
* Correspondence: noureddine_elhaddad@um5.ac.ma; Tel.: +212-5737-7920

Abstract: Climate change has significantly exacerbated the effects of abiotic stresses, particularly high temperatures and drought stresses. This study aims to uncover the mechanisms underlying heat and drought tolerance in lentil accessions. To achieve this objective, twelve accessions were subjected to high-temperature stress (32/20 °C), while seven accessions underwent assessment under drought stress conditions (50% of field capacity) during the reproductive stage. Our findings revealed a significant increase in catalase activity across all accessions under both stress conditions, with ILL7814 and ILL7835 recording the highest accumulations of 10.18 and 9.33 under drought stress, respectively, and 14 μmol H_2O_2 mg protein^{-1} min^{-1} under high temperature. Similarly, ascorbate peroxidase significantly increased in all tolerant accessions due to high temperatures, with ILL6359, ILL7835, and ILL8029 accumulating the highest values with up 50 μmol ascorbate mg protein^{-1} min^{-1}. In contrast, no significant increase was obtained for all accessions subjected to water stress, although the drought-tolerant accessions accumulated more APX activity (16.59 t to 25.08 μmol ascorbate mg protein^{-1} min^{-1}) than the sensitive accessions. The accessions ILL6075, ILL7814, and ILL8029 significantly had the highest superoxide dismutase activity under high temperature, while ILL6363, ILL7814, and ILL7835 accumulated the highest values under drought stress, each with 22 to 25 units mg protein^{-1}. Under both stress conditions, ILL7814 and ILL7835 recorded the highest contents in proline (38 to 45 μmol proline/g FW), total flavonoids (0.22 to 0.77 mg QE g^{-1} FW), total phenolics (7.50 to 8.79 mg GAE g^{-1} FW), total tannins (5.07 to 20 μg CE g^{-1} FW), and total antioxidant activity (60 to 70%). Further, ILL7814 and ILL6338 significantly recorded the highest total soluble sugar content under high temperature (71.57 and 74.24 mg g^{-1}, respectively), while ILL7835 achieved the maximum concentration (125 mg g^{-1}) under drought stress. The accessions ILL8029, ILL6104, and ILL7814 had the highest values of reducing sugar under high temperature with 0.62 to 0.79 mg g^{-1}, whereas ILL6075, ILL6363, and ILL6362 accumulated the highest levels of this component under drought stress with 0.54 to 0.66 mg g^{-1}. Overall, our findings contribute to a deeper understanding of the metabolomic responses of lentil to drought and heat stresses, serving as a valuable reference for lentil stress tolerance breeding.

Keywords: antioxidant activities; catalase; ascorbate peroxidase; superoxide dismutase; proline; flavonoids; sugars; drought stress; heat stress

1. Introduction

Crops are subjected to a wide range of abiotic stresses such as drought, heavy metals, UV rays, heat, and salinity, concurrently throughout their life cycles due to constantly changing environmental conditions, which reduce and limit their growth and productivity [1,2]. Extreme temperatures and drought stress have been identified as two of the most severe environmental risks restricting crop development, productivity, and, consequently, food security in a changing climate [3]. These two stresses contribute significantly to reducing the production of several important staple food crops, which provide 60% of the world's food energy [4–7]. Drought stress and high temperatures have been on the rise in recent decades, and this trend is expected to continue and worsen in the future [8]. Consequently, the resulting conditions are likely to have a significant impact, leading to increased malnutrition and micronutrient deficiencies, particularly in regions such as sub-Saharan Africa, South Asia, Central and South America, and small islands.

Lentil (*Lens culinaris* Medik.) holds a significant position as the third most vital cool-season food legume globally, after chickpea and pea [9], with an area of 5.5 million ha, a production of 5.6 million tons, and a productivity of 1004 Kg ha^{-1} in 2021 [10]. The cultivation of lentil predominantly takes place in rainfed environments, where drought and extreme temperatures are the two most challenging abiotic stresses that reduce its production in arid and semi-arid areas [11,12]. Compared to other leguminous crops, lentil has been found to exhibit greater sensitivity to high temperatures and drought stress, particularly during the reproductive stage, significantly diminishing its yield and productivity [13].

The impacts of extreme temperatures and water stress are receiving increased attention due to the substantial threat to the productivity of leguminous crops. These stresses adversely affect crucial factors such as pollen viability, fertilization, and ultimately, the formation of pods [14]. Elevated temperatures exceeding 32/20 °C (max/min) during the flowering and pod-filling stages have a harmful effect on the growth and grain filling of lentil, and by consequence, negatively impact both yield and nutritional quality [15,16]. Extensive research has been conducted on the effects of high temperatures and drought stress across various plant species, including wheat [17], barley [18], maize [19], chickpea [20], lentil [21], and soybean [22].

Both high temperatures and drought stress induce an imbalance between reactive oxygen species (ROS), such as superoxide radical ($O^{\bullet -}{}_2$), hydrogen peroxide (H_2O_2), hydroxyl radicals (OH•), singlet oxygen ($^1O_2$), and the scavenging activities of antioxidant system components in plants [23]. Abiotic stress causes excessive ROS production, which interferes with photosynthesis, increases lipid peroxidation and electrolyte leakage, damages the structures of cell membranes, and disrupts nucleic acids and proteins. As a result, cells are unable to perform their normal functions [24]. The accumulation of free amino acids, proteins, solutes proline, soluble sugars, and protective heat shock proteins and activation of enzymatic (ascorbate peroxidase, catalase, superoxide dismutase, and peroxidase) and non-enzymatic (glutathione, tocopherols, carotenoids, and ascorbate) antioxidant systems are some of the mechanisms used by plants [25,26]. In addition, secondary metabolites such as phenols, flavonoids, and tannins play a pivotal role in the adaptation and defense of plants against deleterious environmental factors like heat and drought stresses [27]. Accumulation of secondary metabolites in plants during stress conditions contributed to enhance physiological processes and stress tolerance via detoxifying ROS, ultimately promoting plant growth and development [28,29]. Hence, it is important to investigate and understand the enzymatic and non-enzymatic responses to high temperatures and drought stress to improve plants' stress tolerance.

In order to increase lentil production in areas that are severely impacted by climate change, it is essential to investigate the mechanisms linked to tolerance of high temperatures (>32/20 °C) and drought stress. Despite the importance of enzymatic and non-enzymatic responses in the lentil defense system against the increasing frequency of high temperatures and drought stress, there have been limited studies conducted on this aspect [30–32]. In this context, this investigation is the continuation of our previous efforts, and it provides additional insight into the responses of different metabolic traits to drought and heat stresses in lentil accessions. The objectives of the current study were (i) to uncover the metabolic mechanisms underlying heat and drought tolerance in lentil accessions selected from our previous research findings [33,34] and to (ii) identify superior lentil accessions best suited for heat and drought tolerance improvement in lentil germplasm for future programs.

2. Results

2.1. Antioxidant Enzyme Activity (CAT, APX, and SOD)

Seven accessions were subjected to the drought stress (50% field capacity), while twelve accessions were exposed to high-temperature stress (32/20 °C) during the reproductive phase. Our findings showed that stress significantly increased catalase (CAT) activity to higher than in the control plants, especially in the heat-tolerant accessions (6.91 to 16.32 μmol H_2O_2 mg protein^{-1} min^{-1}) and drought-tolerant accessions (4.12 to 10.18 μmol H_2O_2 mg protein^{-1} min^{-1}) compared to the sensitive accessions (Figure 1a,b). The accessions ILL6359, ILL7814, and ILL7835 maintained the highest CAT accumulation with an increase of 56 to 65%; however, ILL5919 and ILL6075 recorded the lowest values under high-temperature conditions with 6.24 and 6.91 μmol H_2O_2 mg protein^{-1} min^{-1}, respectively. In response to drought stress, the accessions ILL6363, ILL7835, and ILL7814 retained the highest CAT activity (9.32 to 10.18 μmol H_2O_2 mg protein^{-1} min^{-1}) with an increase of 27 to 47% compared to the control plants. In contrast, our findings showed that drought stress decreased CAT activity in the sensitive accessions ILL5919 and ILL7820 by 21 and 36%.

Under high temperature, ascorbate peroxidase (APX) significantly increased ($p < 0.05$) in all accessions except ILL5919, ILL6075, ILL6104, and ILL6338, which had values statistically similar to the control plants (Figure 1c). The tolerant accessions ILL6359, ILL8029, and ILL7835 were able to retain the highest APX activity (52.55 to 55.21 μmol ascorbate mg protein^{-1} min^{-1}) in response to high-temperature stress with a 39 to 69% increase compared to the control. In contrast, no significant increase was obtained for all accessions subjected to water stress, although the drought-tolerant accessions accumulated more APX activity (16.59 t to 25.08 μmol ascorbate mg protein^{-1} min^{-1}) than the sensitive accessions. The highest APX activity was recorded for the accessions ILL7835 and ILL7814 with 21.31 and 25.08 μmol ascorbate mg protein^{-1} min^{-1}, respectively (Figure 1d). In addition, our findings indicated that APX activity in the accessions ILL5919, ILL7820, ILL6075, and ILL6363 decreased by 46 to 51% as a result of drought stress.

Superoxide dismutase (SOD) activity significantly increased in all accessions except ILL5919 under heat stress and ILL7820 under drought stress when compared to the control (Figure 1e,f). Under high-temperature stress, the accessions ILL6075, ILL6104, ILL7223, ILL7814, and ILL8029 accumulated the highest SOD activity (20 to 25 units mg protein^{-1}) with a 39 to 65% increase compared to control plants. However, the drought-tolerant accessions ILL6362, ILL6363, ILL7814, and ILL7835 retained higher SOD activity 21.82 to 27.43 units mg protein^{-1} than sensitive accessions (9.79 to 16.02 units mg protein^{-1}) in response to drought stress conditions.

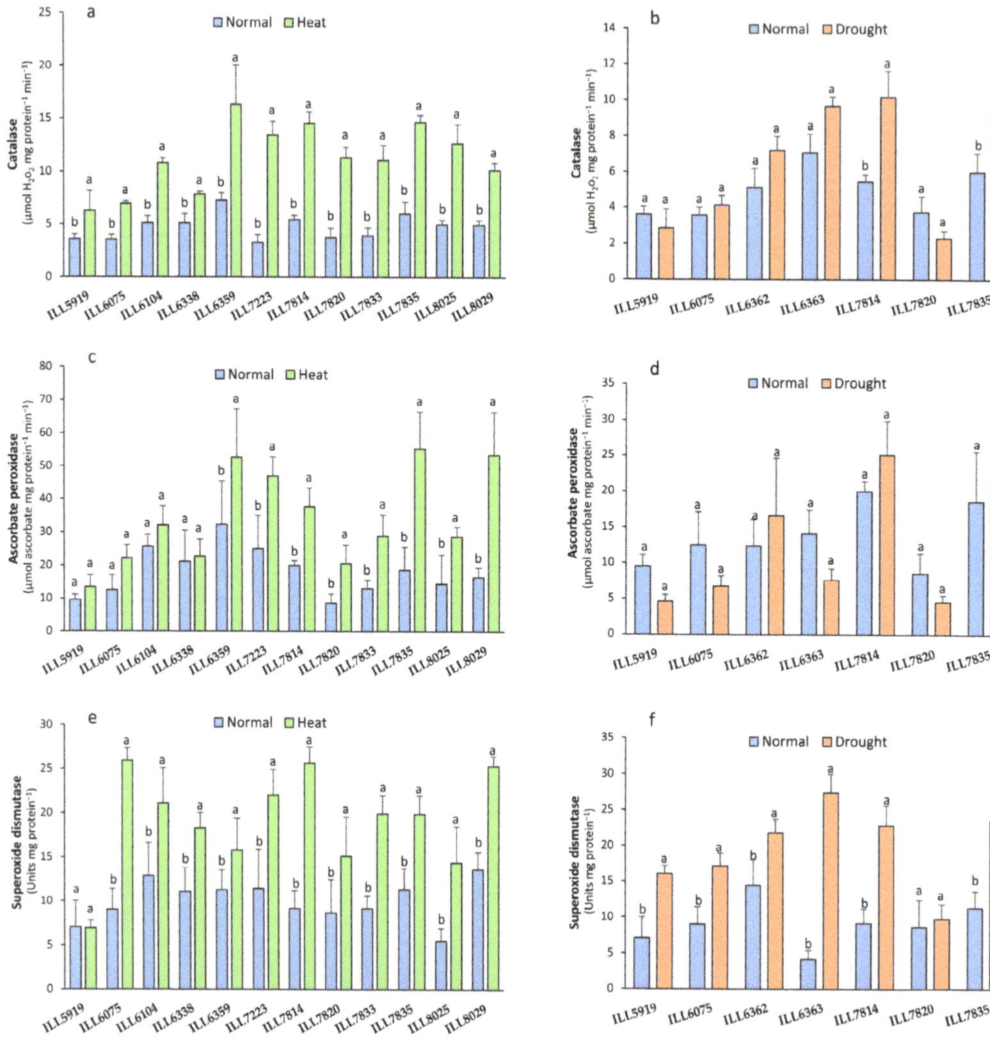

Figure 1. Catalase (CAT), ascorbate peroxidase (APX), and superoxide dismutase (SOD) activities of lentil accessions in response to high-temperature (**a**,**c**,**e**) and drought stress (**b**,**d**,**f**). Significant differences between means of tested accessions under not-stressed and stressed conditions using the LSD test at $p < 0.05$ are indicated by different letters on the bars.

2.2. Effect of High-Temperature and Drought Stress on Proline Content

Proline content (PC) significantly increased ($p < 0.05$) due to high-temperature and drought stress in all accessions except the susceptible accessions (ILL5919 and ILL7820), which had statistically similar PC under both stresses in comparison with the control (Figure 2). Under high-temperature stress, the maximum value of PC was found in ILL7814, ILL7833, and ILL7835 (up to 45 μmol proline/g FW with a 55 to 71% increase compared to control plants). However, ILL6362, ILL7814, and ILL7835 had the highest PC with 34.63, 38.09, and 35.78 μmol proline/g FW, respectively, under drought stress conditions.

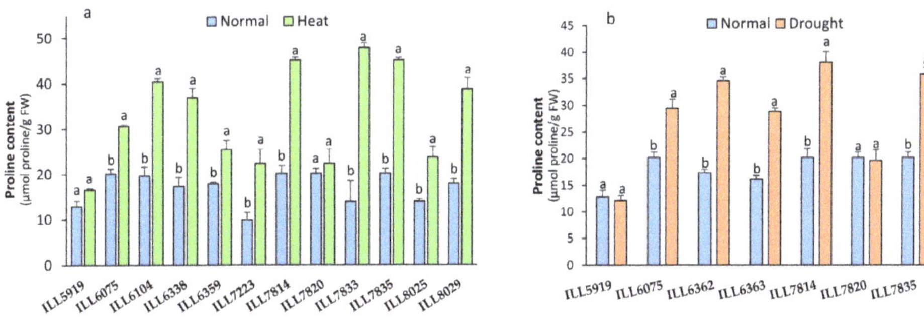

Figure 2. Proline content (PC) response to heat stress (**a**) and drought stress (**b**) in lentil accessions. Significant differences between means of tested accessions under not-stressed and stressed conditions using the LSD test at $p < 0.05$ are indicated by different letters on the bars.

2.3. *Total Flavonoid Content Response to High-Temperature and Drought Stress*

Total flavonoid content (TFC) significantly increased ($p < 0.05$) in five accessions (ILL6075, ILL6104, ILL6338, ILL7833, and ILL7835) in response to high temperature, whereas all accessions except ILL5919 exposed to drought stress increased TFC compared to normal conditions (Figure 3). Under high-temperature stress conditions, ILL6104, ILL6338, and ILL7835 increased TFC by up 70% and accumulated 0.46, 0.64, and 0.77 mg QE g^{-1} FW, respectively. Under drought stress, accessions ILL6362, ILL6363, and ILL7814 had the highest TFC (up to 0.22 mg QE g^{-1} FW) with an increase of 8 to 60% compared to control plants.

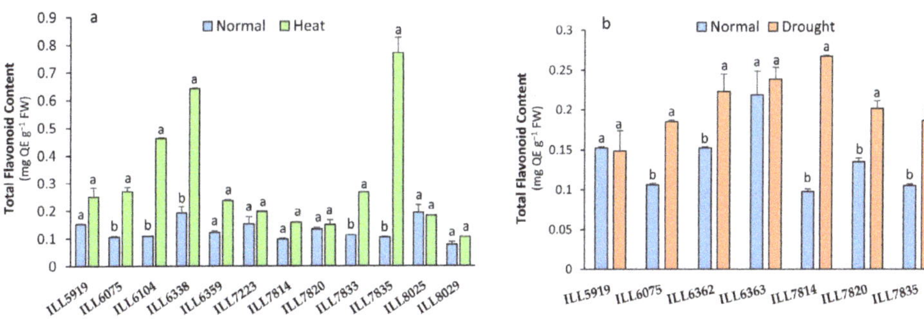

Figure 3. Total flavonoid content (TFC) of lentil accessions in response to heat stress (**a**) and drought stress (**b**). Significant differences between means of tested accessions under not-stressed and stressed conditions using the LSD test at $p < 0.05$ are indicated by different letters on the bars.

2.4. *Total Phenolics Content Response to High-Temperature and Drought Stress*

When subjected to high-temperature stress, all accessions displayed a significant increase in their total phenolics content (TPC), with the exception of ILL5919, which did not exhibit a significant increase compared to the control (Figure 4a). The TPC accumulation was greatest in ILL6104, ILL6338, ILL7814, ILL7835, and ILL8029 (7.50 to 8.79 mg GAE g^{-1} FW) with a 77 to 90% increase compared to control plants. However, only the two tolerant accessions ILL7814 and ILL7835 significantly increased TPC by up to 85% in response to drought stress (Figure 4b). In drought conditions, the average TPCs of ILL7814 and ILL7835 were 7.67 and 8.79 mg GAE g^{-1} FW, respectively.

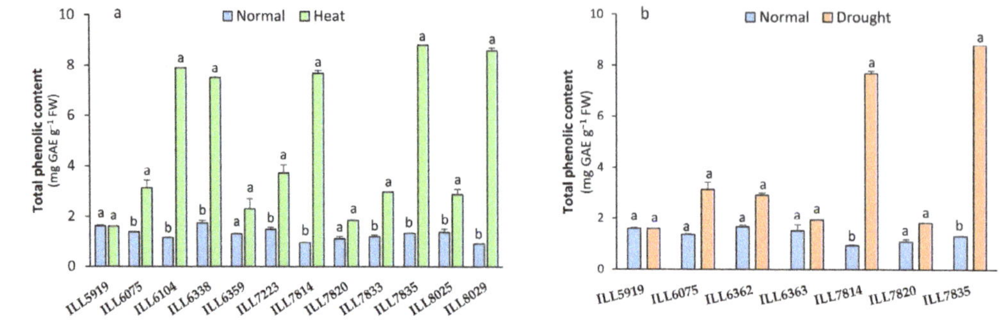

Figure 4. Total phenolic content (TPC) of lentil accessions in response to heat stress (**a**) and drought stress (**b**). Significant differences between means of tested accessions under not-stressed and stressed conditions using the LSD test at $p < 0.05$ are indicated by different letters on the bars.

2.5. Tannin Content Response to High-Temperature and Drought Stress

High-temperature stress caused tannin content (TC) to increase significantly ($p < 0.05$) in comparison to the control in all accessions, while ILL5919 showed no significant difference (Figure 5a). The highest TC accumulation under high-temperature stress was found in accessions ILL6104, ILL7814, ILL7835, and ILL8029 (20.31 to 23.92 µg CE g^{-1} FW), which increased their TC by 80 to 90% compared to control. With drought stress, on the other hand, TC significantly decreased in accessions ILL5919, ILL6075, and ILL6362 by 65 to 78%, while it increased TC in ILL7814 and ILL7835 by up 50% with an average of 5.07 and 8.37 µg CE g^{-1} FW, respectively (Figure 5b).

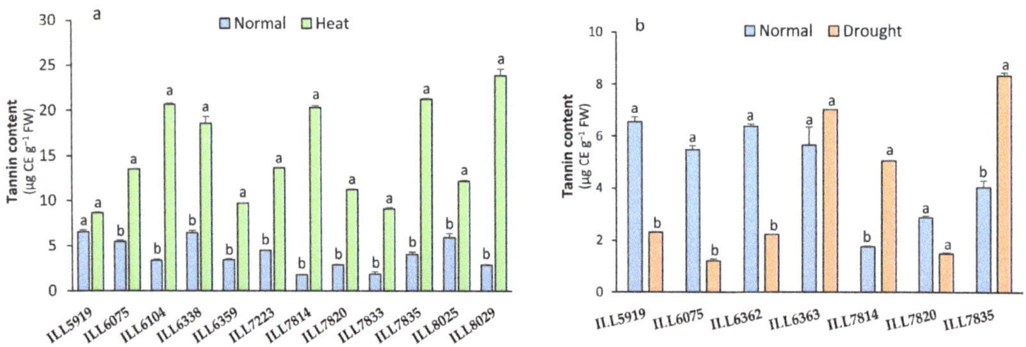

Figure 5. Tannin content (TC) in response to heat stress (**a**) and drought stress (**b**) in lentil accessions. Significant differences between means of tested accessions under not-stressed and stressed conditions using the LSD test at $p < 0.05$ are indicated by different letters on the bars.

2.6. Total Soluble Sugar and Reducing Sugar under Stressed Conditions

Total soluble sugar (TSS) were significantly augmented by 30 to 66% in eight accessions in response to high-temperature conditions, with the highest accumulation recorded in ILL7814 and ILL6338 with 71.57 and 74.24 mg g^{-1}, respectively (Figure 6a). Furthermore, TSS accumulation significantly increased by up 60% in three accessions (ILL5919, ILL6075, and 7835) in response to drought stress (Figure 6b). Among all the accessions, ILL7835 demonstrated the highest TSS content, with an average of 125 mg g^{-1} in response to drought stress conditions.

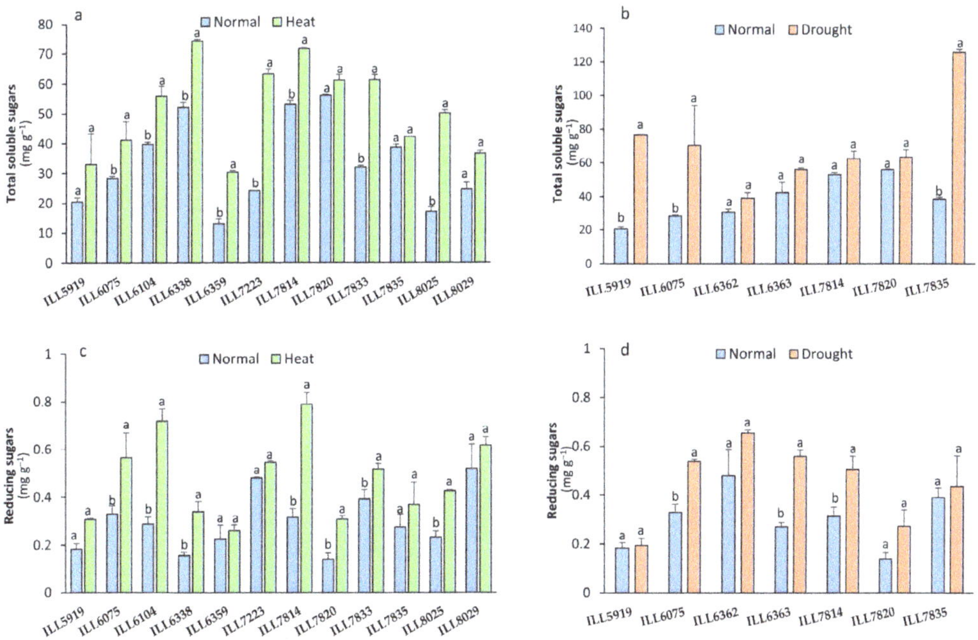

Figure 6. Total soluble sugar (TSS) and reducing sugar (RS) in response to heat stress (**a**,**c**) and drought stress (**b**,**d**) in lentil accessions. Significant differences between means of tested accessions under not-stressed and stressed conditions using the LSD test at $p < 0.05$ are indicated by different letters on the bars. The colored bars represent the treatments as blue for non-stress conditions, green for high temperatures, and orange for drought stress.

Compared to normal conditions, reducing sugar (RS) significantly increased ($p < 0.05$) by 40 to 60% in response to high-temperature stress in seven accessions (Figure 6c). The accessions ILL8029, ILL6104, and ILL7814 accumulated the highest RS under high temperature, with 0.62, 0.72, and 0.79 mg g^{-1}, respectively. On the other hand, the accessions ILL6075, ILL6363, and ILL7814 significantly increased RS by 30 to 50% under drought stress, whereas ILL5919, ILL6362, ILL7820, and ILL7835 had statistically similar RS compared to normal conditions (Figure 6d). Under drought stress, ILL6075, ILL6363, and ILL6362 had the highest RS with 0.54, 0.56, and 0.66 mg g^{-1}, respectively, whereas ILL5919 and ILL7820 had the lowest RS with 0.20 and 0.27 mg g^{-1}, respectively.

2.7. Total Antioxidant Activity Response to High-Temperature and Drought Stress

Total antioxidant activity (TAA) measurements in all 12 accessions under high-temperature stress showed that seven accessions had significantly ($p < 0.05$) higher TAA than the control. The heat-tolerant accessions ILL6104, ILL6338, ILL7814, ILL7835, and ILL8029 had the top TAA with 60 to 74% of DPPH inhibition in high-temperature conditions (Figure 7). However, TAA of accessions ILL7814, ILL6363, and ILL7835 significantly increased by 50 to 67% in response to drought stress, with 60.05, 62.79, and 65.82% of DPPH inhibition, respectively. Under the influence of both high-temperature and drought stress, the susceptible accessions ILL5919 and ILL7820 displayed the lowest accumulation of TAA.

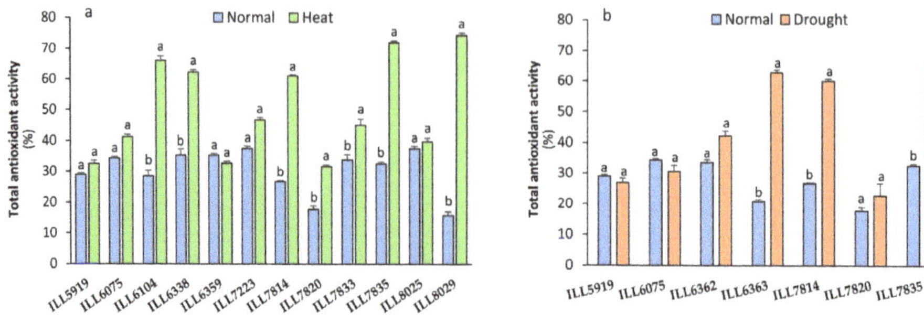

Figure 7. Total antioxidant activity (TAA) in lentil accessions expressed as a percentage of DPPH inhibition in response to heat stress (**a**) and drought stress (**b**). Significant differences between means of tested accessions under not-stressed and stressed conditions using the LSD test at $p < 0.05$ are indicated by different letters on the bars.

2.8. Genotypic Correlation under Normal, High-Temperature, and Drought-Stress Conditions

In normal conditions, TFC exhibited a highly significant positive correlation ($p < 0.01$) with TC and TPC, while it was negatively correlated at $p < 0.05$ with SOD (r = −0.57) (Figure 8a). Furthermore, a remarkably significant positive correlation ($p < 0.001$) between TPC and TC was found (r = 0.88). The heatmap analysis showed that CAT, APX, SOD, and RS were grouped together in the same cluster, while TSS and PC were found in the second cluster (Figure 8b). Under normal conditions, TC, TPC, TFC, and TAA are grouped together. Additionally, genotypic correlation identified three groups. The first group included two heat-tolerant accessions (ILL8029 and ILL6104), two heat- and drought-tolerant accessions (ILL7814 and ILL7835), and one susceptible accession (ILL7820). The second group consisted of two moderately heat- and drought-tolerant accessions (ILL7223 and ILL8025), one drought-tolerant accession (ILL6075), one heat-tolerant accession (ILL7833), and one susceptible accession ILL5919. Three accessions were grouped in one group: two moderately tolerant (ILL6363 and ILL6359), one drought-tolerant (ILL6362), and one heat-tolerant (ILL6338).

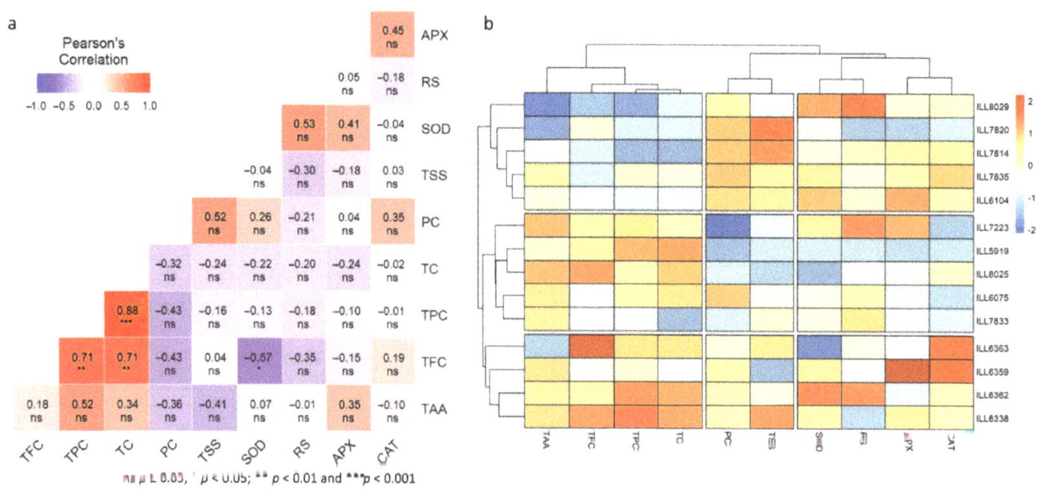

Figure 8. Pearson's correlation coefficients (**a**) and dendrogram along with heatmap (**b**) between various traits of lentil accessions under normal conditions.

In high-temperature conditions, SOD displayed a notable positive correlation with TAA, TC, PC, and RS, whereas CAT had positive correlation ($p < 0.01$) with APX ($r = 0.72$) (Figure 9a). Moreover, TPC, TAA, and TC were found to have a highly positive correlation ($p < 0.001$). The heatmap analysis identified that CAT and APX were clustered together, while TFC was clustered independently (Figure 9b). The remaining components formed the third group. Three heat-tolerant accessions (ILL8029, ILL6104, and ILL6338) and two accessions tolerant to high-temperature and drought stress (ILL7814 and ILL7835) were clustered together in the same group based on genotypic correlation. For almost all the tested components, these accessions had the highest accumulations. The second group contained two moderately tolerant (ILL7223 and ILL8025), one heat-tolerant (ILL7833), one drought-tolerant (ILL6075), and one sensitive (ILL7820) accession. The third group included one susceptible accession (ILL5919) and one moderately tolerant accession (ILL6359).

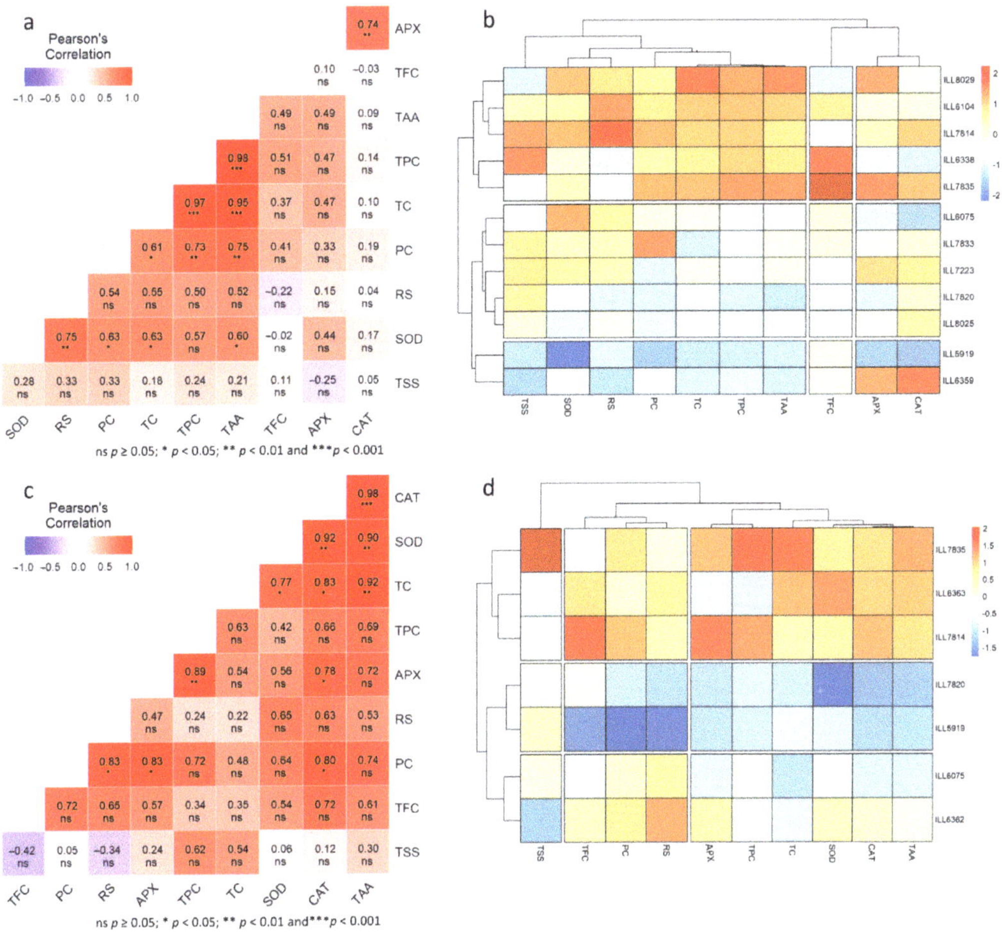

Figure 9. Pearson's correlation coefficients (**left**) and dendrogram along with heatmap (**right**) between various traits for high-temperature (**a**,**b**) and drought-stress (**c**,**d**) conditions.

Under drought stress conditions, SOD was positively correlated with CAT, TAA, and TC (Figure 9c). In addition, APX was found to be positively related to CAT, PC ($p < 0.05$), and TPC ($p < 0.01$). Furthermore, TAA exhibited a positive correlation with TC at $p <$

0.001, whereas PC had a positive correlation with RS (r = 0.83) under drought stress. The heatmap analysis revealed that CAT, SOD, APX, TPC, TAA, and TC were clustered in the same group, whereas RS, PC, and TFC formed the second group (Figure 9d). In the third group, TSS was clustered individually. Furthermore, accessions ILL7835, ILL6363, and ILL7814 were found in the same cluster, while susceptible accessions ILL7820 and ILL5919 were found in a separate group. The accessions ILL6075 and ILL6362, which are drought-tolerant, were found in the third group.

3. Discussion

In the near future, alterations in climatic conditions are anticipated to lead to a frequent occurrences of droughts and extreme temperatures, thereby posing a substantial threat to food security [35]. Both stress factors impose a detrimental effect on metabolic processes, affecting vital aspects like photosynthesis, water relations, nutrient absorption, and consequently, grain yield [36]. Hence, the selection of tolerant genotypes to heat and drought stresses and the comprehension of the underlying mechanisms by plant breeding scientists are extremely important for increasing agricultural productivity in arid and semi-arid environments. In the current study, fourteen lentil accessions were subjected to high-temperature stress (32/20 °C) throughout the critical reproductive stage, while seven accessions were assessed under drought stress by reducing field capacity to 50%. The two sets of accessions were chosen based on their tolerance to high temperatures and drought stress discovered from our previous investigations [33,34]. Under both stressed conditions, the accessions ILL5919 and ILL7820 were used as susceptible controls. The outcomes of our study revealed that the application of high-temperature and drought-stress conditions induced substantial modifications in both enzymatic and biochemical processes in the tested lentil accessions. Remarkably, the concentrations of ascorbate peroxidase, catalase, and superoxide dismutase exhibited a significant increase in the heat-tolerant accessions as compared to the heat-sensitive ones. This suggests a stronger enzymatic response to high-temperature stress in the heat-tolerant accessions. Tolerant accessions ILL7814 and ILL7835 significantly increased CAT and SOD activity in response to heat and drought stresses; however, there was no significant difference in APX across all accessions exposed to drought stress. The moderately tolerant accession ILL6363 accumulated the highest SOD under drought stress, whereas the tolerant accessions ILL6075 and ILL6362 had significant SOD activity. Additionally, it was observed that under both high-temperature and drought-stress conditions, the susceptible accessions ILL5919 and ILL7820 exhibited the lowest antioxidant activity, which suggests their reduced capacity to combat oxidative stress under these challenging conditions. When subjected to high-temperature stress, the moderately heat-tolerant accessions (ILL6075, ILL6359, ILL7223, and ILL8025) displayed a significant increase in the levels of CAT, APX, and SOD enzymes compared to normal conditions. SOD activity was also significantly increased in the ILL6363 accession, which is moderately drought-tolerant.

Both drought and high-temperature stresses can result in elevated production of reactive oxygen species, which hinder various metabolic processes in plants, including those related to photosynthetic components. As a protective mechanism, plants deploy antioxidant enzymes such as CAT, APX, and SOD as their primary defense to scavenge H_2O_2 with different mechanisms and suppress its toxic impacts [37]. Our study observed heightened activities of antioxidant enzymes in nearly all tolerant and moderately tolerant lentil accessions, as a response to the stresses of high temperature and drought, which was in agreement with the outcomes of other investigations [38–42]. Under adverse environmental conditions, complex antioxidant enzyme activities play a critical role in overcoming uncontrolled reactive oxygen species generation and protecting plants from oxidative damage [23]. Through our research, we discovered significant variations in CAT, APX, and SOD activities between tolerant and sensitive accessions. In fact, the tolerant accessions exhibited higher levels of antioxidant enzyme activities compared to the susceptible ones. In lentil, Chakraborty and Pradhan [43] also observed an elevation in APX, SOD, and CAT levels in response to high-temperature stress, with the tolerant accessions demonstrating a more pronounced increase. However, our findings indicate that APX activity did not exhibit a significant increase in lentil accessions under drought stress. Furthermore, the APX activity for four specific accessions (ILL5919, ILL7820, ILL6363, and ILL6075) displayed a non-significant decrease, suggesting a lower level of oxidative stress due to induced antioxidant enzyme activities and detoxified reactive oxygen species, as elucidated by Singh et al. [41]. Reduction in CAT and APX activities was also reported in two grasses and two legume species that were subjected to high temperatures and drought stress [44]. Nonetheless, in comparison to the sensitive controls, the majority of drought-tolerant accessions exhibited notably higher levels of APX and CAT. Similar results regarding antioxidant enzyme activities were reported in other studies in response to heat and drought stresses [45,46]. In addition, our findings confirmed the results of Sarker and Oba [42] in demonstrating that accessions with higher tolerance to drought stress had higher SOD activity contribution compared to accessions with lower tolerance to water stress. Therefore, these enzymatic antioxidants (CAT, APX, and SOD) can be used as economical and effective metabolic indicators for either screening or enhancing lentil germplasm for heat and drought tolerance in the early growth stages.

In lentil, Hosseini et al. [47] identified two groups of SOD genes (Cu/Zn-SOD and Mn-SOD) up-regulated in mitochondria, chloroplast, and cytosol under drought- and heat-stress conditions. Furthermore, previous investigations also showed that SOD activity increased in drought-tolerant genotypes of lentil under drought stress [48], and chloroplastic and cytosolic Cu/Zn-SOD and Mn-SOD were up-regulated under drought stress in soybean [49], wheat [50], and rice [51]. In addition, several cytosolic and chloroplastic APX genes such as lentil homolog of Atperoxidase 1, 12, 52, 67, and 17 were commonly up-regulated under drought and heat stress, and they were found to be important genes used as indicators of tolerance to heat and drought-stress conditions [47]. Rahman et al. [52] also reported that the expressions of APX, Cu/Zn-SOD, and CAT genes are the key candidates in antioxidant defense, which might be useful in molecular breeding strategies for heat and drought tolerance.

Proline stands as one of the most crucial osmoprotectants in plants, serving as a solute that adapts to alterations in the cell's water environment and mitigating the adverse impacts of high temperatures and drought stress on plants [53]. After the removal of stress relief, proline generated during stressful conditions can additionally function as a reservoir for energy and ammonia sources directly influencing plant metabolism [4]. Therefore, higher accumulations of proline are associated with increased stress tolerance. Proline also acts as an antioxidative defense molecule, and it scavenges ROS to activate specific gene functions that are essential for the plant's recovery from stresses [54]. In the current investigation, we observed a significant increase in proline content in all tested accessions subjected to high-temperature and drought stresses, except for the susceptible accessions ILL5919 and ILL7820, which displayed non-significant increases under these stress conditions. These findings align with the observations of Singh et al. [41] who reported higher proline levels in heat-tolerant lentil accessions when compared to sensitive ones. Increasing proline content was also reported by Hosseini et al. [47] in response to heat and drought stresses, and it was identified the hub gene P5CS2 (delta 1-pyrroline-5-carboxylate synthase 2) as the major transcription factor involved in lentil tolerance to these stresses. Furthermore, Hayat et al. [55] highlighted the role of proline in preventing photodamage to chloroplast thylakoid membranes by effectively scavenging and/or reducing the production of reactive oxygen species. Moreover, the utilization of exogenous proline has been shown to regulate drought stress by stimulating plant growth, which is achieved by triggering the antioxidant mechanism, reducing oxidative damage, enhancing the synthesis of compatible solutes, and accelerating proline accumulation, resulting in enhanced photosynthesis and yield attributes [56]. El-Beltagi et al. [57] found that the sequence combination of antioxidant and proline increased the antioxidant defense system and osmolyte synthesis, resulting in accelerated plant growth in chickpea under drought stress. Exogenous proline has the remarkable capacity to interact with enzymes, effectively preserving protein aggregation and preventing thermal denaturation [58].

The soluble sugar and reducing sugar are also among the most important osmotic regulators in plants, serving diverse cellular functions including energy storage, osmoprotectants, and signals for adapting to abiotic stresses [59]. Under the influence of high-temperature and drought stress, the concentrations of total soluble sugar increased in almost all examined accessions within our study. The same trend was also reported by Shah et al. [31] who observed a significant rise in soluble sugar content and proline content in lentil cultivars exposed to water-deficit stress conditions. Furthermore, Sehgal et al. [60] reported similar outcomes, demonstrating a significant increase in soluble sugar in response to heat and drought stresses in lentil. These results confirmed that heat and drought stresses increased the soluble sugar content in plants and thus reduced the osmotic potential and maintained the normal water demand, as described by Li et al. [61]. On the other hand, reducing sugar content was reported to be correlated with grain yield under heat and drought stresses [32]. Sita et al. [62] found that heat-tolerant lentil accessions accumulated more reducing sugar than sensitive ones, which they attributed to an increase in acid invertase activity (hydrolysis of sucrose into fructose and glucose) in the tolerant accessions, which sustained their redox ionic homeostasis.

In addition, our results revealed that tolerant accessions exposed higher concentrations of total flavonoids, total phenolics, and total tannins than sensitive accessions under both high-temperature and drought-stress conditions. A drastic difference has been also observed in the concentrations of phenols, flavonoids, and tannins of control and drought-stress soybean cultivars [49]. These secondary metabolites function as osmoprotectants and play a crucial role in scavenging free radicals against oxidative damage caused by abiotic stress [63]. According to Hassan et al. [64], total phenolic and flavonoid contents are strong antioxidants, and their accumulation in plants has a strong association with drought-stress tolerance. Ghazghazi et al. [54] concluded that polyphenolic compounds serve as subtracts for the synthesis of enzymes in the antioxidant defense network. Further, the second hypothesis suggests that plants use antioxidant enzymes as their primary method of reducing ROS levels rather than phenolic compounds [65]. Total antioxidant activity significantly increased due to both high-temperature and drought stress. Under stressful conditions, tolerant accessions accumulated a greater total antioxidant concentration than susceptible ones. The heat tolerant accessions ILL6104, ILL6338, ILL7814, and ILL8029 had the highest accumulations of total antioxidant activity in high-temperature conditions, while ILL6363, ILL7814, and ILL7835 were the top accumulators under drought stress, inhibiting DPPH by up to 60%. In the presence of high-temperature and drought stress, the increased antioxidant capacity in tolerant accessions can help mitigate oxidative damage to membranes and proteins, thereby preserving the functionality of organelles and promoting overall plant performance [66].

Based on our findings, proline had a significant positive correlation with CAT and APX under drought stress, and with SOD and TAA under high temperatures. These results may be related to the adequate actions of these antioxidants, which can remove ROS and reduce the harmful effects of stress conditions. These outcomes are consistent with the findings of previous reports in lentil [41,67] and other species [38,68,69]. In addition, we observed a significant correlation between TPC and TAA under high-temperature stress conditions. Similar findings were found in several studies [70,71], suggesting that the total antioxidant activity largely depends on the presence of large amounts of phenolic compounds [72]. Our results also revealed a positive correlation between APX and both CAT and TPC under drought stress. Furthermore, TPC accumulation in plants has been shown to have a strong correlation with their resistance to stress [4]. Similarly, TC was positively correlated with TAA, CAT, and SOD under drought stress and with TAA, SOD, PCA, and PC under conditions of heat stress. Total tannin content in plants has been shown to have significant antioxidant activity for removing and preventing ROS formation [73]. Reducing sugar also had a positive correlation with SOD at high-temperature stress and proline content during drought stress. Hussain et al. [4] found that reducing sugar played a significant role in osmoregulation, which protects plants from environmental conditions.

4. Materials and Methods

4.1. Plant Material

A set of 14 diverse lentil accessions were selected based on findings from our previous studies [33,34]. The material included four heat-tolerant accessions, two drought-tolerant accessions, two heat- and drought-tolerant accessions, four moderately heat- and drought-tolerant accessions, and two accessions sensitive to both stresses. The material was obtained from the International Center for Agricultural Research in the Dry Areas (ICARDA). The information regarding these accessions can be found in Table 1.

Table 1. Description of origin, IG number, and classification of lentil accessions used in the current study.

Accession	IG Number	Origin	Classification
ILL5919	69528	Ethiopia	HDS
ILL6075	70130	Pakistan	DT
ILL6104	70159	Pakistan	HT
ILL6338	71270	Pakistan	HT
ILL6359	71291	Pakistan	MT
ILL6362	71294	Pakistan	DT
ILL6363	71295	Pakistan	MT
ILL7223	75942	Nepal	MT
ILL7814	114931	Nepal	HDT
ILL7820	114951	Nepal	HDS
ILL7833	115006	Nepal	HT
ILL7835	115010	Nepal	HDT
ILL8025	117726	Pakistan	MT
ILL8029	117734	Pakistan	HT

HT, heat-tolerant; DT, drought-tolerant; MT, moderately tolerant to heat and drought; HDT, tolerant to heat and drought; HDS, susceptible to heat and drought; IG, ICARDA germplasm.

4.2. Experiments Details

The experiment was conducted in a controlled glasshouse at ICARDA, Rabat, Morocco (latitude 33°58′45.7″ N longitude 6°51′44.9″ W, and altitude 52 m). In plastic pots (20 cm in diameter and 18 cm in height) filled with compost garden soil and sandy-loam soil in a ratio of 2:2 *w*/*w*, seeds were planted, and each filled pot weighed a total of 3 Kg. In each pot, five seeds were initially sown at a depth of 2 cm and were thinned to three per pot after emergence. Three treatments were used: control (C), drought stress (D), and heat stress (H). Randomized complete block design with four replications was used for each treatment. Plants were kept in normal conditions, which included 25/18 °C (day/night) temperature, 90/60% relative humidity (RH), and 450 μmol m^{-2} s^{-1} light intensity. During development, plants were fully irrigated daily between 09:00 and 10:00 h; however, irrigation was restricted during the flower initiation stage for drought treatment to impose water stress. At the first day of anthesis phase (i.e., first flower appearance), plants in heat treatment were exposed to temperatures above 32 °C for four hours in a growth chamber at 65–70% RH for a period of seven days. By maintaining 50% of the relative water content of the control conditions for nine days, plants under drought stress were subjected to severe water stress. Whole plants were collected in liquid nitrogen and kept at −80 °C for various analyses after being subjected to high temperatures and drought treatments during the reproductive period. Concurrently, plants grown under controlled conditions were also collected.

4.3. *Total Phenolic Content*

Total phenolic content was measured using the technique outlined by Waterhouse [74]. In a 2 mL Eppendorf tube, 200 µL of extract from whole plant and 1 mL of Folin–Ciocalteu reagent were added. After five minutes of incubation, 800 µL of aqueous sodium carbonate 20% was added to the mixture. The mixture was incubated in the dark at room temperature for 60 min. At 765 nm, the absorbance was measured with a Thermo scientific spectrophotometer. The total phenolic concentration was calculated using a gallic acid standard curve. The results were expressed as mg gallic acid equivalents (GAE) per gram of fresh weight. Total phenolic content (TPC) was calculated as follows [33]:

$$(\mathrm{TPC})_{(\mathrm{mg}\,\frac{\mathrm{GAE}}{\mathrm{g}}\,\mathrm{dw})} = (\mathrm{C} \times \mathrm{V})/\mathrm{m} \quad (1)$$

where C is the total phenolic concentration (mg L^{-1}), V is the volume of the extraction, and m is the fresh weight of the plant material (g).

4.4. *Estimation of DPPH-Scavenging Activity*

The antioxidant potential of the extracts from whole plants was determined by their ability to scavenge the stable 1,1-diphenyl-2-picrylhydrazyl (DPPH) free radical, according to Williams et al. [75]. Briefly, 200 µL of each extract preparation was added to 1.8 mL of 0.1 mM DPPH dissolved in methanol. The absorbance of the resulting solution was measured at 517 nm after the agitation of the mixture for 30 min at room temperature. A decrease in absorbance indicates an increase in DPPH-scavenging activity. The scavenging effect was characterized using the following equation:

$$\text{scavenging activity} = \left(\frac{\text{Blank absorbance} - \text{Sample absorbance}}{\text{Blank absorbance}}\right) \times 100 \quad (2)$$

4.5. *Total Soluble Sugar Content*

Total soluble sugar concentration from whole plants of each accession was measured using the colorimetric method developed by Dubois et al. [76]. An amount of 2 mL aliquot of a carbohydrate solution was mixed with 1 mL of 5% aqueous solution of phenol in test tubes. Furthermore, a total of 5 mL of concentrated sulfuric acid was rapidly added to the mixture. After allowing the test tubes to stand for 10 min, they were vortexed for 30 s and placed in a water bath at room temperature for 20 min to develop the color. Then, a spectrophotometer was used to record the light absorption at 490 nm. Reference solutions were prepared using the same procedure as described above, except that the 2 mL aliquot of carbohydrate was replaced with bi-distilled water. The phenol used in this procedure was redistilled, and a solution of 5% phenol in water (*w/w*) was prepared immediately before the measurements.

4.6. *Reducing Sugar Content*

The analysis of residual reducing sugar from whole plants of each accession were analyzed using the 3,5-dinitro-salicylic acid (DNS) method described by Miller [77]. In a 25 mL test tube, 1 mL of standard or test sample was added, followed by 3 mL of DNS reagent. The reaction mixture was heated in boiling water for 5 min. The absorbance spectra of the test samples were measured at 540 nm using 2 mm path length quartz cuvettes. The absorbance measurements were compared to a standard mannose curve (180.2 mw) ranging from 0 to 20 mM to identify the presence of reducing sugar released in each extract.

4.7. *Total Flavonoid Content*

Total flavonoid constituents of whole plant extracts were determined using an aluminum chloride colorimetric assay and Quercetine as the standard [78]. An amount of 1 mL of extract was added to a 10 mL flask containing 4 mL of distilled deionized water. A reagent blank using double-distilled H_2O (ddH_2O) instead of the extract was prepared at this stage. After adding 0.3 mL of 5% $NaNO_2$ to the flask, it was vortexed and kept for five minutes. Then, the mixture was treated with 0.3 mL of 10% AlCl3 solution and left for five minutes. At the sixth minute, 2 mL of 1 M NaOH was added, and the total volume was diluted to 10 mL with ddH_2O. The solution was vigorously vortexed, and absorbance was measured at 510 nm against the blank using a spectrophotometer. The procedure was repeated with all standard solutions of Quercetine (20–100 mg/L, Sigma-Aldrich, St. Louis, MO, USA) to obtain a standard curve. The total flavonoid content of the extracts was determined and expressed as milligrams of Quercetine equivalents (QE) per gram of extract. All samples were examined in triplicate.

4.8. Total Tannin Content

Total tannin content was calculated using the vanillin/HCl method developed by Price et al. [79]. The reagent was prepared by mixing equal parts of methanol solutions of 8% HCl and 1% vanillin, just before the reaction. Aliquots of 0.1 mL of lentil leaf extracts and 0.2 mL of vanillin/HCl reagent were mixed and incubated at 30 °C for 20 min. The absorbances were measured at 500 nm. A standard curve was created by using Catechin methanol solutions ranging from 0.1–2.5 mg mL^{-1}. The total tannin content of the extracts was determined and expressed as micrograms of Catechin equivalent (CE) per gram of extracts.

4.9. Total Soluble Protein

The method developed by Lowry [80] was utilized to predict the concentration of soluble proteins. An amount of 100 mg from the whole plant was macerated in 0.1 M phosphate buffer (pH = 7.0) and centrifuged at 513.16× *g* for 15 min to obtain a supernatant. The supernatant was treated with 5 mL of TCA (trichloroacetic acid; 15%) and kept at 4 °C for 24 h. The mixture was then centrifuged at 513.16× *g* for 15 min to separate the precipitates. The supernatant was discarded, and the precipitate was dissolved in 0.1 N NaOH (1 mL), allowed to dissolve completely for 18 h, and then treated as an extract.

4.10. Proline Content

The proline content was extracted and quantified using the methodology described by Bates et al. [81]. A total of 100 mg of leaf was homogenized in 10 mL of 3% sulphosalicyclic acid and centrifuged at 3000 rpm for 10 min. Filtrate was used for estimation by adding 2 mL of acid ninhydrin and 2 mL of glacial acetic acid. After one hour in a 100 °C water bath, the mixture was transferred to a separating funnel. Then, 4 mL of toluene was added. Two distinct layers were formed, and the absorbance of the upper layer was measured at 520 nm using the toluene as the blank reference.

4.11. Antioxidant Enzyme Activity Assays

The activity of catalase (CAT) was determined following the method of Chance and Maehly [82]. Whole plant was ground in mortars using liquid nitrogen. The enzyme was extracted from 0.5 g of tissue using chilled 50 mM sodium phosphate buffer (pH 7.0), containing 1% (*w*/*v*) polyvinylpyrrolidone (PVP), and was centrifuged at 12,000 rpm for 20 min at 4 °C. Enzyme estimation was performed using supernatant. The reaction mixture contained 1.9 mL of 50 mM sodium phosphate buffer (pH 7.5), 0.1 mL enzyme extract, and 1 mL of hydrogen peroxide. The decrease in absorbance after every 30 s interval for 2 min was recorded at 240 nm against a blank. Enzyme activity was calculated using the extinction coefficient of H_2O_2 (0.036 mM^{-1} cm^{-1}) and was expressed as μmol mg $protein^{-1}$ min^{-1}.

Ascorbate peroxidase (APX) activity was measured using the method of Nakano and Asada [83]. A total of 0.5 g of powdered sample was extracted with 2 mL of 50 mM sodium phosphate buffer (pH 7.5) containing 1% PVP and 2 mM ascorbate. The homogenized material was centrifuged at 12,000× *g* for 20 min at 4 °C. Supernatants were used as the crude enzyme source for the enzymatic assays. The reaction mixture contained 50 mM potassium phosphate (pH 7.0), 0.2 mM EDTA, 0.5 mM ascorbic acid, 2% H_2O_2, and 0.1 mL enzyme extract in a final volume of 3 mL. The decrease in absorbance at 290 nm for 1 min was recorded, and the amount of ascorbate oxidized was calculated using the extinction coefficient (e = 2.8 mM^{-1} APX was defined as 1 mmol mL^{-1} per min at 25 °C, cm^{-1}). Enzyme activity was expressed as µmol mg $protein^{-1}$ min^{-1}.

Superoxide dismutase (SOD) activity was assayed by monitoring its ability to inhibit the photochemical reduction of nitro-blue tetrazolium chloride (NBT). One unit of SOD is defined as the amount of enzyme required to inhibit the rate of NBT reduction by 50 percent at 560 nm. In 3 mL of reaction mixture, 75 µM NBT, 1.5 mM riboflavin, 50 mM sodium bicarbonate, 13 mM methionine, 0.1 mM EDTA, 50 mM potassium phosphate buffer (pH 7.5), and 100 µL of enzyme extraction were present. The samples containing the mixture were shaken for 15 min under light at 78 µmol photons s^{-1} m^{-1}, and the absorbance at 560 nm was measured. A control reaction mixture that did not exhibit any color development was used as a reference, and its absorbance at 560 nm was subtracted from the absorbance of the reaction solution [84].

4.12. *Statistical Analysis*

A least significant difference (LSD) test was applied at a significance level of $p < 0.05$ to compare the mean values and determine significant differences. Pearson's correlation coefficient was calculated for the three treatments using metan package in R version 4.1.3 and RStudio version 1.3.31093 [85]. A dendrogram along with a heatmap was performed with the pheatmap R package (Version 1) [86].

5. Conclusions

The current study concluded that tolerant lentil accessions exhibited superior responses to high-temperature and drought-stress conditions than sensitive accessions and produced high enzymatic activities for almost all components. These results align with the classification of these accessions as either tolerant or sensitive to high-temperature or drought stress. In addition, we also observed a moderate response of moderately tolerant accessions under both stresses. The inclusion of differentially sensitive accessions allowed us to delve into the mechanisms associated with tolerance to high-temperature and drought stress, which was one of our primary objectives. Under stress, tolerant accessions maintained higher antioxidant activity, osmolytes, and phenolics metabolism. Consequently, these accessions hold promising potential for maintaining essential physiological processes and achieving high yield levels in regions that are significantly affected by increasing temperatures and drought events. However, more research is needed to investigate the underlying mechanism and identify genes associated with heat or drought tolerance in lentil.

Author Contributions: Conceptualization and methodology, N.E.H., Y.E.-n., A.E.-B., K.H. and S.K.; validation, S.K.; formal analysis, N.E.H.; investigation, N.E.H., Y.E.-n. and K.A.; resources, S.K.; data curation, N.E.H.; writing—original draft preparation, N.E.H.; writing—review and editing, Y.E.-n., H.C., K.A., R.M., A.E.-B., K.H., K.R., A.S., F.M. and S.K.; supervision and project administration, S.K. All authors have read and agreed to the published version of the manuscript.

Funding: Partial funding support for this study was provided by the CGIAR Research Program on Grain Legumes and Dryland Cereals (GLDC).

Data Availability Statement: Datasets generated and/or analyzed during the current study are available from the corresponding author upon reasonable request.

Conflicts of Interest: The authors declare no conflict of interest.

References

1. Kumar, V.; Joshi, S.; Pant, N.C.; Sangwan, P.; Yadav, A.N.; Saxena, A.; Singh, D. Molecular Approaches for Combating Multiple Abiotic Stresses in Crops of Arid and Semi-arid Region. In *Molecular Approaches in Plant Biology and Environmental Challenges*; Springer: Singapore, 2019; pp. 149–170. [CrossRef]
2. Samineni, S.; Mahendrakar, M.D.; Shankar, N.; Hotti, A.; Chand, U.; Rathore, A.; Gaur, P.M. Impact of heat and drought stresses on grain nutrient content in chickpea: Genome-wide marker-trait associations for protein, Fe and Zn. *Environ. Exp. Bot.* **2022**, *194*, 104688. [CrossRef]
3. Çelik, Ö.; Ayan, A.; Atak, Ç. Enzymatic and non-enzymatic comparison of two different industrial tomato (*Solanum lycopersicum*) varieties against drought stress. *Bot. Stud.* **2017**, *58*, 32. [CrossRef] [PubMed]
4. Hussain, H.A.; Men, S.; Hussain, S.; Chen, Y.; Ali, S.; Zhang, S.; Zhang, K.; Li, Y.; Xu, Q.; Liao, C.; et al. Interactive effects of drought and heat stresses on morpho-physiological attributes, yield, nutrient uptake and oxidative status in maize hybrids. *Sci. Rep.* **2019**, *9*, 3890. [CrossRef] [PubMed]
5. Sarkar, S.; Khatun, M.; Era, F.M.; Islam, A.K.M.M.; Anwar, M.P.; Danish, S.; Datta, R.; Islam, A.K.M.A. Abiotic Stresses: Alteration of Composition and Grain Quality in Food Legumes. *Agronomy* **2021**, *11*, 2238. [CrossRef]
6. El Haddad, N.; Sanchez-Garcia, M.; Visioni, A.; Jilal, A.; El Amil, R.; Sall, A.T.; Lagesse, W.; Kumar, S.; Bassi, F.M. Crop Wild Relatives Crosses: Multi-Location Assessment in Durum Wheat, Barley, and Lentil. *Agronomy* **2021**, *11*, 2283. [CrossRef]
7. Zahra, N.; Wahid, A.; Hafeez, M.B.; Ullah, A.; Siddique, K.H.M.; Farooq, M. Grain development in wheat under combined heat and drought stress: Plant responses and management. *Environ. Exp. Bot.* **2021**, *188*, 104517. [CrossRef]
8. Pörtner, H.-O.; Roberts, D.C.; Tignor, M.; Poloczanska, E.; Mintenbeck, K.; Alegría, A.; Craig, M.; Langsdorf, S.; Löschke, S.; Möller, V.; et al. *IPCC Climate Change 2022: Impacts, Adaptation, and Vulnerability. Contribution of Working Group II to the Sixth Assessment Report of the Intergovernmental Panel on Climate Change*; Cambridge University Press: Cambridge, UK; New York, NY, USA, 2022. [CrossRef]
9. Sehgal, A.; Sita, K.; Rehman, A.; Farooq, M.; Kumar, S.; Yadav, R.; Nayyar, H.; Singh, S.; Siddique, K.H. Lentil. In *Crop Physiology Case Histories for Major Crops*; Academic Press: Cambridge, MA, USA, 2021; pp. 408–428. [CrossRef]
10. FAOSTAT. Food and Agriculture Organization Corporate Statistical Database. Available online: http://www.fao.org/ (accessed on 17 January 2020).
11. Kumari, V.V.; Roy, A.; Vijayan, R.; Banerjee, P.; Verma, V.C.; Nalia, A.; Pramanik, M.; Mukherjee, B.; Ghosh, A.; Reja, M. Drought and Heat Stress in Cool-Season Food Legumes in Sub-Tropical Regions: Consequences, Adaptation, and Mitigation Strategies. *Plants* **2021**, *10*, 1038. [CrossRef] [PubMed]
12. Ghanem, M.E.; Guiguitant, J.; Sinclair, T.R. Opportunities to improve the seasonal dynamics of water use in lentil (*Lens culinaris* Medik.) to enhance yield increase in water-limited environments. *Chem. Biol. Technol. Agric.* **2017**, *4*, 22. [CrossRef]
13. Sita, K.; Sehgal, A.; Hanumantha Rao, B.; Nair, R.M.; Vara Prasad, P.V.; Kumar, S.; Gaur, P.M.; Farooq, M.; Siddique, K.H.M.; Varshney, R.K.; et al. Food Legumes and Rising Temperatures: Effects, Adaptive Functional Mechanisms Specific to Reproductive Growth Stage and Strategies to Improve Heat Tolerance. *Front. Plant Sci.* **2017**, *8*, 1658. [CrossRef] [PubMed]
14. Gaur, P.M.; Samineni, S.; Krishnamurthy, L.; Varshney, R.K.; Kumar, S.; Ghanem, M.E.; Beebe, S.E.; Rao, I.M.; Chaturvedi, S.K.; Basu, P.S. High Temperature Tolerance in Grain Legumes. 2014. Available online: https://cgspace.cgiar.org/bitstream/handle/10568/65370/IFLRCVI_ICLGGVII_ELECTRONICProgramBook.pdf?sequence=4 (accessed on 17 January 2022).
15. Bhardwaj, A.; Sita, K.; Sehgal, A.; Bhandari, K.; Kumar, S.; Prasad, P.V.V.; Jha, U.; Kumar, J.; Siddique, K.H.M.; Nayyar, H. Heat Priming of Lentil (*Lens culinaris* Medik.) Seeds and Foliar Treatment with γ-Aminobutyric Acid (GABA), Confers Protection to Reproductive Function and Yield Traits under High-Temperature Stress Environments. *Int. J. Mol. Sci.* **2021**, *22*, 5825. [CrossRef] [PubMed]
16. Choukri, H.; Hejjaoui, K.; El-Baouchi, A.; El Haddad, N.; Smouni, A.; Maalouf, F.; Thavarajah, D.; Kumar, S. Heat and Drought Stress Impact on Phenology, Grain Yield, and Nutritional Quality of Lentil (*Lens culinaris* Medikus). *Front. Nutr.* **2020**, *7*, 596307. [CrossRef] [PubMed]
17. Senapati, N.; Halford, N.G.; Semenov, M.A. Vulnerability of European wheat to extreme heat and drought around flowering under future climate. *Environ. Res. Lett.* **2021**, *16*, 24052. [CrossRef]
18. Naz, R.; Gul, F.; Zahoor, S.; Nosheen, A.; Yasmin, H.; Keyani, R.; Shahid, M.; Hassan, M.N.; Siddiqui, M.H.; Batool, S.; et al. Interactive effects of hydrogen sulphide and silicon enhance drought and heat tolerance by modulating hormones, antioxidant defence enzymes and redox status in barley (*Hordeum vulgare* L.). *Plant Biol.* **2022**, *24*, 684–696. [CrossRef]
19. Liu, X.; Wang, X.; Wang, X.; Gao, J.; Luo, N.; Meng, Q.; Wang, P. Dissecting the critical stage in the response of maize kernel set to individual and combined drought and heat stress around flowering. *Environ. Exp. Bot.* **2020**, *179*, 104213. [CrossRef]
20. Rani, A.; Devi, P.; Jha, U.C.; Sharma, K.D.; Siddique, K.H.M.; Nayyar, H. Developing climate-resilient chickpea involving physiological and molecular approaches with a focus on temperature and drought stresses. *Front. Plant Sci.* **2020**, *10*, 1759. [CrossRef] [PubMed]
21. Choukri, H.; El Haddad, N.; Aloui, K.; Hejjaoui, K.; El-Baouchi, A.; Smouni, A.; Maalouf, F.; Kumar, S. Effect of High Temperature Stress During the Reproductive Stage on Grain Yield and Nutritional Quality of Lentil (*Lens culinaris* Medikus). *Front. Nutr.* **2022**, *9*, 857469. [CrossRef] [PubMed]

22. Cohen, I.; Zandalinas, S.I.; Fritschi, F.B.; Sengupta, S.; Fichman, Y.; Azad, R.K.; Mittler, R. The impact of water deficit and heat stress combination on the molecular response, physiology, and seed production of soybean. *Physiol. Plant.* **2021**, *172*, 41–52. [CrossRef] [PubMed]
23. Mittler, R.; Zandalinas, S.I.; Fichman, Y.; Van Breusegem, F. Reactive oxygen species signalling in plant stress responses. *Nat. Rev. Mol. Cell Biol.* **2022**, *23*, 663–679. [CrossRef] [PubMed]
24. Roy, M.; Niu, J.; Irshad, A.; Kareem, H.A.; Hassan, M.U.; Xu, N.; Sui, X.; Guo, Z.; Amo, A.; Wang, Q. Exogenous melatonin protects alfalfa (*Medicago sativa* L.) seedlings from drought-induced damage by modulating reactive oxygen species metabolism, mineral balance and photosynthetic efficiency. *Plant Stress* **2021**, *2*, 100044. [CrossRef]
25. Sharma, A.; Sharma, P.; Kumar, R.; Sharma, V.; Bhardwaj, R.; Sharma, I. Role of reactive oxygen species in the regulation of abiotic stress tolerance in legumes. In *Abiotic Stress and Legumes*; Academic Press: Cambridge, MA, USA, 2021; pp. 217–243. [CrossRef]
26. Zhao, J.; Wang, X.; Pan, X.; Jiang, Q.; Xi, Z. Exogenous Putrescine Alleviates Drought Stress by Altering Reactive Oxygen Species Scavenging and Biosynthesis of Polyamines in the Seedlings of Cabernet Sauvignon. *Front. Plant Sci.* **2021**, *12*, 767992. [CrossRef] [PubMed]
27. Qaderi, M.M.; Martel, A.B.; Strugnell, C.A. Environmental Factors Regulate Plant Secondary Metabolites. *Plants* **2023**, *12*, 447. [CrossRef] [PubMed]
28. Kumar, K.; Debnath, P.; Singh, S.; Kumar, N. An Overview of Plant Phenolics and Their Involvement in Abiotic Stress Tolerance. *Stresses* **2023**, *3*, 570–585. [CrossRef]
29. Nicolas-Espinosa, J.; Garcia-Ibañez, P.; Lopez-Zaplana, A.; Yepes-Molina, L.; Albaladejo-Marico, L.; Carvajal, M. Confronting Secondary Metabolites with Water Uptake and Transport in Plants under Abiotic Stress. *Int. J. Mol. Sci.* **2023**, *24*, 2826. [CrossRef]
30. Bhandari, K.; Siddique, K.H.M.; Turner, N.C.; Kaur, J.; Singh, S.; Agrawal, S.K.; Nayyar, H. Heat stress at reproductive stage disrupts leaf carbohydrate metabolism, impairs reproductive function, and severely reduces seed yield in lentil. *J. Crop Improv.* **2016**, *30*, 118–151. [CrossRef]
31. Shah, W.; Ullah, S.; Ali, S.; Idrees, M.; Khan, M.N.; Ali, K.; Khan, A.; Ali, M.; Younas, F. Effect of exogenous alpha-tocopherol on physio-biochemical attributes and agronomic performance of lentil (*Lens culinaris* Medik.) under drought stress. *PLoS ONE* **2021**, *16*, e0248200. [CrossRef]
32. Sehgal, A.; Sita, K.; Kumar, J.; Kumar, S.; Singh, S.; Siddique, K.H.M.; Nayyar, H. Effects of drought, heat and their interaction on the growth, yield and photosynthetic function of lentil (*Lens culinaris* Medikus) genotypes varying in heat and drought sensitivity. *Front. Plant Sci.* **2017**, *8*, 1776. [CrossRef]
33. El Haddad, N.; Rajendran, K.; Smouni, A.; Es-Safi, N.E.; Benbrahim, N.; Mentag, R.; Nayyar, H.; Maalouf, F.; Kumar, S. Screening the FIGS set of lentil (*Lens culinaris* Medikus) germplasm for tolerance to terminal heat and combined drought-heat stress. *Agronomy* **2020**, *10*, 1036. [CrossRef]
34. El Haddad, N.; Choukri, H.; Ghanem, M.E.; Smouni, A.; Mentag, R.; Rajendran, K.; Hejjaoui, K.; Maalouf, F.; Kumar, S. High-Temperature and Drought Stress Effects on Growth, Yield and Nutritional Quality with Transpiration Response to Vapor Pressure Deficit in Lentil. *Plants* **2022**, *11*, 95. [CrossRef] [PubMed]
35. Wahab, A.; Abdi, G.; Saleem, M.H.; Ali, B.; Ullah, S.; Shah, W.; Mumtaz, S.; Yasin, G.; Muresan, C.C.; Marc, R.A. Plants' physio-biochemical and phyto-hormonal responses to alleviate the adverse effects of drought stress: A comprehensive review. *Plants* **2022**, *11*, 1620. [CrossRef]
36. Dos Santos, T.B.; Ribas, A.F.; de Souza, S.G.H.; Budzinski, I.G.F.; Domingues, D.S. Physiological Responses to Drought, Salinity, and Heat Stress in Plants: A Review. *Stresses* **2022**, *2*, 113–135. [CrossRef]
37. Gomes, M.P.; Kitamura, R.S.A.; Marques, R.Z.; Barbato, M.L.; Zámocký, M. The role of H_2O_2-scavenging enzymes (ascorbate peroxidase and catalase) in the tolerance of Lemna minor to antibiotics: Implications for phytoremediation. *Antioxidants* **2022**, *11*, 151. [CrossRef] [PubMed]
38. Nikzad, S.; Mirmohammady Maibody, S.A.M.; Ehtemam, M.H.; Golkar, P.; Mohammadi, S.A. Response of seed yield and biochemical traits of *Eruca sativa* Mill. to drought stress in a collection study. *Sci. Rep.* **2023**, *13*, 11157. [CrossRef] [PubMed]
39. Yadav, C.; Bahuguna, R.N.; Dhankher, O.P.; Singla-Pareek, S.L.; Pareek, A. Physiological and molecular signatures reveal differential response of rice genotypes to drought and drought combination with heat and salinity stress. *Physiol. Mol. Biol. Plants* **2022**, *28*, 899–910. [CrossRef] [PubMed]
40. Li, S. Novel insight into functions of ascorbate peroxidase in higher plants: More than a simple antioxidant enzyme. *Redox Biol.* **2023**, *64*, 102789. [CrossRef] [PubMed]
41. Singh, D.; Singh, C.K.; Taunk, J.; Jadon, V.; Pal, M.; Gaikwad, K. Genome wide transcriptome analysis reveals vital role of heat responsive genes in regulatory mechanisms of lentil (*Lens culinaris* Medikus). *Sci. Rep.* **2019**, *9*, 12976. [CrossRef] [PubMed]
42. Sarker, U.; Oba, S. Catalase, superoxide dismutase and ascorbate-glutathione cycle enzymes confer drought tolerance of *Amaranthus tricolor*. *Sci. Rep.* **2018**, *8*, 16496. [CrossRef]
43. Chakraborty, U.; Pradhan, D. High temperature-induced oxidative stress in Lens culinaris, role of antioxidants and amelioration of stress by chemical pre-treatments. *J. Plant Interact.* **2011**, *6*, 43–52. [CrossRef]
44. AbdElgawad, H.; Farfan-Vignolo, E.R.; De Vos, D.; Asard, H. Elevated CO_2 mitigates drought and temperature-induced oxidative stress differently in grasses and legumes. *Plant Sci.* **2015**, *231*, 1–10. [CrossRef] [PubMed]

45. Awasthi, R.; Gaur, P.; Turner, N.C.; Vadez, V.; Siddique, K.H.M.; Nayyar, H. Effects of individual and combined heat and drought stress during seed filling on the oxidative metabolism and yield of chickpea (*Cicer arietinum*) genotypes differing in heat and drought tolerance. *Crop Pasture Sci.* **2017**, *68*, 823–841. [CrossRef]
46. Yadav, R.; Juneja, S.; Kumar, S. Cross priming with drought improves heat-tolerance in chickpea (*Cicer arietinum* L.) by stimulating small heat shock proteins and antioxidative defense. *Environ. Sustain.* **2021**, *4*, 171–182. [CrossRef]
47. Hosseini, S.Z.; Ismaili, A.; Nazarian-Firouzabadi, F.; Fallahi, H.; Rezaei Nejad, A.; Sohrabi, S.S. Dissecting the molecular responses of lentil to individual and combined drought and heat stresses by comparative transcriptomic analysis. *Genomics* **2021**, *113*, 693–705. [CrossRef]
48. Singh, D.; Singh, C.K.; Taunk, J.; Tomar, R.S.S.; Chaturvedi, A.K.; Gaikwad, K.; Pal, M. Transcriptome analysis of lentil (*Lens culinaris* Medikus) in response to seedling drought stress. *BMC Genom.* **2017**, *18*, 206. [CrossRef]
49. Easwar Rao, D.; Viswanatha Chaitanya, K. Changes in the antioxidant intensities of seven different soybean (*Glycine max* (L.) Merr.) cultivars during drought. *J. Food Biochem.* **2020**, *44*, e13118. [CrossRef] [PubMed]
50. Saed-Moucheshi, A.; Sohrabi, F.; Fasihfar, E.; Baniasadi, F.; Riasat, M.; Mozafari, A.A. Superoxide dismutase (SOD) as a selection criterion for triticale grain yield under drought stress: A comprehensive study on genomics and expression profiling, bioinformatics, heritability, and phenotypic variability. *BMC Plant Biol.* **2021**, *21*, 148. [CrossRef] [PubMed]
51. Ubaidillah, M.; Oktaviani, F.; Mufadilah, M.A.; Avivi, S.; Thamrin, N.; Indrawati, A.; Puspito, A.N.; Kim, K.M.; Hartatik, S. Response of regulation of resistance genes, reactive oxygen species, and antioxidant enzymes to salicylic acid treatments in drought tolerant rice. *Agron. Res.* **2023**, *21*, 397–409. [CrossRef]
52. Rahman, M.A.; Woo, J.H.; Song, Y.; Lee, S.-H.; Hasan, M.M.; Azad, M.A.; Lee, K.-W. Heat Shock Proteins and Antioxidant Genes Involved in Heat Combined with Drought Stress Responses in Perennial Rye Grass. *Life* **2022**, *12*, 1426. [CrossRef]
53. Dikilitas, M.; Simsek, E.; Roychoudhury, A. Role of proline and glycine betaine in overcoming abiotic stresses. In *Protective Chemical Agents in the Amelioration of Plant Abiotic Stress: Biochemical and Molecular Perspectives*; John Wiley & Sons: Hoboken, NJ, USA, 2020; pp. 1–23. [CrossRef]
54. Ghosh, U.K.; Islam, M.N.; Siddiqui, M.N.; Cao, X.; Khan, M.A.R. Proline, a multifaceted signalling molecule in plant responses to abiotic stress: Understanding the physiological mechanisms. *Plant Biol.* **2022**, *24*, 227–239. [CrossRef] [PubMed]
55. Hayat, S.; Hayat, Q.; Alyemeni, M.N.; Wani, A.S.; Pichtel, J.; Ahmad, A. Role of proline under changing environments: A review. *Plant Signal. Behav.* **2012**, *7*, 1456–1466. [CrossRef] [PubMed]
56. Anjum, S.A.; Xie, X.; Wang, L.; Saleem, M.F.; Man, C.; Lei, W. Morphological, physiological and biochemical responses of plants to drought stress. *Afr. J. Agric. Res.* **2011**, *6*, 2026–2032.
57. El-Beltagi, H.S.; Mohamed, H.I.; Sofy, M.R. Role of Ascorbic acid, Glutathione and Proline Applied as Singly or in Sequence Combination in Improving Chickpea Plant through Physiological Change and Antioxidant Defense under Different Levels of Irrigation Intervals. *Molecules* **2020**, *25*, 1702. [CrossRef]
58. Ignatova, Z.; Gierasch, L.M. Inhibition of protein aggregation in vitro and in vivo by a natural osmoprotectant. *Proc. Natl. Acad. Sci. USA* **2006**, *103*, 13357–13361. [CrossRef] [PubMed]
59. Ozturk, M.; Turkyilmaz Unal, B.; García-Caparrós, P.; Khursheed, A.; Gul, A.; Hasanuzzaman, M. Osmoregulation and its actions during the drought stress in plants. *Physiol. Plant.* **2021**, *172*, 1321–1335. [CrossRef] [PubMed]
60. Sehgal, A.; Sita, K.; Bhandari, K.; Kumar, S.; Kumar, J.; Vara Prasad, P.V.; Siddique, K.H.M.; Nayyar, H. Influence of drought and heat stress, applied independently or in combination during seed development, on qualitative and quantitative aspects of seeds of lentil (*Lens culinaris* Medikus) genotypes, differing in drought sensitivity. *Plant Cell Environ.* **2019**, *42*, 198–211. [CrossRef] [PubMed]
61. Li, T.; Wang, R.; Zhao, D.; Tao, J. Effects of drought stress on physiological responses and gene expression changes in herbaceous peony (*Paeonia lactiflora* Pall.). *Plant Signal. Behav.* **2020**, *15*, 1746034. [CrossRef] [PubMed]
62. Sita, K.; Sehgal, A.; Bhandari, K.; Kumar, J.; Kumar, S.; Singh, S.; Siddique, K.H.M.; Nayyar, H. Impact of heat stress during seed filling on seed quality and seed yield in lentil (*Lens culinaris* Medikus) genotypes. *J. Sci. Food Agric.* **2018**, *98*, 5134–5141. [CrossRef]
63. Bueno, P.C.P.; Lopes, N.P. Metabolomics to Characterize Adaptive and Signaling Responses in Legume Crops under Abiotic Stresses. *ACS Omega* **2020**, *5*, 1752–1763. [CrossRef] [PubMed]
64. Hassan, M.U.; Aamer, M.; Chattha, M.U.; Haiying, T.; Shahzad, B.; Barbanti, L.; Nawaz, M.; Rasheed, A.; Afzal, A.; Liu, Y.; et al. The critical role of zinc in plants facing the drought stress. *Agriculture* **2020**, *10*, 396. [CrossRef]
65. Ghazghazi, H.; Riahi, L.; Yangui, I.; Messaoud, C.; Rzigui, T.; Nasr, Z. Effect of Drought Stress on Physio-biochemical Traits and Secondary Metabolites Production in the Woody Species *Pinus Halepensis* Mill. At a Juvenile Development Stage. *J. Sustain. For.* **2022**, *41*, 878–894. [CrossRef]
66. Ahanger, M.A.; Ahmad, P. Role of Mineral Nutrients in Abiotic Stress Tolerance: Revisiting the Associated Signaling Mechanisms. In *Plant Signaling Molecules*; Elsevier: Amsterdam, The Netherlands, 2019; pp. 269–285. [CrossRef]
67. Yasmeen, S.; Wahab, A.; Saleem, M.H.; Ali, B.; Qureshi, K.A.; Jaremko, M. Melatonin as a Foliar Application and Adaptation in Lentil (*Lens culinaris* Medik.) Crops under Drought Stress. *Sustainability* **2022**, *14*, 6345. [CrossRef]
68. Wang, G.Y.; Ahmad, S.; Wang, Y.; Wang, B.W.; Huang, J.H.; Jahan, M.S.; Zhou, X.B.; Shi, C.Q. Multivariate analysis compares and evaluates drought and flooding tolerances of maize germplasm. *Plant Physiol.* **2023**, kiad317. [CrossRef] [PubMed]

69. Omar, A.A.; Heikal, Y.M.; Zayed, E.M.; Shamseldin, S.A.M.; Salama, Y.E.; Amer, K.E.; Basuoni, M.M.; Abd Ellatif, S.; Mohamed, A.H. Conferring of Drought and Heat Stress Tolerance in Wheat (*Triticum aestivum* L.) Genotypes and Their Response to Selenium Nanoparticles Application. *Nanomaterials* **2023**, *13*, 998. [CrossRef] [PubMed]
70. Shin, Y.K.; Bhandari, S.R.; Jo, J.S.; Song, J.W.; Lee, J.G. Effect of Drought Stress on Chlorophyll Fluorescence Parameters, Phytochemical Contents, and Antioxidant Activities in Lettuce Seedlings. *Horticulturae* **2021**, *7*, 238. [CrossRef]
71. Popoola, O.O. Phenolic compounds composition and in vitro antioxidant activity of Nigerian *Amaranthus viridis* seed as affected by autoclaving and germination. *Meas. Food* **2022**, *6*, 100028. [CrossRef]
72. Habuš Jerčić, I.; Bošnjak Mihovilović, A.; Matković Stanković, A.; Lazarević, B.; Goreta Ban, S.; Ban, D.; Major, N.; Tomaz, I.; Banjavčić, Z.; Kereša, S. Garlic Ecotypes Utilise Different Morphological, Physiological and Biochemical Mechanisms to Cope with Drought Stress. *Plants* **2023**, *12*, 1824. [CrossRef] [PubMed]
73. Grela, E.R.; Samolińska, W.; Kiczorowska, B.; Klebaniuk, R.; Kiczorowski, P. Content of Minerals and Fatty Acids and Their Correlation with Phytochemical Compounds and Antioxidant Activity of Leguminous Seeds. *Biol. Trace Elem. Res.* **2017**, *180*, 338–348. [CrossRef] [PubMed]
74. Waterhouse, A.L. Determination of total phenolics. *Curr. Protoc. Food Anal. Chem.* **2002**, *6*, I1.1.1–I1.1.8.
75. Brand-Williams, W.; Cuvelier, M.-E.; Berset, C. Use of a free radical method to evaluate antioxidant activity. *LWT-Food Sci. Technol.* **1995**, *28*, 25–30. [CrossRef]
76. Dubois, M.; Gilles, K.A.; Hamilton, J.K.; Rebers, P.A.T.; Smith, F. Colorimetric method for determination of sugars and related substances. *Anal. Chem.* **1956**, *28*, 350–356. [CrossRef]
77. Miller, G.L. Use of dinitrosalicylic acid reagent for determination of reducing sugar. *Anal. Chem.* **1959**, *31*, 426–428. [CrossRef]
78. Chang, C.C.; Yang, M.H.; Wen, H.M.; Chern, J.C. Estimation of total flavonoid content in propolis by two complementary colorimetric methods. *Food Drug Anal.* **2002**, *10*, 178–182.
79. Price, M.L.; Van Scoyoc, S.; Butler, L.G. A critical evaluation of the vanillin reaction as an assay for tannin in sorghum grain. *J. Agric. Food Chem.* **1978**, *26*, 1214–1218. [CrossRef]
80. Lowry, O.H. Protein measurement with the Folin phenol reagent. *J. Biol. Chem.* **1951**, *193*, 265–275. [CrossRef] [PubMed]
81. Bates, L.S.; Waldren, R.P.; Teare, I.D. Rapid determination of free proline for water-stress studies. *Plant Soil* **1973**, *39*, 205–207. [CrossRef]
82. Chance, B.; Maehly, A.C. Assay of catalases and peroxidases. *Methods Enzymol.* **1955**, *2*, 764–775. [CrossRef]
83. Nakano, Y.; Asada, K. Hydrogen Peroxide is Scavenged by Ascorbate-specific Peroxidase in Spinach Chloroplasts. *Plant Cell Physiol.* **1981**, *22*, 867–880. [CrossRef]
84. Giannopolitis, C.N.; Ries, S.K. Superoxide dismutases: I. Occurrence in higher plants. *Plant Physiol.* **1977**, *59*, 309–314. [CrossRef]
85. Olivoto, T.; Lúcio, A.D. metan: An R package for multi-environment trial analysis. *Methods Ecol. Evol.* **2020**, *11*, 783–789. [CrossRef]
86. Kolde, R.; Kolde, M.R. *R Package*, Version 1. Package 'pheatmap'. R Project for Statistical Computing: Vienna, Austria, 2018.

plants

MDPI

Article

Integration of Physiological, Transcriptomic, and Metabolomic Analyses Reveal Molecular Mechanisms of Salt Stress in *Maclura tricuspidata*

Dezong Sui [1], Baosong Wang [1], Yousry A. El-Kassaby [2] and Lei Wang [1,*]

[1] Jiangsu Academy of Forestry, Nanjing 211153, China; jslky@jaf.ac.cn (D.S.); baoswang@163.com (B.W.)
[2] Department of Forest and Conservation Sciences, Faculty of Forestry, University of British Columbia, Vancouver, BC V6T IZ4, Canada; y.el-kassaby@ubc.ca
* Correspondence: sylc@jaf.ac.cn

Abstract: Salt stress is a universal abiotic stress that severely affects plant growth and development. Understanding the mechanisms of *Maclura tricuspidate*'s adaptation to salt stress is crucial for developing salt-tolerant plant varieties. This article discusses the integration of physiology, transcriptome, and metabolome to investigate the mechanism of salt adaptation in *M. tricuspidata* under salt stress conditions. Overall, the antioxidant enzyme system (SOD and POD) of *M. tricuspidata* exhibited higher activities compared with the control, while the content of soluble sugar and concentrations of chlorophyll a and b were maintained during salt stress. KEGG analysis revealed that deferentially expressed genes were primarily involved in plant hormone signal transduction, phenylpropanoid and flavonoid biosynthesis, alkaloids, and MAPK signaling pathways. Differential metabolites were enriched in amino acid metabolism, the biosynthesis of plant hormones, butanoate, and 2-oxocarboxylic acid metabolism. Interestingly, glycine, serine, and threonine metabolism were found to be important both in the metabolome and transcriptome–metabolome correlation analyses, suggesting their essential role in enhancing the salt tolerance of *M. tricuspidata*. Collectively, our study not only revealed the molecular mechanism of salt tolerance in *M. tricuspidata*, but also provided a new perspective for future salt-tolerant breeding and improvement in salt land for this species.

Keywords: *Maclura tricuspidate*; resistance; salt adaptation; multi-omics

1. Introduction

Soil salinization is a major global environmental and socioeconomic issue. Currently, approximately 1.125×10^9 ha of saline–alkali soil exists worldwide [1]. By 2050, it is estimated that over half of the world's arable land will be affected by salinization [2,3]. The total area of salinized and alkaline land in China is widespread, reaching approximately 5.0×10^8 ha [4]. Salt stress caused by neutral salts, such as NaCl and Na_2SO_4, can lead to water deficiency, ion toxicity, decreased nutrient absorption and transport in plant bodies, as well as the accumulation of reactive oxygen species (ROS) that can be toxic to cells [5]. Saline–alkali soil adversely affects plant growth and development, severely restricting the development of sustainable agriculture [6].

Plants develop a variety of physiological and biochemical adaptive mechanisms in response to soil salt stress, known as the plant salt stress response [7]. These mechanisms primarily revolve around regulating ion, osmotic, and nutrient balances, and scavenging ROS [8]. Collectively, these processes enable plants to adapt and survive in environments with high salt concentrations. Studies have shown significant changes in plants in response to salt stress, including oxidative and ion balances and osmotic adjustment [9,10]. ROS is vitally important in responding to stress perception, integrating various stress response signaling networks, activating and domesticating plant defense mechanisms [11]. Under salt stress, plants have the ability to activate enzymatic and non-enzymatic systems to protect cells from ROS toxicity [12,13]. The enzymatic system is composed of antioxidant enzymes, such as catalase (CAT), peroxidase (POD), and superoxide dismutase (SOD) [14]. Non-enzymatic systems generate metabolites that help to eliminate ROS from cells during salt stress, such as ascorbic acid, alkaloids, carotenoids, flavonoids, glutathione, phenolic compounds, and melatonin [7,14–16].

Technological advancements and innovation enable the understanding of changes in genes and metabolites within organisms [17]. High-throughput technologies have facilitated the study of molecular mechanisms and metabolic regulation pathways in plants [18,19]. Although salt tolerance has been proven in certain species, current research primarily centers on model plant species (e.g., *Arabidopsis thaliana, Solanum lycopersicum, Oryza sativa, Triticum aestivum,* and *Populus alba* × *P. glandulosa*) [20–24]. Therefore, it is essential to explore the molecular regulatory mechanisms of plant responses to salt stress, meanwhile, further exploration is needed for the development and utilization of salt-tolerant plant resources.

Maclura tricuspidata (Carr.) Bur. ex Lavallee, Moraceae, is a deciduous broad-leaved spiny tree distributed throughout China, Korea, and Japan [25]. *M. tricuspidata* is widely used in folk medicine to treat various disorders [26]. Additionally, bark fibers are used for papermaking, leaves serve as silkworms feed, and its fruit is consumed [27]. *M. tricuspidata* is a multi-purpose tree species that can survive on moderately saline–alkali soils. In the present study, we employed a combination of multi-omics techniques in *M. tricuspidata* to: (1) elucidate the effects of salt stress on genes and metabolites; (2) identify the key biological pathways associated with regulating salt tolerance; (3) explore the regulatory network of gene–metabolite underlying salt tolerance; and (4) elaborate the molecular mechanisms of salt adaptation. Our study systematically investigated the adaptive mechanisms of *M. tricuspidata* leaves in response to salt stress from three aspects: physiology, transcriptomics, and metabolomics, which are expected to provide new insights into the salt stress adaptation mechanisms of *M. tricuspidata*.

2. Results

2.1. Effects of Salt Stress on Plant Phenotypic Differences and Physiological Characteristics

Compared with the control treatment without salt (CK), *M. tricuspidata* treated with high (h) salt concentration (102 mM NaCl) for 24 h exhibited a few necrotic spot symptoms on the upper surface of the leaves (Figure 1). Furthermore, leaves treated for 48 and 72 h showed symptoms of larger, clearly visible necrotic patches. These symptoms were more obvious under the h treatment than the moderate (m) treatment (68 mM NaCl).

Salt stress can disrupt the balance of ROS scavenging in plants, leading to oxidative stress reactions [28]. Therefore, physiological assessments were conducted to examine the characteristics of *M. tricuspidata* leaves under salt stress. The results showed that there were no significant changes in the levels of soluble sugar among the three different salt stress treatments at the four time points (Figure 2A). CAT activities began to exhibit a significant decrease under 48 h (Figure 2B). SOD activities showed an overall increasing trend, especially under the m and h treatments, with a significant difference compared with the CK (Figure 2C). POD activities significantly increased during most stress treatments, except for 24 h under low-concentration (l) and h treatments, and 72 h under the m treatment (Figure 2D). In comparison to the CK, the MDA content significantly increased under the m and h treatments at 2 h, the m treatment at 24 h, and all three treatments at 48 h (Figure 2E). The concentrations of chlorophyll a and b did not show significant differences among the various treatments (Figure 2F,G). Overall, under salt stress conditions, *M. tricuspidate* was able to maintain an appropriate level of soluble sugar content, and exhibited relatively higher activities of SOD, and POD. Despite experiencing membrane lipid peroxidation, *M. tricuspidate* still maintained stable concentrations of chlorophyll a and b. This physiological adaptation ability may contribute to *M. tricuspidate*'s resistance against salt stress.

Figure 1. Phenotypic differences of *M. tricuspidata* leaves under control (CK), moderate (m), and high (h) salt stress treatments for 24, 48, and 72 h, respectively. Note: CK, 0 mM NaCl; m24, 68 mM NaCl treatment for 24 h; h24, 102 mM NaCl treatment for 24 h; m48, 68 mM NaCl treatment for 48 h; h48, 102 mM NaCl treatment for 48 h; m72, 68 mM NaCl treatment for 72 h; h72, 102 mM NaCl treatment for 72 h.

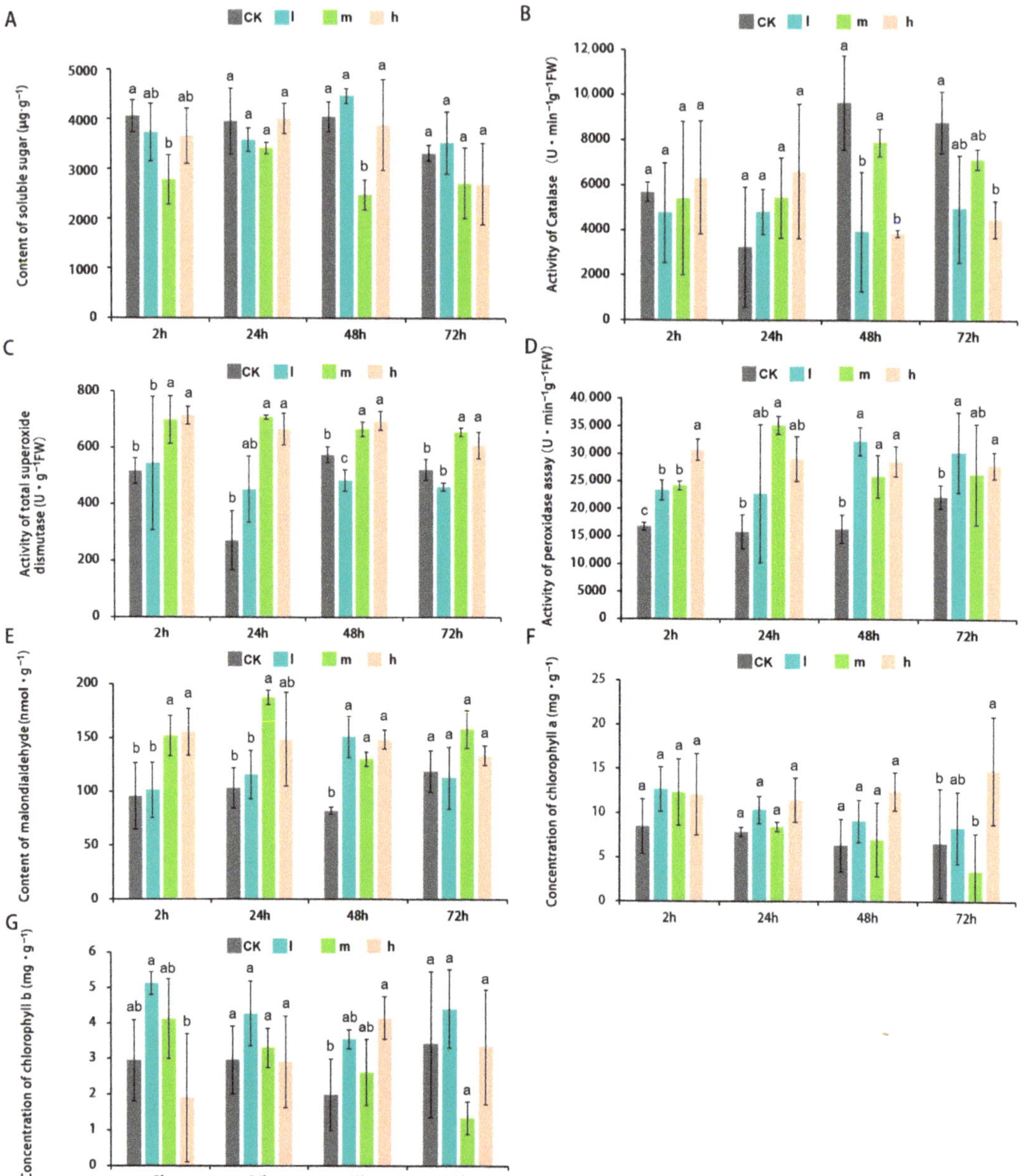

Figure 2. Physiological characteristics of *M. tricuspidata* under salt stress. (**A**) Content of soluble sugar. (**B–D**) Activities of CAT, SOD, and POD. (**E**) Content of MDA. (**F**,**G**) Concentration of chlorophyll a and b. CK, 0 mM NaCl; l, 34 mM NaCl; m, 68 mM NaCl; h, 102 mM NaCl. Different letters indicate a statistically significant difference when analyzed by one-way ANOVA and a multiple comparisons using Duncan's method at $p \leq 0.05$.

2.2. Summary of Transcriptome Sequencing Data

In total, 164.79 Gb of raw data was generated, and the average GC content was 45.52% (Supplementary Table S1). The sequencing quality achieved Q20 and Q30 scores of 97.47

and 92.96%, respectively. PCA results indicated that the biological replicates within each group were clustered together (Supplementary Figure S1A). Moreover, the correlation among samples showed good reproducibility between the biological replicates, which was consistent with the PCA results, indicating a high level of reproducibility (Supplementary Figure S1B).

2.3. Screening and Identification of Differentially Expressed Genes

The present study specifically focused on the analysis of differentially expressed genes (DEGs) in six comparison groups. In the early stages of salt stress (2 h and 24 h) (Figure 3A), the number of up-regulated genes was significantly higher than that of down-regulated genes in four comparison groups, except for the comparison of CK2-vs.-h2, where the number of up-regulated and down-regulated genes were similar. Interestingly, the number of DEGs under the h treatment was lower than that under the m treatment at 2 h, while the number of DEGs under the h treatment was higher than that under the m treatment at 24 h. Until 48 h of stress treatment, the number of DEGs increased significantly, particularly under the h treatment, where the number of DEGs showed a sharp increasing trend. Thus, the salt stress response genes in *M. tricuspidata* were significantly expressed after 48 h of salt stress. In addition, under two different concentrations of salt stress (m and h) at the same treatment time (2, 24, and 48 h), a total of 138, 393, and 1293 commonly DEGs were identified, respectively (Figure 3B–D). Among these, 23 DEGs were found to be common to all six comparison groups (Figure 3E).

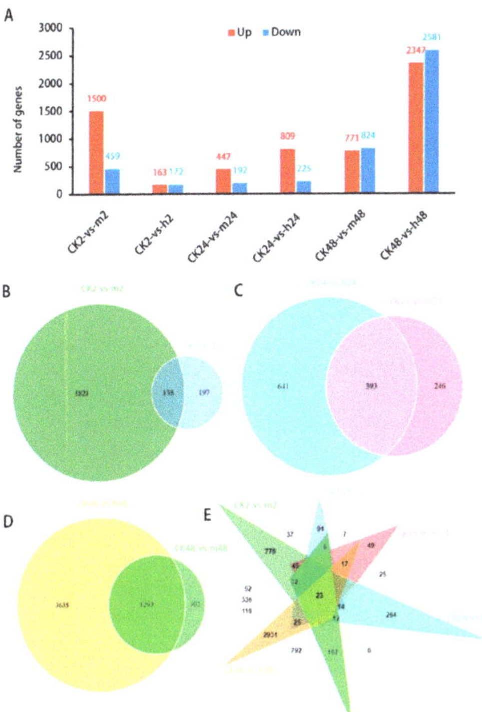

Figure 3. The number of DEGs in different comparison groups (CK2-vs.-m2, CK2-vs.-h2, CK24-vs.-m24, CK24-vs.-h24, CK48-vs.-m48, and CK48-vs.-h48). (**A**) Number of up and down DEGs in six comparison groups. (**B**) Common DEGs under two comparison groups under 2 h treatment. (**C**) Common DEGs under two comparison groups under 24 h treatment. (**D**) Common DEGs under two comparison groups under 48 h treatment. (**E**) Common DEGs among six comparison groups.

2.4. Gene Expression Validation by qRT-PCR

Nine DEGs in the CK48-vs.-h48 comparison were randomly selected for qRT-PCR validation (Figure 4). The expression patterns of these nine genes were consistent with the transcriptomic data, with a correlation coefficient (R) of 0.9932, suggesting that the gene expression patterns detected by RNA-seq are reliable.

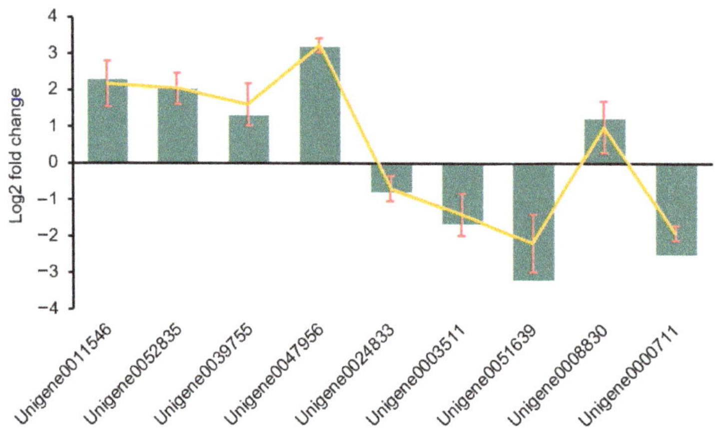

Figure 4. Relative gene expression of DEGs from RNA-seq data verified by qRT-PCR. Note: The green histogram indicates the value of FPKM from RNA-seq and the yellow line indicates the expression level of nine genes detected by qRT-PCR; the error line represents SD calculated from three replicates of qRT-PCR.

2.5. Functional Annotation Analysis of DEGs

To understand the functions of DEGs, GO and KEGG enrichment analysis were performed. The results revealed that 13 GO terms from at least three groups were significantly enriched in six different comparisons (Figure 5A). These terms included phosphorus metabolic process (GO:0006793), response to hormone (GO:0009725), response to organic substance (GO:0010033), small molecule biosynthetic process (GO:0044283), pigment biosynthetic process (GO:0046148), response to stimulus (GO:0050896), and response to other organism (GO:0051707) in biological process ontology. Additionally, transcription regulatory region nucleic acid binding (GO:0001067), protein serine/threonine kinase activity (GO:0004674), transferase activity (GO:0016740), phosphotransferase activity, alcohol group as acceptor (GO:0016773), transmembrane transporter activity (GO:0022857), and chlorophyll synthetase activity (GO:0046408) were identified in the molecular function ontology. Moreover, KEGG enrichment results showed that 9 pathways from at least three groups were significantly enriched (Figure 5B). These pathways included phenylpropanoid biosynthesis (ko00940), flavonoid biosynthesis (ko00941), tropane, piperidine and pyridine alkaloid biosynthesis (ko00960), plant–pathogen interaction (ko04626), metabolic pathways (ko01100), biosynthesis of secondary metabolites (ko01110), ubiquinone and other terpenoid-quinone biosynthesis (ko00130), mitogen-activated protein kinase (MAPK) signaling pathway-plant (ko04016), and plant hormone signal transduction (ko04075).

Figure 5. Enrichment analysis of DEGs in response to salt stress. (**A**) GO functional annotation of DEGs. (**B**) Enrichment analysis of DEGs in KEGG pathways. Darker red color indicates higher degree of enrichment. NA is referred to No Abundance.

2.5.1. Plant Hormone Signal Transduction

Plant hormones play a critical role in plant responses to environmental stresses. In this study, we found that the plant hormone signal transduction pathway was significantly enriched in three comparison groups. Specifically, auxin influx carrier (*AUX1*), auxin/indole-3-acetic acid (*AUX/IAA*), and auxin response factor (*ARF*), which are involved in auxin signaling, were more significantly upregulated under salt stress (Figure 6). Gretchen hagen3 (*GH3*), an early responsive genes of plant auxin, was upregulated only in the first (CK2-vs.-m2) and sixth (CK48-vs.-h48) comparison groups. Small auxin-up RNAs (*SAUR*), another early responsive gene of plant auxin, was quickly and specifically upregulated after salt stress treatment. In the cytokinine (CTK) signal transduction pathway, cytokinin receptor (*CRE1*) was significantly upregulated under 48 h treatment, and *AHP* and *ARR* were found to be upregulated in various comparison groups. In the gibberellin (GA) signaling pathway, gibberellin-insensitive dwarf 1 (*GID1*) and *DELLA* were significantly upregulated in various comparison groups. In contrast, in the abscisic acid (ABA) pathway, pyrabactin resistance/-likes (*PYR/PYL*), protein phosphatase type II C (*PP2C*), sucrose non-fermenting 1-related protein kinase 2 (*SnRK2*), and ABA responsive element binding factor (*ABF*), which are the main proteins involved in the ABA receptor signaling pathway, exhibited a downward trend in expression in most comparison groups, except for nine *PP2C*, one *SnRK2*, and three *ABF*, which were upregulated in the CK48-vs.-h48 comparison. In the ethylene pathway, ethylene receptor (*ETR*) and EIN3-binding F box protein 1 and 2 (*EBF1/2*) were mainly downregulated, except for CK48-vs.-h48. In the brassinosteroid (BR) pathway, BR insensitive 1 (*BRI1*), BR signaling kinase (*BSK*), touch 4 (*TCH4*), and D-type cyclin 3 (*CYCD3*) were upregulated in various comparison groups. In the jasmonic acid (JA) pathway, jasmonate ZIM domain-containing protein (*JAZ*) and myelocytomatosis protein 2 (*MYC2*) were downregulated in several comparison groups but significantly upregulated under CK2-vs.-m2. Finally, one DEG encoding the transcription factor *TGA* and three pathogenesis-related protein 1 (*PR-1*) in the salicylic acid (SA) signaling pathway were downregulated in the early treatment (2 h) but upregulated in the 24 and 48 h treatments.

2.5.2. Phenylpropanoid Biosynthesis

Phenylpropanoids play an important role in plants' response to both biotic and abiotic stresses. In our study, we observed that DEGs involved in the phenylpropanoid biosynthesis pathway were significantly affected by salt stress treatment across six comparison groups (Figure 7A). Key enzymes in the phenylpropanoid biosynthesis pathway, including phenylalanine ammonialyase (*PAL*), 4-coumarate-CoA ligase (*4CL*), caffeate/5-hydroxyferulate 3-O-methyltransferase (*COMT*), caffeoyl shikimate esterase (*CSE*), cinnamoyl-CoA reductase (*CCR*), and cinnamyl alcohol dehydrogenase (*CAD*), exhibited differential expression patterns among the six comparison groups, particularly in the CK2-vs.-m2 group, where most of these enzymes showed an upregulated expression. Interestingly, the expression trend of these enzymes in the CK48-vs.-h48 group was opposite to that in the CK2-vs.-m2 group. However, it is worth noting that, as the salt concentration increased, the change in gene expression at the same treatment time became more consistent.

2.5.3. Flavonoid Biosynthesis

Genes involved in flavonoid biosynthesis, including chalcone synthase (*CHS*), flavonol synthase (*FLS*), dihydroflavonol 4-reductase (*DFR*), and leucoanthocyanidin reductase (*LAR*), were significantly enriched in all six comparison groups (Figure 7B). Interestingly, these four genes exhibited completely opposite expression patterns between the CK2-vs.-m2 (upregulated change) group and the CK48-vs.-h48 group (downregulated change).

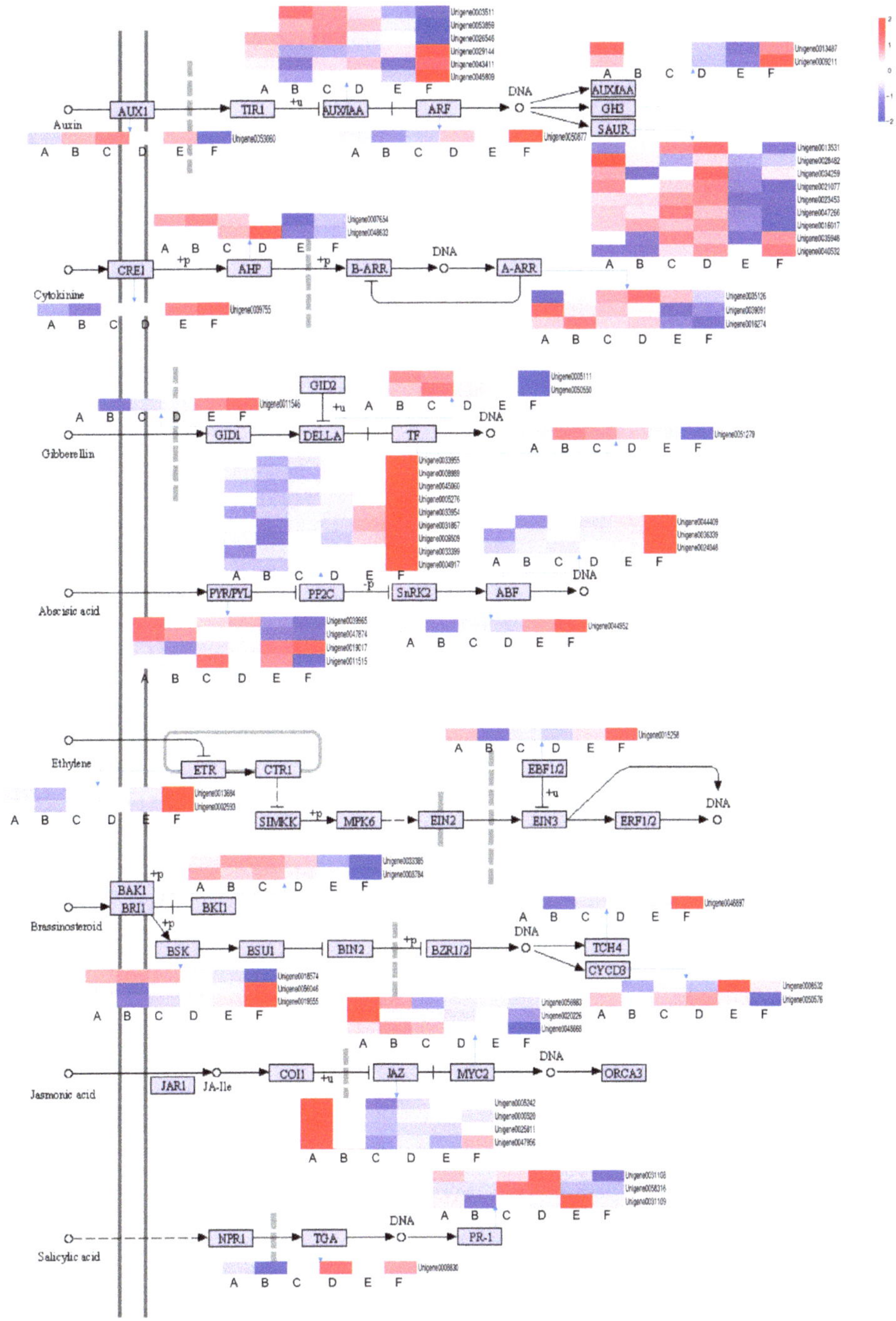

Figure 6. Expression of DEGs involved in the plant hormone signal transduction pathway of *M. tricuspidata* under salt stress. (A) The comparison of CK2-vs.-m2. (B) CK2-vs.-h2. (C) CK24-vs.-m24. (D) CK24-vs.-h24. (E) CK48-vs.-m48. (F) CK48-vs.-h48. The redder the heatmap color, the higher the level of gene expression.

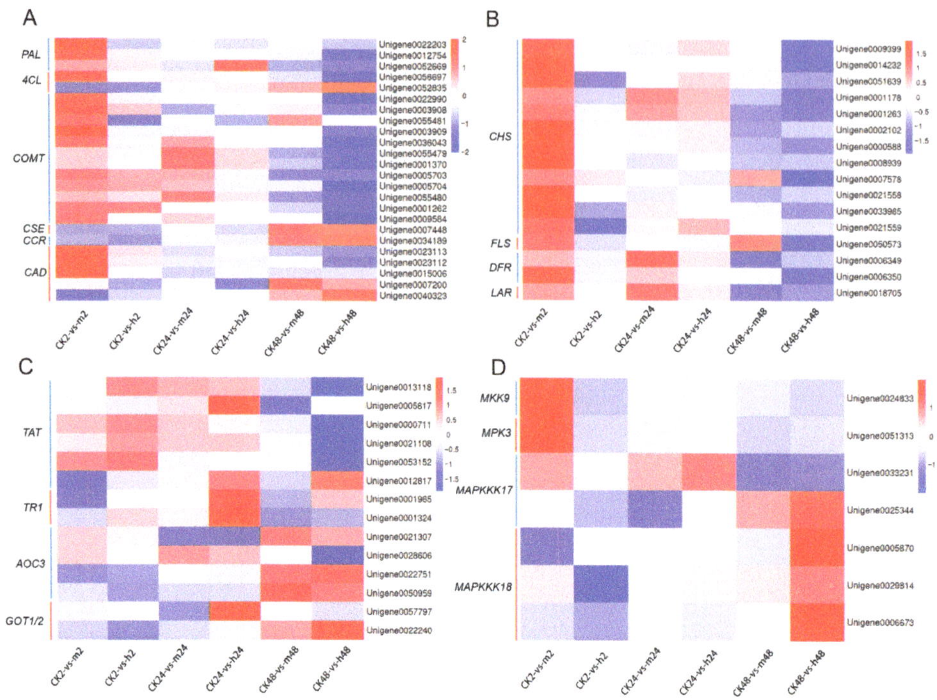

Figure 7. Expression of DEGs in four important metabolic pathways under salt stress. (**A**) Phenylpropanoid biosynthesis. (**B**) Flavonoid biosynthesis. (**C**) Tropane, piperidine and pyridine alkaloid biosynthesis. (**D**) MAPK signaling pathway plant. The redder the heatmap color, the higher the level of gene expression.

2.5.4. Tropane, Piperidine, and Pyridine Alkaloid Biosynthesis

In the tropane, piperidine, and pyridine alkaloid biosynthesis pathway, four types of gene transcripts were identified as tyrosine aminotransferase (*TAT*), tropinone reductase 1 (*TR1*), allene oxide cyelase 3 (*AOC3*), and glutamic-oxaloacetic transaminase (*GOT1/2*). Most of these genes exhibited upregulated expression levels across the six comparison groups (Figure 7C).

2.5.5. MAPK Signaling Pathway Plant

MAPK signaling pathway showed significant enrichments in all six comparison groups (Figure 7D). This pathway included important genes such as mitogen-activated protein kinase kinase (*MKK9*), mitogen-activated protein kinase (*MPK3*), and mitogen-activated protein kinase kinase kinase (*MAPKKK17/18*). In the early stage of treatment, *MKK9* and *MPK3* showed upregulation, but as the concentration and time increased, their expression levels decreased. Conversely, one *MAPKKK17* and three *MAPKKK18* genes displayed downregulation in the early stage, but their expression levels increased with a higher concentration and longer treatment time.

2.6. Summary of the LC-MS/MS Metabolomics

In the current study, both positive and negative ion mode (POS and NEG) were employed to achieve a wider coverage of metabolites and obtain more desirable detection results. A total of 1506 and 1721 known metabolites were successfully identified under the POS and NEG modes, respectively. The number of differentially accumulated metabolites (DAMs) is shown in Table 1.

Table 1. The number of DAMs between the different comparison groups.

Ion Mode	Group	Up	Down	Total
POS	CK2-vs.-m2	18	38	56
	CK2-vs.-h2	57	59	116
	CK24-vs.-m24	4	16	20
	CK24-vs.-h24	54	36	90
	CK48-vs.-m48	44	35	79
	CK48-vs.-h48	39	36	75
NEG	CK2-vs.-m2	48	65	113
	CK2-vs.-h2	70	63	133
	CK24-vs.-m24	10	31	41
	CK24-vs.-h24	32	44	76
	CK48-vs.-m48	44	29	73
	CK48-vs.-h48	26	32	58

2.7. Differential Metabolites Respond to Salt Stress

The KEGG pathway enrichment analysis revealed a significant enrichment of several pathways including alanine, aspartate and glutamate metabolism (ko00250), glycine, serine, and threonine metabolism (ko00260), butanoate metabolism (ko00650), biosynthesis of plant hormones (ko01070), and 2-oxocarboxylic acid metabolism (ko01210) in at least two comparison groups. Among these, the pathway of glycine, serine, and threonine metabolism and 2-oxocarboxylic acid metabolism were significantly enriched in four comparison groups (Figure 8A). Consequently, the subsequent analysis focuses on the detailed exploration of these five pathways (Figure 8B).

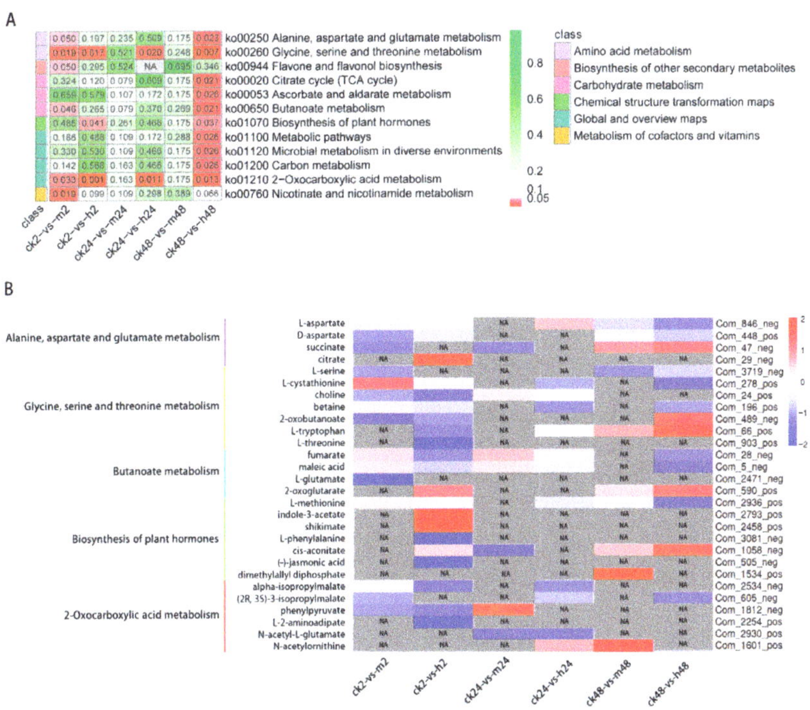

Figure 8. Differential metabolites of *M. tricuspidata* in response to salt stress between different comparisons: (**A**) KEGG analysis of differential metabolites; and (**B**) Abundance of DAMs in five pathways. NA is referred to No Abundance.

2.7.1. Alanine, Aspartate, and Glutamate Metabolism

Across the six comparison groups under stress treatment, four differential metabolites (L-aspartate, D-aspartate, succinate, and citrate) were found to be significantly enriched in the alanine, aspartate, and glutamate metabolism pathway. Among these metabolites, L-aspartate exhibited a significant increase within 24 h under high concentration stress, succinate showed a significant increase within 48 h under stress, and citrate displayed a significant increase within 2 h under the h treatment. However, the metabolites in other parts either significantly decreased or were not detected.

2.7.2. Glycine, Serine, and Threonine Metabolism

Among the six comparison groups analyzed, seven metabolites (L-serine, L-cystathionine, choline, betaine, 2-oxobutanoate, L-tryptophan, and L-threonine) involved in glycine, serine, and threonine metabolism were found to be enriched in four comparison groups. Specifically, at three time points under the h treatment, L-cystathionine under the m treatment for 2 h, 2-oxobutanoate under the h treatment for 48 h, and L-tryptophan for 48 h, the abundance of these metabolites significantly increased during these specific treatment conditions.

2.7.3. Butanoate Metabolism

Four metabolites (fumarate, maleic acid, L-glutamate, and 2-oxoglutarate) involved in butanoate metabolism were found to be significantly enriched in the first and sixth comparison groups. Fumarate and maleic acid showed a slight increase at 2 h and 24 h under the m treatment, while the abundance of 2-oxoglutarate increased under the h treatment at 2 h and 48 h.

2.7.4. Biosynthesis of Plant Hormones

Plant hormones play a crucial role in regulating various metabolic processes to adapt to changes in external environmental conditions. Seven metabolites (L-methionine, indole-3-acetate, shikimate, L-phenylalanine, cis-aconitate, (−)-jasmonic acid, and dimethylallyl diphosphate) involved in the biosynthesis of plant hormone pathways were significantly enriched under the h treatment at 2 h and 48 h. The abundance of indole-3-acetate and shikimate significantly increased under the h treatment at 2 h. Cis-aconitate abundance increased under the h treatment at 2 h, as well as under the m and h treatments at 48 h. Additionally, the abundance of dimethylallyl diphosphate significantly increased under the m treatment at 48 h.

2.7.5. 2-Oxocarboxylic Acid Metabolism

Six metabolites (alpha-isopropylmalate, (2R,3S)-3-isopropylmalate, phenylpyruvate, L-2-aminoadipate, N-acetyl-L-glutamate, and N-acetylornithine) involved in 2-oxycarboxylic acid metabolism were significantly enriched in four out of the six comparison groups, including the m and h treatment at 2 h, and the h treatment at 24 h and 48 h. Phenylpyruvate abundance showed an increasing trend under the m at 24 h, while N-acetylornithine abundance increased under the h treatment at 24 h and the m treatment at 48 h.

2.8. Conjoint Analysis of Transcriptomic and Metabolomic Data

The establishment and visualization of gene–metabolite interactions were based on KEGG pathway maps. We conducted a comprehensive analysis of omics data under 2, 24, and 48 h of salt treatment, resulting in the identification of 10,388 DEGs and 650 DAMs across all comparisons. Two pathways, namely metabolic pathways, glycine, serine, and threonine metabolism, were found to be highly significantly enriched in the sixth comparison groups (Supplementary Tables S2 and S3). Furthermore, the glycine, serine, and threonine metabolism pathway, which was observed in both the metabolome and transcriptome–metabolome correlation analyses, may be the essential biological processes in response to salinity.

Pearson correlation coefficient was used to evaluate the relationship between DEGs and DMAs. Ultimately, the top 250 DEGs were selected based on their correlation coefficients with DMAs. The expression heatmap of these DEGs and DMAs was shown in Figure 9A. It can be observed that the DMA of Com_2378_pos and Com_7892_pos were significantly positively correlated with other DEGs (except Unigene0047966), while Unigene0047966 was positively correlated with other DMAs, and Unigene0046388 was positively correlated with all DMAs. Moreover, a network diagram of the correlation coefficients between these DEGs and DMAs is presented in Figure 9B. The biosynthesis of secondary metabolites, tryptophan metabolism, and metabolic pathways were prominent, indicating that these three pathways may play a pivotal part in the transcriptional and metabolic response to salt stress.

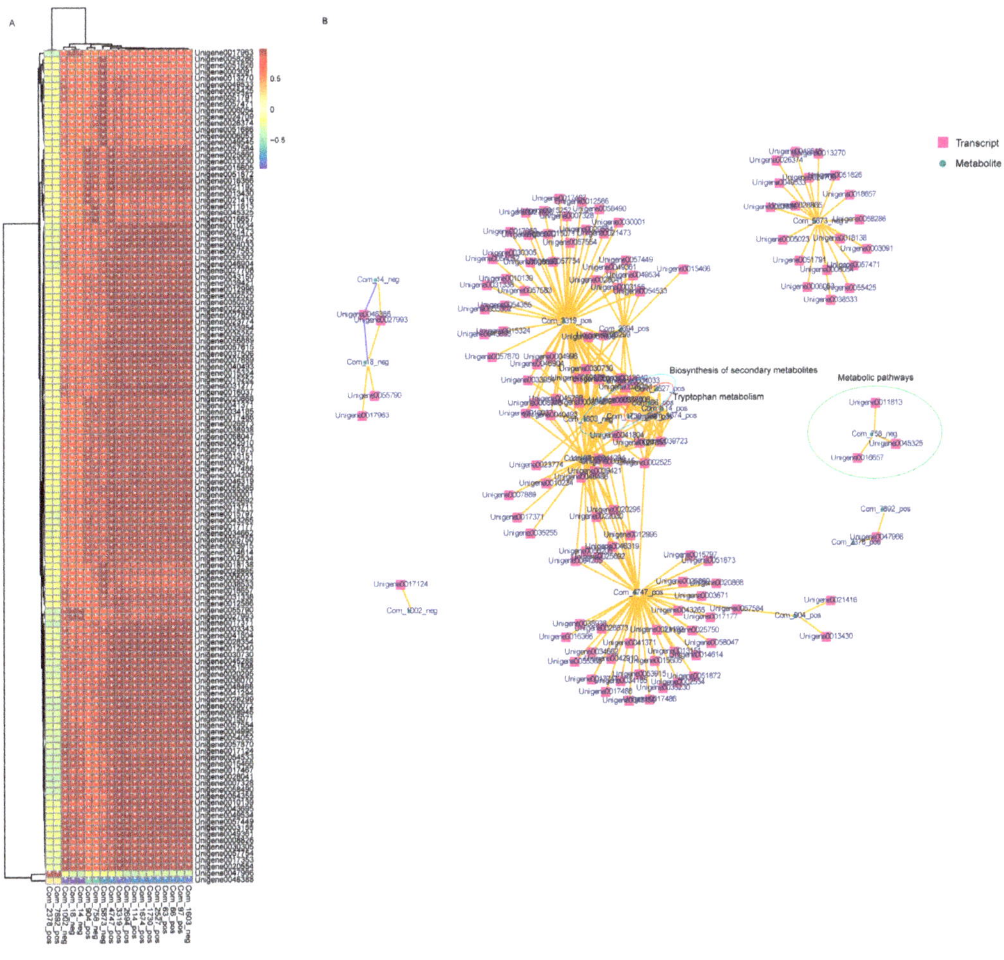

Figure 9. Correlation analysis of DEGs and DAMs in *M. tricuspidate* under salt stress. (**A**) Heatmap of the top 250 DEGs and 19 DAMs. (**B**) Cytoscape network of the top DEGs and top DAMs within the same KEGG pathways. The asterisk (*) represents positive correlation.

3. Discussion

3.1. Salinity-Activated Hormone Signaling to Participate in Stress Response

Plants optimize the balance between growth and stress by integrating exogenous salt stress and endogenous development signals [29,30]. Generally, significant changes occurring in plants under stress conditions can affect hormone flux and signal transduction [31].

Auxin participates in plants' adaptation to environmental stress by regulating developmental processes and morphological structures of plant leaves [32]. Salt stress commonly leads to a reduction in IAA levels and alters the localization of *AUX1* and *Pin1/2* [33], resulting in the downregulation of auxin receptor-encoding genes *TIR1* and *AFB2*, inhibiting plant growth [34]. Our transcriptome analysis revealed that the gene expression of *AUX1*, *AUX/IAA*, and *SAUR* in the IAA signal transduction pathway was mostly upregulated initially, but started to downregulate under 48 h of stress treatment. This suggests that the initial salt stress may promote the auxin signal transduction process but inhibits it during the later stages of stress. In cotton, an increase in IAA levels and the activation of the auxin signal transduction pathway contribute to salt stress adaptation [35]. Similarly, in *Arabidopsis thaliana*, an increase in IAA levels also enhances salt tolerance [36]. These findings are consistent with our current results, indicating that salinity activates the transcription of key genes involved in IAA signal transduction and enhances salt adaptation.

The appropriate modulation of CTK metabolism and signal transduction are beneficial for plant survival in saline environments [37]. Under salt stress conditions, the expression of CTK signal components, such as *AHK*, *AHP*, *CRF*, and *ARR*, undergoes significant changes [29]. In the '84 K' poplar clone, salt stress was found to activate the cytokinin signaling pathway, and 59 *ARRs* showed transcriptional changes during salt treatment, indicating their potential involvement in salt response [38]. Consistent with these findings, we observed that the transcription of *AHP* and *A-ARR* was inhibited in the later stage of the CTK signaling pathway under salt stress. However, *CRE1* exhibited an upregulated expression pattern.

The dynamic balance mechanism of active GA content in plants relies on the core components of the GA signaling pathway, including DELLA proteins (such as RGA, GAI, RGL1/2/3), F-box protein SLEEPY1, and soluble GA receptor protein GID1 [39,40]. When exposed to salt stress, *Arabidopsis* experiences a decrease in GA content, while the accumulation of DELLA proteins can enhance plant resistance to salt stress [41]. Additionally, DELLA proteins can counteract the production of ROS induced by abiotic stress, thereby increasing the plant's resilience [42]. In our observations, we observed an increase in *GID1* expression level under 48 h treatment, while *DELLA* showed the opposite trend.

ABA plays an essential role in defending against salt stress [43,44]. Under salt stress, ABA activates downstream response genes, forming a PYL–ABA–PP2C complex that removes the PP2C inhibition of SnRK2. This leads to SnRK2 phosphorylating TFs, which helps maintain ion balance and clears ROS, thereby mitigating the damage caused by stress [8]. In the '84 K' poplar, researchers constructed a subnetwork primarily consisting of *ABF* and *PP2C* in the ABA transduction pathway [38]. In our study, we observed the upregulation of four different *PYL* genes in different comparison groups, but the expression levels of *PP2C*, *SnRK2*, and *ABF* were the highest under 48 h of stress treatment. These results differ from those found by Zhao et al. [38], which might be attributed to the fact that they studied six different time points in poplar, whereas we only examined three time points in our study.

Ethylene signal transduction plays an important role in plant's response to salt stress [29]. The exogenous application of ethylene increased the sensitivity of rice seedlings to salt, while the application of the ethylene inhibitor 1-methylcyclopropene (1-MCP) could enhance rice's tolerance to salt [45]. The lectin receptor-like kinase *OsSIT1* activated *MPK3/6*, leading to increased ethylene biosynthesis, accumulation of ROS, and salt sensitivity [46].

In *Arabidopsis* subjected to high salt treatment, the degradation of two F-box proteins, EBF1/ EBF 2, was promoted, facilitating the accumulation of two core TFs, *EIN3* and *EIL1* [47]. However, our study found that the expression of *ETR* and *EBF1/2* genes was upregulated after 48 h of high salt stress, which is inconsistent with the findings of Peng et al. [47]. This discrepancy suggests that the different regulatory mechanisms may exist in various plant species.

BRs play a vital role in regulating plant salt tolerance; however, their effect is not always positive. Liu et al. (2020) proposed that appropriately increasing the level or enhancing the signaling of BRs can improve plant salt tolerance, but excessive or insufficient levels of BRs may be detrimental to plant salt tolerance [48]. The membrane-bound receptor *BRI1* recognizes the presence of BR, inducing the formation of a heterodimer between *BRI1* and BRI1-associated kinase 1 (*BAK1*), which in turn activates downstream components such as *BSKs* and BR-insensitive 2 (*BIN2*), ultimately leading to the activation of BR-responsive genes [49]. *ZmBES1/BZR1-5* has been reported to have increased expression in maize shoots, while its expression is consistently suppressed in roots, indicating that salt stress may activate distinct BR responses in different tissues/organs within the same plant [50]. Our research found that the expression pattern of one of the three *BSK* genes was consistent with that of *BRI1*, and the expression levels of these genes increased at 2 h and 24 h of salt treatment, indicating that the *BRI1* gene plays a regulatory role in downstream *BSK* genes.

JA is one of the key hormones involved in plant responses to abiotic stress. Previous studies have shown that JA enhances tomato's salt tolerance by maintaining the balance of ROS [21]. JAZ proteins act as transcriptional repressors, inhibiting the activity of key TFs (such as basic helix–loop–helix (*bHLH*), *MYC2*, and its homologues *MYC3/4/5*) in the JA pathway [51]. Herein, we observed that the expression of *JAZ* and the major regulatory factor *MYC2* was upregulated in the early stages of salt stress and downregulated in the later stages in the JA signaling pathway, indicating that salt stress activates JA signal transduction.

SA, as a key defensive hormone, can alleviate salt stress in plants [52]. SA enhances the antioxidant system, promotes the synthesis of osmotic substances, and accelerates plant photosynthesis under salt stress [53]. The transcriptomic analysis of cotton has shown that important *TGA* elements involved in the SA signaling pathway were downregulated [35]. In this study, we found that the expression levels of TGA genes decreased in the early stages of stress treatment but increased in the later stages, and this effect was relieved in the later stages. These results suggest that SA participates in the salt tolerance response of *M. tricuspidata*.

3.2. Phenylpropane and Flavonoid Biosynthesis Pathways Involved in the Response to Salt Stress

Our transcriptomic analysis revealed that the majority of genes involved in phenylpropane biosynthesis were significantly upregulated during the early stage of salt treatment (2 and 24 h) (Figure 7A). However, in the later stage (48 h), the expression level of *4CL* in the general phenylpropane biosynthetic pathway, as well as key genes in the specific lignin biosynthetic pathway, such as *CCR*, *CAD*, and *CSE*, were significantly upregulated. This suggests that lignin synthesis was not inhibited under salt stress conditions. Cinnamates and flavonoids in the mulberry 'Guisangyou 12′ were induced and expressed under salt stress, and studies have shown that phenylpropane may be involved in salt stress response and enhance mulberry's salt tolerance [54]. Lignin plays a protective role in cells against abiotic stresses and serves as a structural component that helps perceive and respond to environmental stresses [55,56]. The upregulation of these genes may protect plants from salt stress, and the upregulation of lignin biosynthesis could be an important defense mechanism in *M. tricuspidata* against salt stress.

The present study identified DEGs in the flavonoid biosynthesis pathway that were upregulated and downregulated in the first and last comparison groups, respectively (Figure 7B), exhibiting an opposite trend under different stress times. This is consistent with previous research on *Belamcanda chinensis*, where the levels of flavonoids varied with the degree and duration of drought stress [57]. Zhang et al. found that, as salt stress increased, the flavonoid content in *Cyclocarya paliurus* also showed an increasing trend [58]. Under single and combined stress of NaCl and $CdCl_2$, flavonoids were found to be involved in maintaining the normal state of *Tamarix hispida* [59]. Research on *Glycyrrhiza uralensis* has found that *PAL*, cinnamate 4-hydroxylase (*C4H*), *4CL*, *CHS*, chalcone-isomerase (*CHI*), and *FLS* exhibited higher expression levels under salt treatment [60]. Salt treatment significantly increased the flavonoid biosynthesis pathway of *Cyclocarya paliurus* seedlings at the metabolic level to cope with salt stress [61]. Consistently, Wanichthanarak et al. found that the KEGG pathways were enriched in plant hormone signal transduction, phenylpropanoid, and flavonoid biosynthesis in rice to respond to salinity stress [62]. Therefore, our transcriptomic analysis revealed significant changes in genes related to flavonoid biosynthesis, and the dynamic changes in flavonoids may play a role in alleviating salt stress in *M. tricuspidata.*

3.3. Alkaloids and Plant MAPK Signaling Pathway Enhance the Salt Tolerance

Alkaloids, as secondary metabolites of tryptophan and glutamic acid synthesis, are also involved in defense against salt stress [63]. In this study, the tropane, piperidine, and pyridine alkaloid biosynthesis pathway revealed that four important genes, namely *TAT*, *TR1*, *AOC3*, and *GOT1/2*, were upregulated in most of the comparison groups (Figure 7C). Especially under 48 h treatment, the upregulated expression of *AOC3* and *GOT1/2* was more pronounced, suggesting that the accumulation of these genes may be a defensive response to salt stress. Jia et al. found that the changes in alkaloid levels in *Malus halliana* can alleviate the damage caused by ROS, thereby enhancing its resistance to salt stress [64]. Triple mutants of *Atahp2,3,5* and *Atarr1,10,12* in *Arabidopsis* exhibited greater salt tolerance, and transcriptomic analysis revealed that the DEGs were mainly enriched in phenylpropanoid, flavonoid, terpenoid, piperidine, and pyridine alkaloid biosynthesis, as well as plant hormone signaling transduction [65].

MAPK is generally considered to play a crucial role in the biological response to environmental stress by phosphorylating downstream signaling pathways [66]. In this study, four types of MAPK signaling transduction genes were identified, including *MKK9*, *MPK3*, and *MAPKKK17/18*. The early stage (2 h) of salt treatment induced upregulated expression of *MKK9* and *MPK3*, while the expression of *MAPKKK17/18* was significantly upregulated in the later stage (48 h) (Figure 7D). Yan et al. reported that *MPK3/6* interact with each other and phosphorylate the key cytokinin signaling components of *ARR1/10/12* under salt stress [67]. Additionally, *OsMAPKKK63* (*Arabidopsis* ortholog *AtMPKKK15/16/17/18*) was induced by high salt and interacted with *OsMKK1/6*, the mediators of salt stress response, and mutants of *OsMAPKKK63* indeed demonstrated the necessity of the normal response to high salt in rice [68]. The genome-wide identification of MAPK has revealed its potential in improving salt tolerance in *Gossypium hirsutum* [69]. MAPK signal transduction may induce a large number of antioxidants and free radical scavenging enzymes in *Dendrobium huoshanense*, thereby enhancing its salt tolerance [70]. Therefore, we speculate that the MAPK signaling pathway may play a vital role in enhancing the salt tolerance of *M. tricuspidata* in response to salt stress.

3.4. Amino Acid Metabolism Plays a Role in Enhancing the Salt Tolerance of *M. tricuspidata*

Amino acids accumulate in response to various abiotic stress tolerances and serve a variety of functions in plant growth and development [71]. The present study's metabolomic analysis revealed that the DAMs under salt treatments were mainly associated with alanine, aspartate, and glutamate metabolism, as well as the glycine, serine, and threonine metabolism pathways (Figure 8B). Research on *M. halliana* has found that the upregulation of the MhASP3 protein involved in the synthesis of aspartate and glutamate initiates amino acid defense responses to counteract salt stress [64]. The integration analysis of transcriptome and metabolome under salt stress revealed a strong enhancement of amino acid metabolism, including alanine, aspartate, glutamate, glycine, serine, and threonine, contributing to salt tolerance in *Sesamum indicum* [72]. A metabolomic analysis of *Glycine max* revealed that salt-tolerant soybean enhances salt tolerance by enhancing amino acid and organic acid metabolism [73]. Moreover, research has found that the accumulation of several amino acids, including serine, phenylalanine, and threonine, can enhance plant salt tolerance by mediating the ROS clearance and controlling damage repair processes [74,75]. The above metabolites were also present in our study, indicating that amino acid metabolism plays an essential role in the salt tolerance response in *M. tricuspidata*. Notably, in our integrated analysis, we also identified glycine, serine, and threonine metabolism as an important one that showed a significant correlation, indicating its crucial role in the salt stress response process. However, the butanoate metabolism and 2-oxocarboxylic acid metabolism pathways were unique in this study, suggesting that their presence indicates different mechanisms of salt tolerance in different plants.

4. Materials and Methods

4.1. Plant Material

M. tricuspidata seedings were collected from the Dafeng provenance in Yancheng, Jiangsu province (120°85′ E; 32°98′ N). These seedlings were cultivated in the greenhouse of Jiangsu Academy of Forestry (118°46′ E; 31°51′ N) for a growth period of 150 d, serving as the plant materials for this study. Under natural outdoor conditions, healthy seedlings were cultivated in culture containers filled with a mixture of peat soil and perlite in a ratio of 3:1 (v/v). The cultivation container used was a black plastic box measuring 500 cm × 450 cm × 380 cm.

4.2. Salt Treatment and Sample Collection

The seedlings were treated as follows: (a) control (CK: 0 mM NaCl); (b) low concentration (l: 34 mM NaCl); (c) moderate concentration (m: 68 mM NaCl); and (d) high concentration (h: 102 mM NaCl). These concentration levels were selected and adjusted based on previous research on salt tolerance in plants of the Moraceae family [76]. Meanwhile, four different time points (2, 24, 48, and 72 h) were chosen to observe the physiological and molecular responses to salt stress. Each treatment had 3 biological repeats, with each replicate consisting of 9 plants, resulting in a total of 108 plants (3 × 4 × 9). On September 26th, 2022, a 2 L NaCl solution of the corresponding concentration was irrigated once into each culture container, while the control treatment received an equal volume of pure water. Container trays were used to collect any exudate, which was then poured back into the container to prevent any outflow.

After each time point of the stress treatments, three samples were collected. Each sample consisted of three randomly selected plants combined to create a mixed sample. Each of these mixed samples contained 27 functional leaves from the same part of each plant, all with the same age. This provides the necessary tissue for the 3 replicates of physiological characteristics and transcriptome analysis, as well as 6 replicates for metabolomics. The collected samples were deposited in a −80 °C refrigerator until further use.

4.3. Physiological Characteristics Measurements

Plant soluble sugar (Product model: A145-1-1) and malondialdehyde (MDA) (Colorimetric method) (A003-3-1) contents, CAT (Visible light) (A007-1-1) and POD (A084-3-1) activities, total SOD (Hydroxylamine method) (A001-1-2), and the chlorophyll a and chlorophyll b (A147-1-1) concentrations were determined following the instructions provided in the kits obtained from Nanjing Jiancheng Bioengineering Institute (Nanjing, Jiangsu, China). Plant-soluble sugars, enzyme activities, and chlorophyll were extracted by homogenizing plant leaf tissue.

4.4. Transcriptome Analysis

4.4.1. RNA Extraction and Library Construction

Total RNA was extracted using a FastPure plant total RNA isolation kit (Polysaccharides & Polyphenolics-rich) from VazymeBiotech, Nanjing, China, following the manufacturer's protocol. The quality, concentration, and integrity of RNA were assessed using

NanoPhotometer spectrophotometer (Implen, Calabasas, CA, USA), Qubit2.0 fluorometer (Invitrogen, Carlsbad, CA, USA), and Agilent 2100 bioanalyzer (Agilent Technologies, Palo Alto, CA, USA), respectively. RNA-seq libraries were generated using the NEBNext® Ultra™ RNA Library Prep Kit for Illumina® NEB, #E7530L, Ipswich, MA, USA) according to the manufacturer's instruction, and then sequenced using an Illumina NovaSeq6000 (Illumina, San Diego, CA, USA).

4.4.2. Data Analysis and Functional Annotation

The sequencing data were filtered using Fastp (version 0.23.2) (https://github.com/OpenGene/fastp, accessed on 6 September 2022) [77]. After filtering out reads containing adapter sequences, over 10% N bases, all A bases, and low-quality ($Q \leq 20$) readings of more than 50 bases, the resulting clean reads were obtained. The cleaned reads were then de novo assembled into transcripts using Trinity software (version 2.13.2) (https://github.com/trinityrnaseq/trinityrnaseq/, accessed on 10 September 2022) [78]. The integrity of the assembly sequence was assessed using BUSCO software (version 3.0.2) [79]. The expression abundance of the unigenes was calculated using RSEM software (version 3.0.2) in terms of fragments per kilobase of exon per million mapped reads (FPKM) values [80]. DEGs were identified using the DESeq2 software (version 1.6.3) based on the criteria of false detection rate (FDR) < 0.05 and (|log2 fold change (FC)| > 1) [81]. The DEGs were functionally annotated using the Gene Ontology (GO) and Kyoto Encyclopedia of Genes and Genomes (KEGG) pathways [82]. The significantly enriched GO terms and KEGG pathways were determined based on an FDR cutoff of 0.05.

4.5. Metabolomics Analysis

4.5.1. Metabolites Extraction and Ultra-High-Performance Liquid Chromatography (UHPLC)-MS/MS Analyses

The ground leaf tissue samples (100 mg) were placed in an Eppendorf (EP) tube and resuspended in 500 µL of prechilled 80% methanol. The mixture was incubated in an ice water bath for 5 min, followed by centrifugation at 4 °C, 15,000× *g* for 20 min. Subsequently, a portion of the supernatant was diluted to a final concentration of 53% methanol. The diluted samples were transferred to new EP tube and centrifuged again. Finally, the supernatant was injected for analysis using an LC-MS/MS system [83].

The UHPLC-MS/MS analysis was performed using the Vanquish UHPLC system (Thermo Fisher, Dreieich, Germany) connected to the Orbitrap Q ExactiveTM HF-X mass spectrometer (Thermo Fisher, Dreieich, Germany). The samples were injected onto a Hypesil Gold column (100 × 2.1 mm, 1.9 µm) and eluted with a linear gradient at a flow rate of 0.2 mL/min for 17 min. For the positive polarity mode, the eluent A consisted of 0.1% formic acid in water, while the eluent B was methanol. For the negative polarity mode, the eluent A was 5 mM ammonium acetate (pH = 9.0), and the eluent B was methanol. The solvent gradient settings were as follows: 2% B for 1.5 min; 2–100% B for 12 min; 100% B for 14 min; 100–2% B for 14.1 min; and 2% B for 17 min. During the analysis, the spray voltage was 3.2 kV, the capillary temperature was 320 °C, and the flow rate of the sheath gas and aux gas were 40 Arb and 10 Arb, respectively.

4.5.2. Data Processing, Metabolite Identification, and Multivariate Statistical Analysis

Compound discoverer 3. 1 (CD3.1, Thermo Fisher, Hercules, CA, USA) was used to process the raw data files generated by UHPLC-MS/MS, and peak alignment, peak picking, and quantitation were performed for each metabolite. The main characteristics were as follows: retention time tolerance, 0.2 min; actual mass tolerance, 5 ppm; signal intensity tolerance, 30%; signal/noise ratio, 3; and minimum intensity threshold, 100,000. Afterwards, the peak intensities were normalized to the total spectral intensity. Molecular formulas were predicted using the normalized data of additive ions, molecular ion peaks, and fragment ions. The obtained peaks were matched with the mzCloud (https://www.mzcloud.org/, accessed on 12 January 2023). Accurate qualitative and relative quantitative results were obtained from mz Vaultand Mass Listdatabase. Quality control samples were used to assess the system's stability, and blank samples were employed to remove the background ions. Statistical analysis was performed using R software (version 3.4.3), Python (version 2.7.6) and CentOS (version 6.6).

4.5.3. Identification of Differential Metabolites and Analysis of Metabolic Pathway

The variable importance in the projection (VIP) scores of orthogonal partial least squares discriminant analysis (OPLS-DA) and the *p*-values from *t*-tests were utilized to identify DAMs [84]. Metabolites were defined as differential whenever VIP $\geq$ 1 and *p*-value < 0.05, and were mapped to the KEGG metabolic pathways for annotation and enrichment analysis.

4.6. Integrative Analysis of Omics Data

In the association analysis, a shared KEGG pathway analysis was performed to identify a common pathway involving differential genes and metabolites between groups [85]. For the integrative analysis, a Pearson correlation coefficient model was employed. The correlation was established by calculating the expression of all differential genes and the abundance of all differential metabolites using a union approach across the comparison groups.

4.7. Statistical Analysis

IBM SPSS statistics (version 21) (IBM Inc., Armonk, NY, USA) was used to perform the analysis of variance (ANOVA), and significant variations were determined using Duncan's multiple range test method at a significance level of $p = 0.05$. The Cytoscape program (version 3.3.0) (NIGMS, Bethesda, MD, USA) was used to perform the network regulation diagram.

4.8. Quantitative Real-Time PCR (qRT-PCR) Validation of DEGs

qRT-PCR analysis was used to verify the accuracy of transcriptome profiles. The gene expression levels of 9 randomly selected genes were analyzed. Relative expression levels were calculated by the $2^{-\Delta\Delta Ct}$ method [86]. Ubiquitin (UBQ) was selected as an internal reference, and PCR primers for the tested genes were designed based on the RNA-Seq data (Supplementary Table S4).

5. Conclusions

In the current study, we conducted an integrated analysis of physiology, transcriptome, and metabolome in *M. tricuspidata* under salt treatment. Our findings revealed that the plant's salt adaptability is associated with maintaining appropriate levels of soluble sugar, increased activities of SOD and POD, as well as the maintained concentrations of chlorophyll a and chlorophyll b. Through KEGG pathway enrichment analysis, we identified key DEGs directly involved in plant hormone signal transduction, phenylpropanoid and flavonoid biosynthesis, alkaloids, and the MAPK signaling pathway. Furthermore, our differential metabolite analysis unveiled the involvement of the main pathways related to amino acid metabolism (alanine, aspartate and glutamate; glycine, serine, and threonine), butanoate and 2-oxocarboxylic acid metabolism, and the biosynthesis of plant hormones. Notably, the glycine, serine, and threonine metabolism pathway showed significance in both the metabolome and transcriptome–metabolome correlation analyses, highlighting its crucial role in response to salinity. Based on our findings, we constructed a regulatory mechanism model depicting salt stress adaptation in *M. tricuspidata* (Figure 10). This study will contribute to our understanding of the molecular mechanism underlying salt tolerance in *M. tricuspidata*, and provide valuable insights for the future development of salt-tolerant breeding and the improvement in salt land.

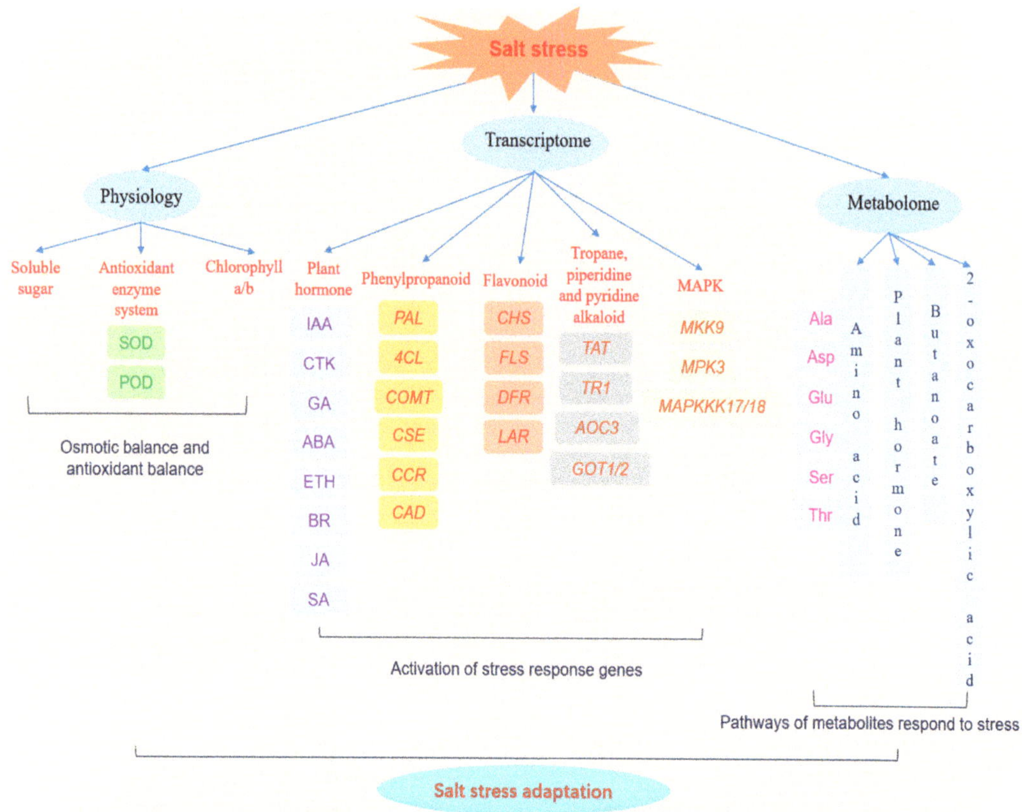

Figure 10. Schematic diagram of the mechanism of salt stress adaptation in *M. tricuspidate*.

Supplementary Materials: The following supporting information can be downloaded at: https://www.mdpi.com/article/10.3390/plants13030397/s1. Figure S1: Overview of *M. tricuspidate* transcriptome responses to salt stress. (A) Principal component analysis (PCA) plots of transcripts under salt stress. (B) The Pearson correlation coefficient between the samples shown in the heat map. Table S1: Summary of the RNA sequencing data. Table S2: Information of metabolites involved in metabolic pathways. Table S3: Information of metabolites involved in glycine, serine, and threonine metabolism. Table S4: Primers used in qRT-PCR analyses.

Author Contributions: D.S. conceived the project, prepared the data analysis, and wrote the manuscript. B.W. prepared data analysis and provided the sample. Y.A.E.-K. implemented the language editing, revised the manuscript. L.W. supervised the project and revised the manuscript. All authors have read and agreed to the published version of the manuscript.

Funding: This work was supported by Forestry Science and Technology Innovation and Extension Project in Jiangsu Province (LYKJ[2021]26); and Independent Scientific Research Project of Jiangsu Academy of Forestry (ZZKY202103).

Data Availability Statement: The transcriptome sequence data for *Maclura tricuspidate* were deposited at NCBI under BioProject code PRJNA1065092. The original contributions presented in the study are included in the article/Supplementary Materials.

Conflicts of Interest: The authors declare no conflict of interest.

References

1. Hassani, A.; Azapagic, A.; Shokri, N. Global predictions of primary soil salinization under changing climate in the 21st century. *Nat. Commun.* **2021**, *12*, 6663. [CrossRef]
2. Kumar, P.; Sharma, P.K. Soil salinity and food security in India. *Front. Sustain. Food Syst.* **2020**, *4*, 533781. [CrossRef]
3. Zhang, H.; Yu, F.; Xie, P.; Sun, S.; Qiao, X.; Tang, S.; Chen, C.; Yang, S.; Mei, C.; Yang, D.; et al. A G_{γ} protein regulates alkaline sensitivity in crops. *Science* **2023**, *379*, eade8416. [CrossRef]
4. Zhu, G.; Lu, H.; Shi, X.; Wang, Y.; Zhi, W.; Chen, X.; Liu, J.; Ren, Z.; Shi, Y.; Ji, Z.; et al. Nitrogen management enhanced plant growth, antioxidant ability, and grain yield of rice under salinity stress. *Agron. J.* **2020**, *112*, 550–563. [CrossRef]
5. Kerbab, S.; Silini, A.; Bouket, A.C.; Cherif-Silini, H.; Eshelli, M.; El Houda Rabhi, N.; Belbahri, L. Mitigation of NaCl stress in wheat by rhizosphere engineering using salt habitat adapted PGPR halotolerant bacteria. *Appl. Sci.* **2021**, *11*, 1034. [CrossRef]
6. Shahid, S.A.; Zaman, M.; Heng, L. Soil salinity: Historical perspectives and a world overview of the problem. In *Guideline for Salinity Assessment, Mitigation and Adaptation Using Nuclear and Related Techniques*; Springer: Cham, Switzerland, 2018; pp. 43–53.
7. Zhang, H.; Han, B.; Wang, T.; Chen, S.; Li, H.; Zhang, Y.; Dai, S. Mechanisms of plant salt response: Insights from proteomics. *J. Proteome Res.* **2012**, *11*, 49–67. [CrossRef] [PubMed]
8. Liu, C.; Mao, B.; Yuan, D.; Chu, C.; Duan, M. Salt tolerance in rice: Physiological responses and molecular mechanisms. *Crop J.* **2022**, *10*, 13–25. [CrossRef]
9. Xu, Y.; Magwanga, R.O.; Cai, X.; Zhou, Z.; Wang, X.; Wang, Y.; Zhang, Z.; Jin, D.; Guo, X.; Wei, Y.; et al. Deep transcriptome analysis reveals reactive oxygen species (ROS) network evolution, response to abiotic stress, and regulation of fiber development in cotton. *Int. J. Mol. Sci.* **2019**, *20*, 1863. [CrossRef] [PubMed]
10. Fan, Y.; Lu, X.; Chen, X.; Wang, J.; Wang, D.; Wang, S.; Guo, L.; Rui, C.; Zhang, Y.; Cui, R.; et al. Cotton transcriptome analysis reveals novel biological pathways that eliminate reactive oxygen species (ROS) under sodium bicarbonate ($NaHCO_3$) alkaline stress. *Genomics* **2021**, *113*, 1157–1169. [CrossRef] [PubMed]
11. Mittler, R.; Zandalinas, S.I.; Fichman, Y.; Van Breusegem, F. Reactive oxygen species signaling in plant stress responses. *Nat. Rev. Mol. Cell Biol.* **2022**, *23*, 663–679. [CrossRef]
12. Racchi, M.L. Antioxidant defenses in plants with attention to *Prunus* and *Citrus* spp. *Antioxidants* **2013**, *2*, 340–369. [CrossRef] [PubMed]
13. Mostofa, M.G.; Hossain, M.A.; Fujita, M. Trehalose pretreatment induces salt tolerance in rice (*Oryza sativa* L.) seedlings: Oxidative damage and co-induction of antioxidant defense and glyoxalase systems. *Protoplasma* **2015**, *252*, 461–475. [CrossRef] [PubMed]
14. Bandehagh, A.; Salekdeh, G.H.; Toorchi, M.; Mohammadi, A.; Komatsu, S. Comparative proteomic analysis of canola leaves under salinity stress. *Proteomics* **2011**, *11*, 1965–1975. [CrossRef] [PubMed]
15. Huang, C.; He, W.; Guo, J.; Chang, X.; Su, P.; Zhang, L. Increased sensitivity to salt stress in an ascorbate-deficient *Arabidopsis* mutant. *J. Exp. Bot.* **2005**, *56*, 3041–3049. [CrossRef] [PubMed]
16. Khan, M.N.; Khan, Z.; Luo, T.; Liu, J.; Rizwan, M.; Zhang, J.; Zhang, J.; Xu, Z.; Wu, H.; Hu, L. Seed priming with gibberellic acid and melatonin in rapeseed: Consequences for improving yield and seed quality under drought and non-stress conditions. *Ind. Crops Prod.* **2020**, *156*, 112850. [CrossRef]
17. Todd, E.V.; Black, M.A.; Gemmell, N.J. The power and promise of RNA-seq in ecology and evolution. *Mol. Ecol.* **2016**, *25*, 1224–1241. [CrossRef]
18. Goodwin, S.; McPherson, J.D.; McCombie, W.R. Coming of age: Ten years of next-generation sequencing technologies. *Nat. Rev. Genet.* **2016**, *17*, 333–351. [CrossRef]
19. Pirih, N.; Kunej, T. Toward a taxonomy for multi-omics science? Terminology development for whole genome study approaches by omics technology and hierarchy. *OMICS J. Integr. Biol.* **2017**, *21*, 1–16. [CrossRef]
20. Rasmussen, S.; Barah, P.; Suarez-Rodriguez, M.C.; Bressendorff, S.; Friis, P.; Costantino, P.; Bones, A.M.; Nielsen, H.B.; Mundy, J. Transcriptome responses to combinations of stresses in *Arabidopsis*. *Plant Physiol.* **2013**, *161*, 1783–1794. [CrossRef]
21. Abouelsaad, I.; Renault, S. Enhanced oxidative stress in the jasmonic acid-deficient tomato mutant *def-1* exposed to NaCl stress. *J. Plant Physiol.* **2018**, *226*, 136–144. [CrossRef]
22. Wang, W.S.; Zhao, X.Q.; Li, M.; Huang, L.Y.; Xu, J.L.; Zhang, F.; Cui, Y.R.; Fu, B.Y.; Li, Z.K. Complex molecular mechanisms underlying seedling salt tolerance in rice revealed by comparative transcriptome and metabolomic profiling. *J. Exp. Bot.* **2016**, *67*, 405–419. [CrossRef]
23. Li, L.; Peng, Z.; Mao, X.; Wang, J.; Li, C.; Chang, X.; Jing, R. Genetic insights into natural variation underlying salt tolerance in wheat. *J. Exp. Bot.* **2021**, *72*, 1135–1150. [CrossRef]
24. Xu, W.; Wang, Y.; Xie, J.; Tan, S.; Wang, H.; Zhao, Y.; Liu, Q.; El-Kassaby, Y.A.; Zhang, D. Growth-regulating factor 15-mediated gene regulatory network enhances salt tolerance in poplar. *Plant Physiol.* **2023**, *191*, 2367–2384. [CrossRef]
25. Hu, Y.; Li, Z.; Wang, L.; Deng, L.; Sun, J.; Jiang, X.; Zhang, Y.; Tian, L.; Wang, Y.; Bai, W. Scandenolone, a natural isoflavone derivative from *Cudrania tricuspidata* fruit, targets EGFR to induce apoptosis and block autophagy flux in human melanoma cells. *J. Funct. Foods.* **2017**, *37*, 229–240. [CrossRef]
26. Kim, D.W.; Kim, J.K.; Gebru, Y.A.; Kim, Y.H.; Choi, H.S.; Kim, M.K. Identification of novel parishin compounds from the Twig of *Maclura tricuspidata* and comparative analysis of parishin derivatives in different parts. *Molecules* **2022**, *28*, 7. [CrossRef]
27. Wu, Z.Y.; Raven, P.H.; Hong, D.Y. *Flora of China Vol. 5, Moraceae through Maclura*; Missouri Botanical Garden Press: St. Louis, MO, USA; Science Press: Beijing, China, 2003; p. 16.

28. Liu, J.; Fu, C.; Li, G.; Khan, M.N.; Wu, H. ROS homeostasis and plant salt tolerance: Plant nanobiotechnology updates. *Sustainability* **2021**, *13*, 3552. [CrossRef]
29. Yu, Z.P.; Duan, X.B.; Luo, L.; Dai, S.J.; Ding, Z.J.; Xia, G.M. How plant hormones mediate salt stress responses. *Trends Plant Sci.* **2020**, *25*, 1117–1130. [CrossRef] [PubMed]
30. Xiang, Y.H.; Yu, J.J.; Liao, B.; Shan, J.X.; Ye, W.W.; Dong, N.Q.; Guo, T.; Kan, Y.; Zhang, H.; Yang, Y.B.; et al. An α/β hydrolase family member negatively regulates salt tolerance but promotes flowering through three distinct functions in rice. *Mol. Plant* **2022**, *15*, 1908–1930. [CrossRef] [PubMed]
31. Kohli, A.; Sreenivasulu, N.; Lakshmanan, P.; Kumar, P.P. The phytohormone crosstalk paradigm takes center stage in understanding how plants respond to abiotic stresses. *Plant Cell Rep.* **2013**, *32*, 945–957. [CrossRef]
32. Martín-Rejano, E.M.; Camacho-Cristóbal, J.J.; Herrera-Rodríguez, M.B.; Rexach, J.; Navarro-Gochicoa, M.T.; González-Fontes, A. Auxin and ethylene are involved in the responses of root system architecture to low boron supply in *Arabidopsis* seedlings. *Physiol. Plant.* **2011**, *142*, 170–178. [CrossRef]
33. Jiang, K.; Moe-Lange, J.; Hennet, L.; Feldman, L.J. Salt stress affects the redox status of *Arabidopsis* root meristems. *Front. Plant Sci.* **2016**, *7*, 81. [CrossRef] [PubMed]
34. Iglesias, M.J.; Terrile, M.C.; Windels, D.; Lombardo, M.C.; Bartoli, C.G.; Vazquez, F.; Bartoli, C.G.; Vazquez, F.; Estelle, M.; Casalongué, C.A. MiR393 regulation of auxin signaling and redox-related components during acclimation to salinity in *Arabidopsis*. *PLoS ONE* **2014**, *9*, e107678. [CrossRef] [PubMed]
35. Ju, F.; Pang, J.; Sun, L.; Gu, J.; Wang, Z.; Wu, X.; Ali, S.; Wang, Y.; Zhao, W.; Wang, S.; et al. Integrative transcriptomic, metabolomic and physiological analyses revealed the physiological and molecular mechanisms by which potassium regulates the salt tolerance of cotton (*Gossypium hirsutum* L.) roots. *Ind. Crops Prod.* **2023**, *193*, 116177. [CrossRef]
36. Krishnamurthy, A.; Rathinasabapathi, B. Auxin and its transport play a role in plant tolerance to arsenite-induced oxidative stress in *Arabidopsis thaliana*. *Plant Cell Environ.* **2013**, *36*, 1838–1849. [CrossRef]
37. Keshishian, E.A.; Cliver, B.R.; McLaughlin, W.F.; Hallmark, H.T.; Plačková, L.; Goertzen, L.R.; Novák, O.; Cobine, P.A.; Leisner, C.P.; Rashotte, A.M. CYTOKININ RESPONSE FACTOR2 is involved in modulating the salt stress response. *Plant J.* **2022**, *110*, 1097–1110. [CrossRef]
38. Zhao, Y.; Jia, K.; Tian, Y.; Han, K.; El-Kassaby, Y.A.; Yang, H.; Si, H.; Sun, Y.; Li, Y. Time-course transcriptomics analysis reveals key responses of populus to salt stress. *Ind. Crops Prod.* **2023**, *194*, 116278. [CrossRef]
39. Olszewski, N.; Sun, T.P.; Gubler, F. Gibberellin signaling: Biosynthesis, catabolism, and response pathways. *Plant Cell* **2002**, *14*, S61–S80. [CrossRef]
40. Sun, T.P.; Gubler, F. Molecular mechanism of gibberellin signaling in plants. *Annu. Rev. Plant Biol.* **2004**, *55*, 197–223. [CrossRef]
41. Yang, D.L.; Li, Q.; Deng, Y.W.; Lou, Y.G.; Wang, M.Y.; Zhou, G.X.; Zhang, Y.Y.; He, Z.H. Altered disease development in the eui mutants and Eui overexpressors indicates that gibberellins negatively regulate rice basal disease resistance. *Mol. Plant* **2008**, *1*, 528–537. [CrossRef]
42. Achard, P.; Renou, J.P.; Berthomé, R.; Harberd, N.P.; Genschik, P. Plant DELLAs restrain growth and promote survival of adversity by reducing the levels of reactive oxygen species. *Curr. Biol.* **2008**, *18*, 656–660. [CrossRef]
43. Niu, M.; Xie, J.; Chen, C.; Cao, H.; Sun, J.; Kong, Q.; Shabala, S.; Shabala, L.; Huang, Y.; Bie, Z. An early ABA-induced stomatal closure, Na^+ sequestration in leaf vein and K^+ retention in mesophyll confer salt tissue tolerance in *Cucurbita* species. *J. Exp. Bot.* **2018**, *69*, 4945–4960. [CrossRef]
44. Chen, K.; Li, G.J.; Bressan, R.A.; Song, C.P.; Zhu, J.K.; Zhao, Y. Abscisic acid dynamics, signaling, and functions in plants. *J. Integr. Plant Biol.* **2020**, *62*, 25–54. [CrossRef] [PubMed]
45. Yang, C.; Ma, B.; He, S.J.; Xiong, Q.; Duan, K.X.; Yin, C.C.; Chen, H.; Lu, X.; Chen, S.Y.; Zhang, J.S. *MAOHUZI6/ETHYLENE INSENSITIVE3-LIKE1* and *ETHYLENE INSENSITIVE3-LIKE2* regulate ethylene response of roots and coleoptiles and negatively affect salt tolerance in rice. *Plant Physiol.* **2015**, *169*, 148–165. [CrossRef] [PubMed]
46. Li, C.H.; Wang, G.; Zhao, J.L.; Zhang, L.Q.; Ai, L.F.; Han, Y.F.; Sun, D.Y.; Zhang, S.W.; Sun, Y. The receptor-like kinase SIT1 mediates salt sensitivity by activating MAPK3/6 and regulating ethylene homeostasis in rice. *Plant Cell* **2014**, *26*, 2538–2553. [CrossRef] [PubMed]
47. Peng, J.; Li, Z.; Wen, X.; Li, W.; Shi, H.; Yang, L.; Zhu, H.; Guo, H. Salt-induced stabilization of EIN3/EIL1 confers salinity tolerance by deterring ROS accumulation in *Arabidopsis*. *PLoS Genet.* **2014**, *10*, e1004664. [CrossRef] [PubMed]
48. Liu, J.; Yang, R.; Jian, N.; Wei, L.; Ye, L.; Wang, R.; Gao, H.; Zheng, Q. Putrescine metabolism modulates the biphasic effects of brassinosteroids on canola and *Arabidopsis* salt tolerance. *Plant Cell Environ.* **2020**, *43*, 1348–1359. [CrossRef]
49. Planas-Riverola, A.; Gupta, A.; Betegón-Putze, I.; Bosch, N.; Ibañes, M.; Caño-Delgado, A.I. Brassinosteroid signaling in plant development and adaptation to stress. *Development* **2019**, *146*, dev151894. [CrossRef] [PubMed]
50. Sun, F.; Yu, H.; Qu, J.; Cao, Y.; Ding, L.; Feng, W.; Khalid, M.H.B.; Li, W.; Fu, F. Maize ZmBES1/BZR1-5 decreases ABA sensitivity and confers tolerance to osmotic stress in transgenic *Arabidopsis*. *Int. J. Mol. Sci.* **2020**, *21*, 996. [CrossRef]
51. Song, S.; Huang, H.; Wang, J.; Liu, B.; Qi, T.; Xie, D. MYC5 is involved in jasmonate-regulated plant growth, leaf senescence and defense responses. *Plant Cell Physiol.* **2017**, *58*, 1752–1763. [CrossRef]
52. Ahanger, M.A.; Aziz, U.; Alsahli, A.A.; Alyemeni, M.N.; Ahmad, P. Influence of exogenous salicylic acid and nitric oxide on growth, photosynthesis, and ascorbate-glutathione cycle in salt stressed *Vigna angularis*. *Biomolecules* **2019**, *10*, 42. [CrossRef]

53. Torun, H.; Novák, O.; Mikulík, J.; Strnad, M.; Ayaz, F.A. The Effects of exogenous salicylic acid on endogenous phytohormone status in *Hordeum vulgare* L. under salt stress. *Plants* **2022**, *11*, 618. [CrossRef] [PubMed]
54. Gan, T.; Lin, Z.; Bao, L.; Hui, T.; Cui, X.; Huang, Y.; Wang, H.; Su, C.; Jiao, F.; Zhang, M.; et al. Comparative proteomic analysis of tolerant and sensitive varieties reveals that phenylpropanoid biosynthesis contributes to salt tolerance in mulberry. *Int. J. Mol. Sci.* **2021**, *22*, 9402. [CrossRef] [PubMed]
55. Li, X.; Bonawitz, N.D.; Weng, J.K.; Chapple, C. The growth reduction associated with repressed lignin biosynthesis in *Arabidopsis thaliana* is independent of flavonoids. *Plant Cell* **2010**, *22*, 1620–1632. [CrossRef] [PubMed]
56. Parrotta, L.; Guerriero, G.; Sergeant, K.; Cai, G.; Hausman, J.F. Target or barrier? The cell wall of early-and later-diverging plants vs cadmium toxicity: Differences in the response mechanisms. *Front. Plant Sci.* **2015**, *6*, 133. [CrossRef] [PubMed]
57. Zhu, Y.; Pu, B.Q.; Xie, G.Y.; Tian, M.; Xu, F.Y.; Qin, M.J. Dynamic changes of flavonoids contents in the different parts of rhizome of *Belamcanda chinensis* during the thermal drying process. *Molecules* **2014**, *19*, 10440–10454. [CrossRef] [PubMed]
58. Zhang, J.; Li, B.; Yang, Y.; Hu, W.; Chen, F.; Xie, L.; Fan, L. Genome-wide characterization and expression profiles of the superoxide dismutase gene family in *Gossypium*. *Int. J. Genomics* **2016**, *2016*, 8740901. [CrossRef] [PubMed]
59. Xie, Q.; Liu, B.; Dong, W.; Li, J.; Wang, D.; Liu, Z.; Gao, C. Comparative transcriptomic and metabolomic analyses provide insights into the responses to NaCl and Cd stress in *Tamarix hispida*. *Sci. Total Environ.* **2023**, *884*, 163889. [CrossRef] [PubMed]
60. Wang, C.; Chen, L.; Cai, Z.C.; Chen, C.; Liu, Z.; Liu, X.; Zou, L.; Chen, J.; Tan, M.; Wei, L.; et al. Comparative proteomic analysis reveals the molecular mechanisms underlying the accumulation difference of bioactive constituents in *Glycyrrhiza uralensis* Fisch under salt stress. *J. Agric. Food Chem.* **2020**, *68*, 1480–1493. [CrossRef]
61. Zhang, L.; Zhang, Z.; Fang, S.; Liu, Y.; Shang, X. Integrative analysis of metabolome and transcriptome reveals molecular regulatory mechanism of flavonoid biosynthesis in *Cyclocarya paliurus* under salt stress. *Ind. Crops Prod.* **2021**, *170*, 113823. [CrossRef]
62. Wanichthanarak, K.; Boonchai, C.; Kojonna, T.; Chadchawan, S.; Sangwongchai, W.; Thitisaksakul, M. Deciphering rice metabolic flux reprograming under salinity stress via in silico metabolic modeling. *Comput. Struct. Biotechnol. J.* **2020**, *18*, 3555–3566. [CrossRef]
63. Hussein, M.M.; Faham, S.Y.; Alva, A.K. Role of foliar application of nicotinic acid and tryptophan on onion plants response to salinity stress. *J. Agric. Sci.* **2014**, *6*, 41–51. [CrossRef]
64. Jia, X.M.; Zhu, Y.F.; Hu, Y.; Zhang, R.; Cheng, L.; Zhu, Z.L.; Zhao, T.; Zhang, X.Y.; Wang, Y.X. Integrated physiologic, proteomic, and metabolomic analyses of *Malus halliana* adaptation to saline–alkali stress. *Hortic. Res.* **2019**, *6*, 91. [CrossRef] [PubMed]
65. Abdelrahman, M.; Nishiyama, R.; Tran, C.D.; Kusano, M.; Nakabayashi, R.; Okazaki, Y.; Matsuda, F.; Chávez Montes, R.A.; Mostofa, M.G.; Li, W.Q.; et al. Defective cytokinin signaling reprograms lipid and flavonoid gene-to-metabolite networks to mitigate high salinity in *Arabidopsis*. *Proc. Natl. Acad. Sci. USA* **2021**, *118*, e2105021118. [CrossRef] [PubMed]
66. Rohila, J.S.; Yang, Y. Rice mitogen-activated protein kinase gene family and its role in biotic and abiotic stress response. *J. Integr. Plant Biol.* **2007**, *49*, 751–759. [CrossRef]
67. Yan, Z.; Wang, J.; Wang, F.; Xie, C.; Lv, B.; Yu, Z.; Dai, S.; Liu, X.; Xia, G.; Tian, H.; et al. MPK3/6-induced degradation of ARR1/10/12 promotes salt tolerance in *Arabidopsis*. *EMBO Rep.* **2021**, *22*, e52457. [CrossRef]
68. Na, Y.J.; Choi, H.K.; Park, M.Y.; Choi, S.W.; Xuan Vo, K.T.; Jeon, J.S.; Kim, S.Y. OsMAPKKK63 is involved in salt stress response and seed dormancy control. *Plant Signal. Behav.* **2019**, *14*, e1578633. [CrossRef]
69. Sadau, S.B.; Mehari, T.G.; Ahmad, A.; Tajo, S.M.; Ibrahim, S.; Iqbal, M.S.; Elasad, M.; Zhang, J.; Wei, H.; Yu, S. Genome wide identification and characterization of MAPK genes reveals their potential in enhancing drought and salt stress tolerance in *Gossypium hirsutum*. *J. Cotton Res.* **2022**, *5*, 23. [CrossRef]
70. Song, C.; Zhang, Y.; Chen, R.; Zhu, F.; Wei, P.; Pan, H.; Chen, C.; Dai, J. Label-free quantitative proteomics unravel the impacts of salt stress on *Dendrobium huoshanense*. *Front. Plant Sci.* **2022**, *13*, 874579. [CrossRef]
71. Batista-Silva, W.; Heinemann, B.; Rugen, N.; Nunes-Nesi, A.; Araújo, W.L.; Braun, H.P.; Hildebrandt, T.M. The role of amino acid metabolism during abiotic stress release. *Plant Cell Environ.* **2019**, *42*, 1630–1644. [CrossRef]
72. Zhang, Y.; Li, D.; Zhou, R.; Wang, X.; Dossa, K.; Wang, L.; Zhang, Y.; Yu, J.; Gong, H.; Zhang, X.; et al. Transcriptome and metabolome analyses of two contrasting sesame genotypes reveal the crucial biological pathways involved in rapid adaptive response to salt stress. *BMC Plant Biol.* **2019**, *19*, 66. [CrossRef]
73. Yang, D.S.; Zhang, J.; Li, M.X.; Shi, L.X. Metabolomics analysis reveals the salt-tolerant mechanism in *Glycine soja*. *J. Plant Growth Regul.* **2017**, *36*, 460–471. [CrossRef]
74. Yang, Y.; Guo, Y. Unraveling salt stress signaling in plants. *J. Integr. Plant Biol.* **2018**, *60*, 796–804. [CrossRef]
75. Ke, Y.Z. Muberry Adaptability to Salinity and Its Salt-Tolerant Mechanism and Application to Saline-Alkali Soils. Ph.D. Thesis, Chinese Academy of Forestry, Beijing, China, 2008; p. 19.
76. Sanchez, D.H.; Pieckenstain, F.L.; Escaray, F.; Erban, A.; Kraemer, U.T.E.; Udvardi, M.K.; Kopka, J. Comparative ionomics and metabolomics in extremophile and glycophytic *Lotus* species under salt stress challenge the metabolic pre-adaptation hypothesis. *Plant Cell Environ.* **2011**, *34*, 605–617. [CrossRef]
77. Chen, S.; Zhou, Y.; Chen, Y.; Gu, J. fastp: An ultra-fast all-in-one FASTQ preprocessor. *Bioinformatics* **2018**, *34*, i884–i890. [CrossRef]
78. Grabherr, M.G.; Haas, B.J.; Yassour, M.; Levin, J.Z.; Thompson, D.A.; Amit, I.; Adiconis, X.; Fan, L.; Raychowdhury, R.; Zeng, Q.; et al. Full-length transcriptome assembly from RNA-Seq data without a reference genome. *Nat. Biotechnol.* **2011**, *29*, 644–652. [CrossRef]

79. Simão, F.A.; Waterhouse, R.M.; Ioannidis, P.; Kriventseva, E.V.; Zdobnov, E.M. BUSCO: Assessing genome assembly and annotation completeness with single-copy orthologs. *Bioinformatics* **2015**, *31*, 3210–3212. [CrossRef]
80. Li, B.; Dewey, C.N. RSEM: Accurate transcript quantification from RNA-Seq data with or without a reference genome. *BMC Bioinf.* **2011**, *12*, 323. [CrossRef]
81. Love, M.I.; Huber, W.; Anders, S. Moderated estimation of fold change and dispersion for RNA-seq data with DESeq2. *Genome Biol.* **2014**, *15*, 550. [CrossRef] [PubMed]
82. Kanehisa, M.; Araki, M.; Goto, S.; Hattori, M.; Hirakawa, M.; Itoh, M.; Katayama, T.; Kawashima, S.; Okuda, S.; Tokimatsu, T.; et al. KEGG for linking genomes to life and the environment. *Nucleic Acids Res.* **2007**, *36*, D480–D484. [CrossRef] [PubMed]
83. Want, E.J.; Masson, P.; Michopoulos, F.; Wilson, I.D.; Theodoridis, G.; Plumb, R.S.; Shockcor, J.; Loftus, N.; Holmes, E.; Nicholson, J.K. Global metabolic profiling of animal and human tissues via UPLC-MS. *Nat. Protoc.* **2013**, *8*, 17–32. [CrossRef] [PubMed]
84. Saccenti, E.; Hoefsloot, H.C.; Smilde, A.K.; Westerhuis, J.A.; Hendriks, M.M. Reflections on univariate and multivariate analysis of metabolomics data. *Metabolomics* **2014**, *10*, 361–374. [CrossRef]
85. Cho, K.; Cho, K.S.; Sohn, H.B.; Ha, I.J.; Hong, S.Y.; Lee, H.; Kim, Y.; Nam, M.H. Network analysis of the metabolome and transcriptome reveals novel regulation of potato pigmentation. *J. Exp. Bot.* **2016**, *67*, 1519–1533. [CrossRef] [PubMed]
86. Schmittgen, T.D.; Livak, K.J. Analyzing real-time PCR data by the comparative C_T method. *Nat. Protoc.* **2008**, *3*, 1101–1108. [CrossRef] [PubMed]

plants

MDPI

Article

Effect of UV-A Irradiation on Bioactive Compounds Accumulation and Hypoglycemia-Related Enzymes Activities of Broccoli and Radish Sprouts

Gongheng Che [1], Mingmei Chen [1], Xiaodan Li [1,2,3], Junxia Xiao [1,2,3], Liang Liu [1,2,3] and Liping Guo [1,2,3,*]

1 College of Food Science and Engineering, Qingdao Agricultural University, Qingdao 266109, China; chegongheng1204@163.com (G.C.); cmm1954154242@163.com (M.C.); lixd1231@163.com (X.L.); xjxfood@qau.edu.cn (J.X.); 201901104@qau.edu.cn (L.L.)
2 Key Laboratory of Special Food Processing (Co-construction by Ministry and Province), Ministry of Agriculture Rural Affairs, Qingdao Agricultural University, Qingdao 266109, China
3 Shandong Technology Innovation Center of Special Food, Qingdao 266109, China
* Correspondence: happyglp@qau.edu.cn; Tel.: +86-0532-58957771

Abstract: In the present study, different intensities of UV-A were applied to compare their effects on growth, bioactive compounds and hypoglycemia-related enzyme activities in broccoli and radish sprouts. The growth of sprouts was decreased after UV-A irradiation. A total of 12 W of UV-A irradiation resulted in the highest content of anthocyanin, chlorophyll, polyphenol and ascorbic acid in broccoli and radish sprouts. The highest soluble sugar content was recorded in sprouts under 8 W of UV-A irradiation, while no significant difference was obtained in soluble protein content among different UV-A intensities. Furthermore, 12 W of UV-A irradiation induced the highest glucosinolate accumulation, especially glucoraphanin and glucoraphenin in broccoli and radish sprouts, respectively; thus, it enhanced sulforaphane and sulforaphene formation. The α-amylase, α-glucosidase and pancrelipase inhibitory rates of two kinds of sprouts were enhanced significantly after UV-A irradiation, indicating UV-A-irradiation-treated broccoli and radish sprouts have new prospects as hypoglycemic functional foods.

Keywords: UV-A irradiation; bioactive compounds; hypoglycemia-related enzymes; broccoli sprouts; radish sprouts

1. Introduction

Cruciferous sprouts are rich in glucosinolates, phenolic compounds, anthocyanins, ascorbic acid, etc., which are related to the prevention of chronic diseases. Especially, broccoli and radish sprouts have attracted extensive attention from researchers due to their various and prominent biological activities. It has been reported that broccoli sprouts are effective against various cancers and novel coronaviruses, as well as possessing antioxidant, anti-inflammatory, antibacterial, analgesic, antidepressant and anti-obesity properties, etc. [1–3]. On the other hand, radish sprouts also have positive antioxidant, anti-cancer, anti-inflammatory and anti-radical properties [4,5]. In addition, radish sprouts are thought to have greater anti-cancer potential than broccoli sprouts due to the absence of epithiospecifier protein (ESP), resulting in a higher yield of isothiocyanates with anti-cancer properties [6]. Hence, the enhancement of accumulation of bioactive compounds and bioactivities is important in developing the corresponding functional foods of broccoli and radish sprouts and their industrial production.

Plant photoreceptors detect light with a variety of wavelengths, such as blue (400–500 nm) and ultraviolet (280–315 nm), and mount responses that are specific to the wavelength [7]. Ultraviolet light (UV) as an abiotic stressor, can cause different responses in plants, including damage to DNA and proteins, the generation of reactive oxygen species (ROS), changes in plant growth and development and the enhancing of secondary metabolites [8]. Moreira-Rodríguez et al. [9] reported that UV-A treatment was able to enhance the content of phenolic compounds in broccoli sprouts, such as gallic acid and kaempferol glucoside, compared to UV-B treatment. Kang et al. [8] found that the phenolic compound content and phenylalanine ammonia lyase (PAL) activity of sweet basil sprouts were highest at a UV-A intensity of 20 $W{\cdot}m^{-2}$, but showed an inhibitory effect at a UV-A intensity of 30 $W{\cdot}m^{-2}$. The results of Jiang et al. [10] showed that supplementation with 12 $\mu mol{\cdot}m^{-2}\ s^{-1}$ UV-A significantly increased the content of total glucosinolates in kale sprouts; meanwhile, 4-methoxyglucobrassicin and neoglucobrassicin content increased by 31.29% and 79.02%, respectively, compared to the control. Flavonoids are a kind of polyphenols, and anthocyanins are a kind of flavonoids; they are widely found in plants. Studies have shown that polyphenol intake is associated with a reduced risk of chronic disease, and the contents of polyphenol, flavonoid and anthocyanin in "red butter" lettuce sprouts under UV-A treatment were higher than the control with 17.24%, 11.57% and 14.11%, respectively [11]. However, there is limited information on the effects of UV-A irradiation on the enrichment of secondary metabolites of broccoli and radish sprouts.

Since the 21st century, the number of patients with type II diabetes mellitus (T2DM) has increased dramatically, and has become a serious public health problem worldwide. Given the association between diabetes and premature death, effective measures must be taken to prevent it in our daily life. Recently, the nutritional and medicinal values of sprouts have been extensively researched in many countries, and it is generally accepted that sprouts have some potential in the prevention and treatment of diabetes. It has been reported that fava bean sprouts showed the highest improvement effect in lowering blood glucose levels in diabetic rats with a reduction from 399.2 mg/dL to 139.6 mg/dL, as well as reduced complications, including lipid metabolism and oxidative stress associated with diabetes [12]. Additionally, Laila et al. [13] found that fenugreek sprouts exhibited the best α-glucosidase inhibitory activity (95.24%) and also showed a moderate inhibitory effect on α-amylase (41.64%) and invertase (49.24%) at the vitro level. As we all know, broccoli and radish sprouts are rich in phenolic compounds and glucosinolates metabolite, isothiocyanates, which have a blood glucose-lowering effect. It deserves to be investigated whether broccoli and radish sprouts have an effect on lowering blood glucose levels.

Therefore, the aim of the present study was to investigate the growth and the content of secondary metabolites, such as the glucosinolates, isothiocyanates, phenolic compounds, anthocyanins and ascorbic acid of broccoli and radish sprouts under UV-A irradiation, and also to compare the hypoglycemic enzyme activities, including the α-amylase, α-glucosidase and pancrelipase of broccoli and radish sprouts in vitro level.

2. Results

2.1. Effect of UV-A Irradiation on Growth and Anthocyanin, Chlorophyll Content in Broccoli and Radish Sprouts

As shown in Figure 1A,B, UV-A irradiation inhibited the growth of broccoli and radish sprouts. With the increase in light intensity, the length and weight of broccoli and radish sprouts showed different degrees of reduction. At a light intensity of 16 W, the length of the broccoli and radish sprouts were lower than the control by 44.05% and 42.82%, respectively; the plant weight was reduced by 34.19% and 36.30% compared to the control, respectively. From Figure 1C, the highest anthocyanin content was 7.27 mg/100 g at a 12 W light intensity in broccoli sprouts and 2.52 mg/100 g at 16 W light intensity in radish sprouts, which was 3.97 and 2.36 times of that in the control, respectively. In Figure 1D, UV-A irradiation significantly increased the chlorophyll enrichment of both sprouts. The highest chlorophyll content in broccoli sprouts was recorded at 12 W light intensity, which was 13.45 times of control, and showed a decreasing trend at 16 W. The highest chlorophyll content in radish sprouts was recorded at 16 W and was 31.42 times that of the control.

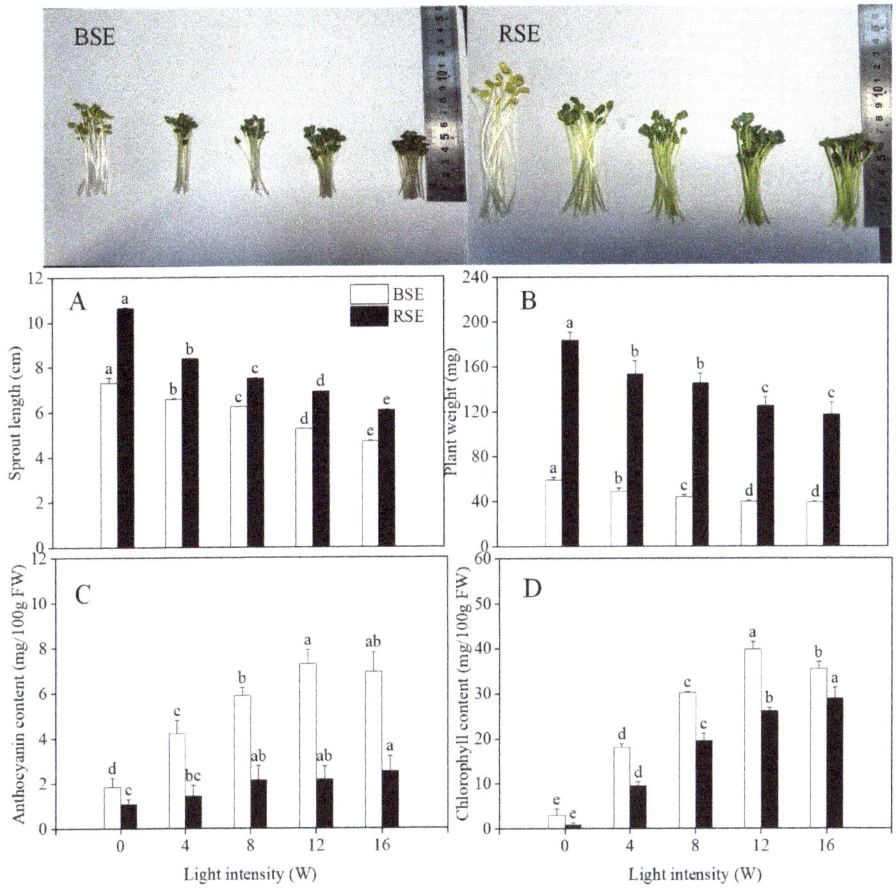

Figure 1. Effect of UV-A irradiation on sprout length (**A**), plant weight (**B**), anthocyanin content (**C**) and chlorophyll content (**D**) in broccoli and radish sprouts. BSE: broccoli sprout extracts; RSE: radish sprout extracts. 0 W: seeds treated with dark; 4 W, 8 W, 12 W, 16 W: seeds treated with UV-A, respectively. The letters of a–e are significantly different at $p < 0.05$.

2.2. *Effect of UV-A Treatment on Soluble Sugar and Soluble Protein Content in Broccoli and Radish Sprouts*

Soluble sugar content in radish sprouts was higher than that in broccoli sprouts, and UV-A irradiation significantly enhanced its content, compared to the control (Figure 2A). A UV-A intensity of 8 W resulted in the highest soluble sugar content in broccoli and radish sprouts, which was 2.27 and 1.96 times of that in the control, respectively. As can be seen from Figure 2B, no significant difference was found in the soluble protein content in broccoli sprouts among different UV-A intensities; however, it first increased and then decreased in radish sprouts with the increasing of UV-A intensity.

2.3. *Effect of UV-A Irradiation on Polyphenol and Ascorbic Acid Content in Broccoli and Radish Sprouts*

As shown in Figure 3, the total polyphenol content and ascorbic acid content in broccoli and radish sprouts was first increased and then decreased with the increasing of UV-A intensity. The polyphenol content in broccoli and radish sprouts was 540.26 and 443.55 mg/100 g at a 12 W light intensity, which was 1.31 and 1.23 times of that in the control samples, respectively (Figure 3A). On the other hand, the highest ascorbic acid content in the broccoli and radish sprouts was recorded at the light intensity of 8 W, which was 2.12 and 2.13 times in comparison with the control, respectively (Figure 3B).

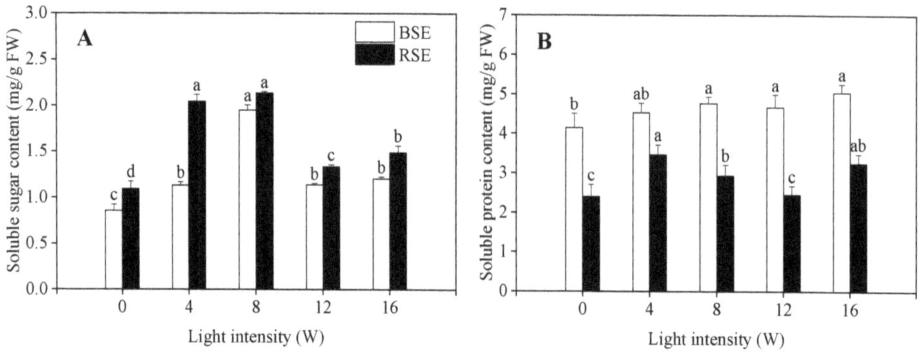

Figure 2. Effect of UV-A irradiation on soluble sugar content (**A**) and soluble protein content (**B**) in broccoli and radish sprouts. BSE: broccoli sprout extracts; RSE: radish sprout extracts. 0 W: seeds treated with dark; 4 W, 8 W, 12 W, 16 W: seeds treated with UV-A, respectively. The letters of a–d are significantly different at $p < 0.05$.

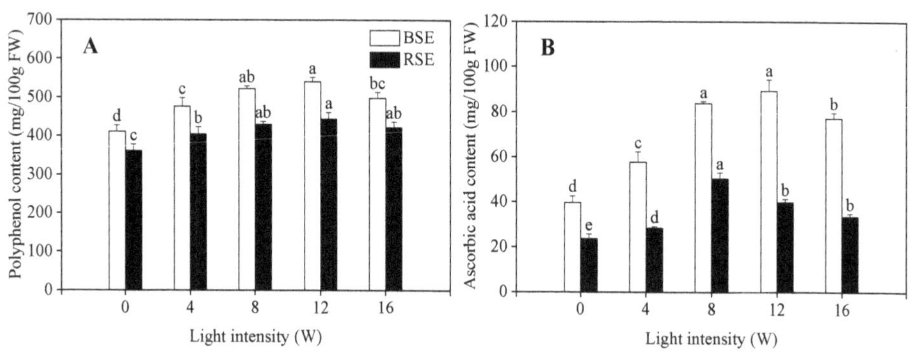

Figure 3. Effect of UV-A irradiation on polyphenol content (**A**) and ascorbic acid content (**B**) in broccoli and radish sprouts. BSE: broccoli sprouts extracts; RSE: radish sprouts extracts. 0 W: seeds treated with dark; 4 W, 8 W, 12 W, 16 W: seeds treated with UV-A respectively. The letters of a–e are significantly different at $p < 0.05$.

2.4. *Effect of UV-A Irradiation on Glucosinolate Content in Broccoli and Radish Sprouts*

UV-A irradiation significantly induced the accumulation of aliphatic and indole glucosinolates in broccoli and radish sprouts (Table 1). It was found that GRA and RAE, the most abundant glucosinolate in broccoli and radish sprouts, respectively; showed a trend of increasing, then decreasing with the increasing of UV-A intensity, and it reached the highest level at 12 W UV-A intensity, which was 2.99 and 2.84 times that of the control. Moreover, other individual glucosinolates in the broccoli and radish sprouts were also significantly enhanced after UV-A irradiation. The highest total glucosinolate content in broccoli and radish sprouts was found at a 12 W UV-A intensity, which was 2.57 and 2.20 times that of the control.

Table 1. Effect of UV-A irradiation on glucosinolate content in broccoli and radish sprouts. BSE: broccoli sprout extracts; RSE: radish sprout extracts. 0 W: seeds treated with dark; 4 W, 8 W, 12 W, 16W: seeds treated with UV-A, respectively. GRA: Glucoraphanin; RAE: Glucoraphenin; GNA: Gluconapin; GER: Glucoerucin; RSA: Glucoraphasatin; PRO: Progoitrin; 4HGB: 4-Hydroxyglucobrassicin; GB: Glucobrassicin; 4MGB: 4-Methoxyglucobrassicin; NGB: Neoglucobrassicin. The letters of a–e are significantly different at $p < 0.05$.

Light Intensity/W		µmol/g										
		Aliphatic Glucosinolate						Indole Glucosinolate				Total Glucosinolate
		GRA	RAE	GNA	GER	RSA	PRO	4HGB	GB	4MGB	NGB	
BSE	0	3.61 ± 0.07 e	—	0.99 ± 0.01 e	0.35 ± 0.02 e	—	—	0.15 ± 0.01 d	0.36 ± 0.03 e	0.51 ± 0.01 d	0.11 ± 0.01 e	6.08 ± 0.16 e
	4	5.86 ± 0.21 d	—	1.37 ± 0.01 d	0.52 ± 0.01 d	—	—	0.26 ± 0.02 c	0.39 ± 0.02 d	0.52 ± 0.02 c	0.15 ± 0.02 d	9.07 ± 0.31 d
	8	8.31 ± 0.28 c	—	1.73 ± 0.01 c	0.74 ± 0.00 c	—	—	0.27 ± 0.02 b	0.34 ± 0.04 c	0.53 ± 0.01 b	0.17 ± 0.02 c	12.09 ± 0.38 c
	12	10.83 ± 0.19 a	—	2.17 ± 0.02 a	1.17 ± 0.01 a	—	—	0.33 ± 0.01 a	0.40 ± 0.04 b	0.54 ± 0.01 a	0.21 ± 0.05 a	15.62 ± 0.33 a
	16	9.78 ± 0.21 b	—	1.86 ± 0.02 b	1.05 ± 0.01 b	—	—	0.33 ± 0.00 a	0.42 ± 0.02 a	0.54 ± 0.01 a	0.18 ± 0.03 b	14.16 ± 0.30 b
RSE	0	—	1.47 ± 0.03 e	0.59 ± 0.01 e	—	0.16 ± 0.03 e	0.22 ± 0.02 c	0.05 ± 0.00 c	0.22 ± 0.04 d	0.31 ± 0.01 b	—	3.02 ± 0.14 e
	4	—	1.89 ± 0.04 d	0.68 ± 0.01 d	—	0.25 ± 0.02 d	0.25 ± 0.04 b	0.08 ± 0.01 b	0.23 ± 0.04 c	0.31 ± 0.01 b	—	3.69 ± 0.17 d
	8	—	2.78 ± 0.04 c	0.75 ± 0.00 c	—	0.33 ± 0.02 c	0.27 ± 0.02 a	0.08 ± 0.00 b	0.24 ± 0.04 b	0.31 ± 0.02 b	—	4.76 ± 0.14 c
	12	—	4.18 ± 0.07 a	1.04 ± 0.03 a	—	0.56 ± 0.04 a	0.19 ± 0.01 d	0.09 ± 0.00 a	0.25 ± 0.04 a	0.32 ± 0.01 a	—	6.63 ± 0.20 a
	16	—	3.92 ± 0.07 b	0.82 ± 0.02 b	—	0.45 ± 0.05 b	0.19 ± 0.02 d	0.08 ± 0.02 b	0.24 ± 0.00 b	0.32 ± 0.02 a	—	6.02 ± 0.20 b

2.5. Effect of UV-A Irradiation on Myrosinase Activity of Broccoli and Radish Sprouts

It was noted that UV-A did not affect reduced myrosinase activity in broccoli and radish sprouts. In addition, no significant difference was observed in myrosinase activity between the two kinds of sprouts. The lowest myrosinase activity was found in two sprouts treated with a UV-A intensity of 16 W, which was decreased by 46.09% and 24.08% compared to the control, respectively (Figure 4).

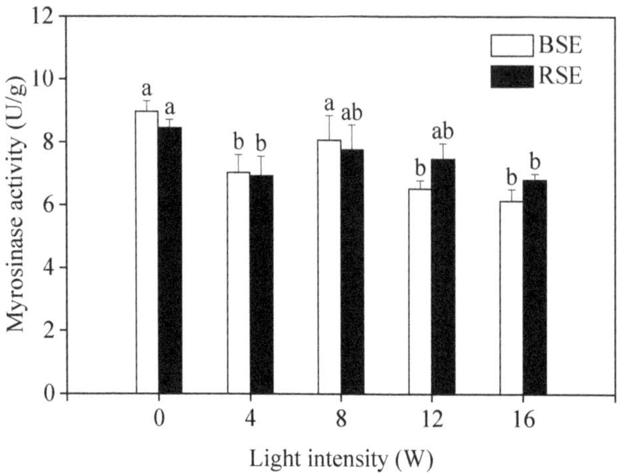

Figure 4. Effect of UV-A irradiation on myrosinase activity in broccoli and radish sprouts. BSE: broccoli sprout extracts; RSE: radish sprout extracts. 0 W: seeds treated with dark; 4 W, 8 W, 12 W, 16 W: seeds treated with UV-A, respectively. The letters of a–b are significantly different at $p < 0.05$.

2.6. Effect of UV-A Irradiation on Sulforaphane and Sulforaphene Formation in Broccoli and Radish Sprouts

As can be seen from Figure 5A,B, all UV irradiation significantly induced sulforaphane and sulforaphene formation in broccoli and radish sprouts compared to the control. Meanwhile, sulforaphane formation was significantly higher than that of sulforaphene; both reached the highest level at a UV-A intensity of 12 W, which was 3.89 and 3.37 times of that in the control, respectively.

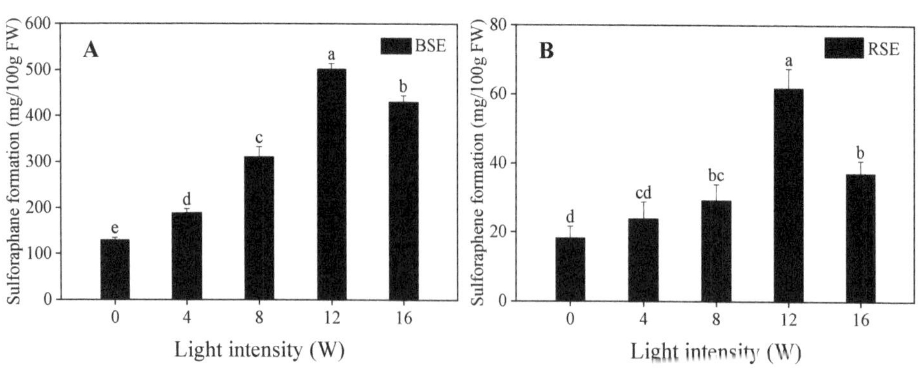

Figure 5. Effect of UV-A irradiation on sulforaphane formation (**A**) and sulforaphane formation (**B**) in broccoli and radish sprouts. BSE: broccoli sprout extracts; RSE: radish sprout extracts. 0 W: seeds treated with dark; 4 W, 8 W, 12 W, 16 W: seeds treated with UV-A, respectively. The letters of a–e are significantly different at $p < 0.05$.

2.7. Effect of UV-A Irradiation on α-Amylase Inhibition Rate in Broccoli and Radish Sprouts

According to Figure 6, it was easy to see that UV-A irradiation showed a positive effect on the α-amylase activity inhibition rate between the two kinds of sprouts. As the concentration of the broccoli and radish sprouts increased, the α-amylase activity inhibition rate also increased. At a concentration of 0.6 g/mL, the inhibition rate of α-amylase activity in broccoli and radish sprouts under UV-A irradiation was 9.06 and 7.94 times in comparison with the control, respectively.

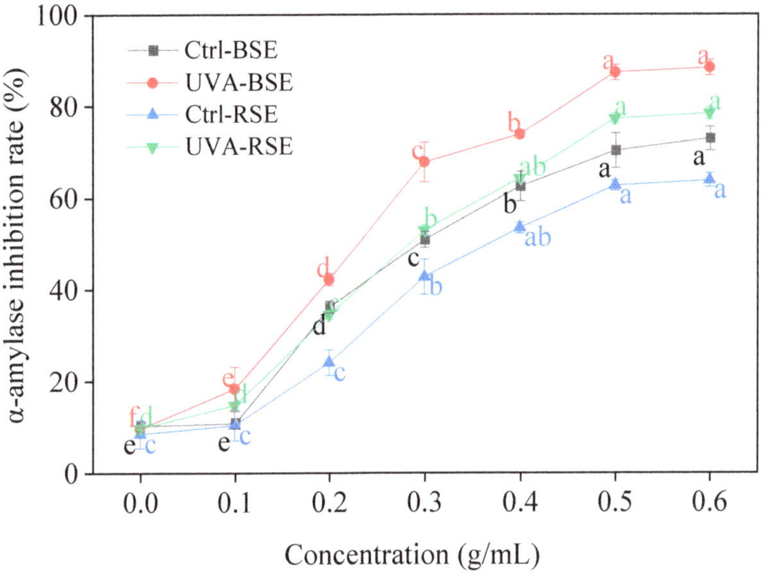

Figure 6. Effect of UV-A irradiation on α-amylase inhibition rate in broccoli and radish sprouts. Ctrl-BSE (black line): broccoli seeds treated with dark; UVA-BSE (red line): broccoli seeds treated with 12 W; Ctrl-RSE (blue line): radish seeds treated with dark; UVA-RSE (green line): radish seeds treated with 12 W. 0: water; 0.1, 0.2, 0.3, 0.4, 0.5, 0.6: 0.2, 0.4, 0.6, 0.8, 1.0, 1.2 g sprouts treated with 2 mL water, respectively. The letters of a–f are significantly different at $p < 0.05$.

2.8. Effect of UV-A Irradiation on α-Glucosidase Inhibition Rate in Broccoli and Radish Sprouts

With the sprout concentration increasing, the inhibitory rate of broccoli and radish sprouts against α-glucosidase enhanced continuously (Figure 7). The sprout concentration of 0.6 g/mL under UV-A irradiation resulted in the highest inhibition rate of α-glucosidase, which increased by 79.84% and 68.99% in comparison with the control, respectively.

2.9. Effect of UV-A Irradiation on Pancrelipase Inhibition Rate in Broccoli and Radish Sprouts

UV-A irradiation-treated broccoli and radish sprouts showed significant pancrelipase inhibitory activity (Figure 8). At a concentration of 0.6 g/mL, the inhibition rate of broccoli sprouts on pancrelipase activity was up to 90.31%, and that of radish sprouts was 74.42%, which increased by 79.84% and 64.74%, compared to the control, respectively.

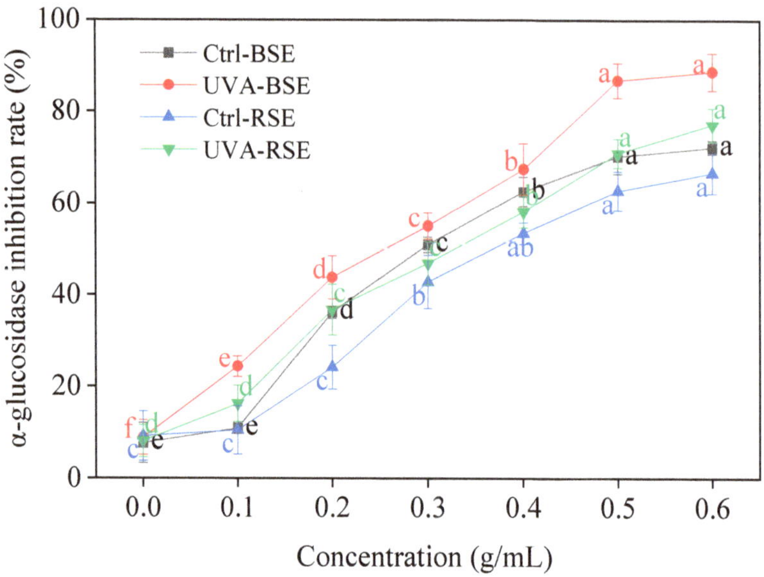

Figure 7. Effect of UV-A irradiation on α-glucosidase inhibition rate in broccoli and radish sprouts. Ctrl-BSE (black line): broccoli seeds treated with dark; UVA-BSE (red line): broccoli seeds treated with 12 W; Ctrl-RSE (blue line): radish seeds treated with dark; UVA-RSE (green line): radish seeds treated with 12 W. 0: water; 0.1, 0.2, 0.3, 0.4, 0.5, 0.6: 0.2, 0.4, 0.6, 0.8, 1.0, 1.2 g sprouts treated with 2 mL water, respectively. The letters of a–f are significantly different at $p < 0.05$.

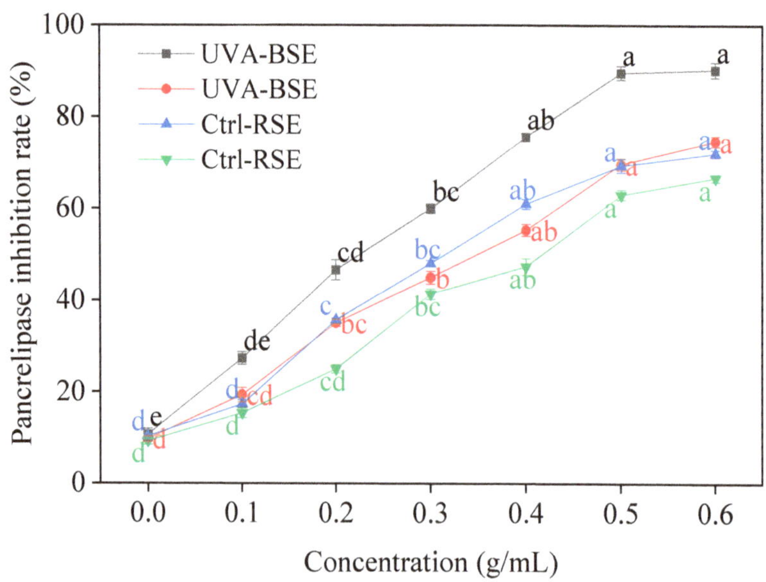

Figure 8. Effect of UV-A irradiation on pancrelipase inhibition rate in broccoli and radish sprouts. Ctrl-BSE (black line): broccoli seeds treated with dark; UVA-BSE (red line): broccoli seeds treated with 12 W; Ctrl-RSE (blue line): radish seeds treated with dark; UVA-RSE (green line): radish seeds treated with 12 W. 0: water; 0.1, 0.2, 0.3, 0.4, 0.5, 0.6: 0.2, 0.4, 0.6, 0.8, 1.0, 1.2 g sprouts treated with 2 mL water, respectively. The letters of a–e are significantly different at $p < 0.05$.

3. Discussion

Broccoli and radish sprouts as new functional vegetables have attracted widespread attention from scholars all around the world. It has been reported that light plays an important role in plant growth, morphology and development. Tsormpatsidis et al. [14] reported that UV-A irradiation reduced the height of lettuce sprouts, while it increased secondary metabolites. In addition, the inhibition of stem elongation by UV-A irradiation has been confirmed in several plants, such as kale and basil [15,16]. This is consistent with the results of the present study, in which UV-A irradiation reduced the sprout length and weight of the broccoli and radish sprouts, but the radish sprouts were significantly higher than the broccoli sprouts; which might be due to photosynthetic characteristics and auxin levels among different UV-A intensities during germination [17].

In addition to growth, UV-A irradiation also regulates the accumulation of anthocyanin and chlorophyll in plants. According to Zhang et al. [18], UV-A irradiation significantly enhanced anthocyanin accumulation in radish sprouts by upregulating the expression of genes related to anthocyanin biosynthesis, including CRY, PHOT and COP1. Moreover, Nguyen et al. [19] also reported UV-A irradiation increased anthocyanin accumulation in "green and red perilla". Similar results were also obtained in the present study. The anthocyanin content in broccoli and radish sprouts was enhanced after UV-A irradiation, and the likely explanation for this is the fact that the expression of genes and enzyme activity are related to anthocyanin biosynthesis under UV-A irradiation. Chlorophyll is an important product of photosynthesis. The results of Johnson et al. [20] showed that in the chlorophyll UV-absorption spectra, the absorbance of chlorophyll a at 340 and 389 nm was 51% and 72% of the absorbance peak in Sorghum bicolor, respectively. A previous study reported that the chlorophyll content in *Rosa hybrida* and *Fuchsia hybrida* was enhanced slightly by UV-A irradiation [21]. It is obvious that a similar phenomenon was detected in the present study. UV-A irradiation increased the chlorophyll content in the broccoli and radish sprouts, which might mean that the two kinds of sprouts directly absorbed specific wavelengths of UV-A, which then promoted the electron transport chain in the photosynthetic system, and thus promoted photosynthesis to produce more chlorophyll [22].

Plants produce soluble sugars during photosynthesis, which are the material basis for the synthesis of macromolecular compounds, such as polysaccharides and fats, and which also play an important role in participating in the carbon metabolism of plants. Mao et al. [22] found that the soluble sugar content in red and green pak choi was decreased after different UV-A intensities treatments. On the contrary, in the present study, the soluble sugar content in broccoli and radish sprouts increased significantly with the increasing of UV-A intensity. This might be due to high UV-A irradiation possibly damaging the integrity of the plant tissue, and thus hindering the normal metabolism of carbohydrates in the plant body [23]. However, this discrepancy needs further study. Soluble protein is one of the main components of plant cells and plays an important physiological function. No significant effect on soluble protein content under different UV-A intensities was observed in broccoli sprouts; however, in the present study it was found to be enhanced in radish sprouts. Interestingly, the phenomenon was similar to Tezuk [24], where UV-A irradiation was useful in promoting the accumulation of soluble protein content in radish plants. The likely explanation for this was that the soluble protein content might be related to the growth speed of two kinds of sprouts, but the exact mechanism needs to be further investigated.

Phenolic compounds and ascorbic acid play an important role in the antioxidant process of plants. Galactono-lactone dehydrogenase is a key enzyme in the ascorbic acid metabolic pathway, especially in the germination stage of the embryo [25]. Jia et al. [26] found that UV-A irradiation significantly increased the expression of ascorbic acid biosynthesis genes such as GME, VTC2, VTC4, GDH and GLDH levels in bean sprouts, thus significantly enhancing the ascorbic acid content. In the present study, the changes of ascorbic acid showed dramatic improvement in the broccoli and radish sprouts under

UV-A irradiation, implying that UV-A irradiation also enhanced the ascorbic acid content by elevating the expression level and enzyme activity of the ascorbic acid biogenesis of two kinds of sprouts. Phenolic compound synthesis is influenced by multiple metabolic pathways [27]. A previous study reported that UV-A irradiation enhanced the phenolic compound content and PAL enzyme activity of sweet basil, especially with an intensity of 20 W/m^2, resulting in a PAL enzyme activity and phenolic compound content of 38.4% and 36.3% higher than that of the control, respectively [8]. In the present study, the phenolic compound content in the two kinds of sprouts increased significantly under UV-A irradiation. The possible explanation for this was that UV-A radiation triggered oxidative stress and disrupted reactive oxygen homeostasis, thereby allowing phenolic synthases to function in stress-induced defense reactions that were directly involved in phenolic compound biogenesis [28].

UV-A as a signal could simultaneously stimulate different metabolic networks, and in the present study it also regulated glucosinolate metabolism in broccoli and radish sprouts. Zhuang et al. [29] reported that yellow and purple LED lights induced the glucoraphanin accumulation of broccoli sprouts by upregulating aliphatic glucosinolate biosynthetic genes, especially CYP79F1, CYP83A1, UGT74B1 and $FMO_{GS\text{-}OX1}$. In addition, UV-A irradiation significantly increased the aliphatic glucosinolate content of Chinese kale by upregulating the expression of glucosinolate biosynthesis-related genes, including BCAT4, GGP1, SUR1, AOP2 and AOP3 [30]. Meanwhile, UV-A irradiation could also enhance the indolic glucosinolate accumulation of kale [10]. In the present study, the aliphatic and indolic glucosinolate content in broccoli and radish sprouts showed dramatic enhancement after UV-A irradiation. This might be because UV-A irradiation could induce gene expression in aliphatic and indole glucosinolate biosynthesis. Nevertheless, the specific mechanism of aliphatic and indole glucosinolate biosynthesis needs further study.

Sulforaphane and sulforaphene are the abundant isothiocyanates in broccoli and radish sprouts, respectively; which are the products formed by glucoraphanin and glucoraphenin under the catalysis of myrosinase. A previous study demonstrated that blue, purple and red light were beneficial to sulforaphane formation in broccoli florets [31]. O'Hare et al. [6] reported that the presence of ESP led to the hydrolysis of glucosinolate toward nitrile and thus decreased the isothiocyanate formation; broccoli sprouts contain ESP, but radish sprouts have no ESP. Therefore, in the present study, only 33.73% of the glucoraphanin translated to sulforaphane, but 77.99% of the glucoraphanin translated to sulforaphane. In addition to the glucosinolate content, myrosinase activity is also related to isothiocyanate formation. In the present study, myrosinase activity did not change or decrease after UV-A irradiation. Hence, the enhancement of sulforaphane and sulforaphene formation after UV-A irradiation was attributed to the increased content of glucoraphanin and glucoraphenin, rather than myrosinase activity.

α-amylase, α-glucosidase and pancrelipase are able to promote the breakdown of food in the mouth, stomach and intestines, releasing glucose into the bloodstream and raising blood sugar levels; thus, restraining the enzyme activity can predict the potential for the treatment of various disease such as diabetes [32,33]. The current approach to control hyperglycemia is to inhibit the activity of α-amylase, α-glucosidase and pancrelipase, thereby slowing down the digestion of carbohydrates in resorbable monosaccharides. In the present study, what was noteworthy was that UV-A irradiation showed excellent α-amylase, α-glucosidase and pancrelipase enzyme inhibition activity, and its effect on broccoli sprouts was better than that on radish sprouts. Therefore, two kinds of sprouts treated with UV-A radiation had the potential to be used as functional food and raw materials to reduce blood sugar.

4. Materials and Methods

4.1. Plant Materials, Cultivation and Treatment

Broccoli (*Brassica oleracea* L.) and radish (*Raphanus sativus* L.) seeds were purchased from Weifang Shouhe Seed Co., (Shandong, China). The varieties of broccoli were "Excellent broccoli", and the varieties of radish were "Weixian radish". They were disinfected with sodium hypochlorite and rinsed to neutral, then soaked in a water bath at 30 °C for 4 h. After that, the seeds were evenly spread in vermiculite-filled incubation trays at a density of 100 seeds per germination tray (60 cm × 30 cm × 4 cm), then germinated in the dark at 25 °C for 1 d [34,35] and incubated in UV-A incubators with different intensities of UV light (0, 4, 8, 12, 16 W) with a 16/8 h light/dark photoperiod. The model number of UV-A was F4T5BL and it was purchased from Beijing Ruilaixing Technology Co., Ltd., (Beijing, China). The plants were sprayed with deionized water twice a day, and samples were harvested at 5 d for measurement.

4.2. Determination of Sprout Length, Weight, Anthocyanin and Chlorophyll Content

A sample of 20 fresh sprouts was randomly selected from the sprouts treated with different UV light intensities for measuring sprout length and weight.

The content of anthocyanin was measured according to the method of Su et al. [36]. A total of 0.2 g of fresh sample was homogenized with 4 mL acidified ethanol. The homogenates were centrifuged at 10,000× *g* for 15 min, the supernatants were collected and the absorbance was measured at 535 nm.

The chlorophyll content was determined according to a previously published method [11]. Approximately 0.5 g of fresh sprouts were homogenized with 10 mL of 95% ethanol, and the homogenates were then filtered and transferred to a 25 mL volumetric flask. The absorbance measured at the wavelengths of 665 and 649 nm were chlorophyll a and chlorophyll b, respectively. The chlorophyll content in this study was the sum of chlorophyll a and chlorophyll b. The units of anthocyanin and chlorophyll content were expressed as mg/100 g FW.

4.3. Soluble Sugar and Soluble Protein Content Determination

The measurement of soluble sugar and soluble protein content was performed according to the method described by He et al. [11], though with some modifications. A total of 0.5 g of fresh sprouts were homogenized with 15 mL distilled water and then boiled for 20 min. The homogenates were allowed to cool and filter, then 1 mL of filtrate was taken and diluted by adding 6 mL distilled water. After that, 5 mL of anthrone solution was added to the extract and reacted in a boiling water bath for 10 min; finally, the absorbance value was measured at a 620 nm wavelength.

Approximately 0.2 g of fresh sprouts were ground into a homogenate using 4 mL of distilled water, then centrifuged at 10,000× *g* for 5 min. The supernatant was added to 5 mL of G-250 Coomassie brilliant blue solution and the reaction was carried out at room temperature for 5 min. The absorbance value was measured at a 595 nm wavelength. The soluble sugar and soluble protein content were expressed as mg/g FW.

4.4. Ascorbic Acid and Total Phenols Content Analysis

The method for determination of ascorbic acid content was referenced by Guo et al. [37], with some modifications. Approximately 0.3 g of fresh sprouts were extracted by grinding with 5 mL of 2% oxalic acid, and the homogenates were centrifuged at 8000× *g* for 5 min. The supernatant was filtered through a 0.45 μm membrane and then analyzed by HPLC.

A previously described method for total phenol content analysis was used in this study [37]. Fresh broccoli and radish sprouts (200 mg) were homogenized with 5 mL of 50% methanol, which was then centrifuged at 10,000× *g* for 15 min at room temperature. The supernatant was mixed with 1 mL Folin's phenol solution (0.2 mol/L) and 2 mL Na_2CO_3 solution (2 g/100 mL), and the absorbance was measured at 765 nm after the reaction for 1.5 h at room temperature in the dark. The results of ascorbic acid and total phenol content were expressed as mg/100 g FW.

4.5. Glucosinolate Content Determination

The extraction and determination of glucosinolates were conducted by Guo et al. [38]. Approximately 0.5 g of fresh sprouts were homogenized with 4 mL of boiled 75% methanol solution and extracted at 80 °C for 15 min. After the homogenates were centrifuged at 10,000× *g* for 10 min, the supernatant was flowed through DEAE Sephadex A-25 column to purity glucosinolates, then analyzed by HPLC system with a UV detection at 226 nm. The unit of glucosinolate content was expressed as μmol/g FW.

4.6. Myrosinase Activity Assay

Determination was performed according to the experimental method of Guo et al. [37]. Fresh broccoli and radish sprouts (500 mg) were homogenized with 3 mL of 0.1 mol/L phosphate buffer (pH 6.5) under ice bath conditions, and centrifuged at 10,000× *g* for 15 min at 4 °C. The supernatant obtained was the crude enzyme. The crude enzyme and sinigrin (1 mmol/L) were mixed thoroughly, then reacted in a constant temperature water bath at 37 °C for 10 min. The enzyme was inactivated at 100 °C for 5 min and then measured using a glucose kit. One unit of enzyme activity was taken as 1 nmol of glucose per minute. The unit of myrosinase activity was expressed as U/g FW.

4.7. Sulforaphane and Sulforaphene Assay

The determination of sulforaphane and sulforaphene was referred to Guo et al. [39]. Approximately 0.5 g of frozen broccoli and radish sprouts were ground and mixed with 4 mL distilled water, then added to 4 mL dichloromethane. The dichloromethane phase was evaporated and dissolved with 2 mL of 10% acetonitrile, then analyzed by HPLC (Thermo Fisher Scientific Dionex Ultimate 3000, Thermo Field Biological Co. Ltd., Chelmsford, MA, USA). The units of sulforaphane and sulforaphene formation were expressed as mg/100 g FW.

4.8. Hypoglycemia-Related Enzyme Activity Assay

4.8.1. Sample Treatments

Fresh broccoli and radish sprouts (0, 0.1, 0.2, 0.3, 0.4, 0.5 and 0.6 g) under 12 W irradiation were homogenized with 4 mL 50% ethanol, then the homogenates were concentrated with a nitrogen blow-drying apparatus, and 2 mL distilled water was added to re-dissolve the homogenates.

4.8.2. α-Amylase Inhibitory Rate

The α-amylase inhibitory activity was determined by reference to Chen et al. [40], with slight modifications. The α-amylase solution (1 U/mL) and different concentrations of sample solution were added into a test tube, incubated for 10 min in a water bath at 37 °C, and finally, 1 mL DNS reagent was added. The enzyme was inactivated in a boiling water bath for 5 min, cooled and diluted by adding 5 mL of distilled water, then the absorbance value was measured at 520 nm. The α-amylase inhibitory rate was calculated using the following equation:

$$R/\% = \left(1 - \frac{OD_A - OD_a}{OD_B - OD_b}\right) \times 100\%$$

where R is α-amylase inhibitory activity, %; OD_A is the absorption value of the sample, enzyme and substrate; ODa is the absorption value of the sample, buffer and substrate; OD_B is the absorption value of the buffer, enzyme and substrate; ODb is the absorption value of the buffer, buffer and substrate.

4.8.3. α-Glucosidase Inhibitory Rate

The determination of α-glucosidase inhibitory activity was slightly adjusted with the method of Ke et al. [41]. The α-glucosidase solution (0.5 U/mL) and different concentrations of the sample solution were incubated in a water bath at 37 °C for 15 min, before 0.5 mL of PNPG solution was added as substrate and finally 5 mL of 0.1 mol/L Na_2CO_3 was added to terminate the reaction. The absorbance value was recorded at 400 nm and the calculation method was consistent with α-amylase inhibitory activity.

4.8.4. Pancreatic Lipase Inhibitory Rate

A previously described method for pancreatic lipase inhibitory activity was used in this study [42]. The sprouts were treated as above. The sample solution of different concentrations, pancreatic lipase solution and Tris-HCL buffer solution were mixed, incubated at 37 °C for 10 min, before 6 mL 4-nitrophenylaurate was added as the substrate. The solution was reacted at 37 °C for 20 min and terminated in a boiling water bath for 10 min. The absorbance value was measured at 405 nm, and the calculation method was consistent with α-amylase inhibitory activity.

4.9. Statistical Analyses

All data are presented as the mean ± standard deviation of three repeats. The data were statistically analyzed with SPSS 26 (SPSS Inc., Chicago, IL, USA) and mapped with Origin 2021. Data were statistically analyzed by one-way ANOVA and Duncan's multiple-range tests. Differences at $p < 0.05$ were considered significant.

5. Conclusions

In conclusion, our results acquired in the present study demonstrated that UV-A irradiation reduced the growth of broccoli and radish sprouts. Nevertheless, UV-A irradiation, especially 12 W of UV-A irradiation, significantly induced health compound accumulation, including anthocyanin, chlorophyll, polyphenol and ascorbic acid. Furthermore, 12 W of UV-A intensity treatment also resulted in the highest aliphatic and indole glucosinolate content, especially glucoraphanin and glucoraphenin, and contributed to the increase of sulforaphane and sulforaphane formation, respectively. On the other hand, broccoli and radish sprouts treated with UV-A irradiation of 12 W showed a significant inhibitory effect on α-amylase, α-glucosidase and pancrelipase, and the broccoli sprouts were even better. Hence, UV-A irradiation under 12 W might be applied to improve the quality of broccoli and radish sprouts so as to achieve the purpose of alleviating diabetes. In addition, the next study will compare the hypoglycemic activity of the two kinds of sprouts in vivo to provide a basis for the application of the sprouts.

Author Contributions: Investigation, data curation, writing—original draft, G.C.; investigation, data curation, M.C.; formal analysis, software, X.L.; software, project administration, J.X.; methodology, writing—review and editing, L.L.; funding acquisition, writing—review and editing, validation L.G. All authors have read and agreed to the published version of the manuscript.

Funding: This study was supported by the National Natural Science Foundation of China (No. 31801457).

Data Availability Statement: The data is contained within the manuscript.

Conflicts of Interest: The authors declare that they have no known competing financial interests or personal relationships that could have appeared to influence the work reported in this paper.

References

1. Bousquet, J.; Le Moing, V.; Blain, H.; Czarlewski, W.; Zuberbier, T.; de la Torre, R.; Pizarro Lozano, N.; Reynes, J.; Bedbrook, A.; Cristol, J.P.; et al. Efficacy of broccoli and glucoraphanin in COVID-19: From hypothesis to proof-of-concept with three experimental clinical cases. *World Allergy Organ. J.* **2021**, *14*, 100498. [CrossRef] [PubMed]
2. Patel, B.; Mann, G.E.; Chapple, S.J. Concerted redox modulation by sulforaphane alleviates diabetes and cardiometabolic syndrome. *Free Radic. Biol. Med.* **2018**, *122*, 150–160. [CrossRef]
3. Xing, J.J.; Cheng, Y.L.; Chen, P.; Shan, L.; Ruan, R.; Li, D.; Wang, L.J. Effect of high-pressure homogenization on the extraction of sulforaphane from broccoli (*Brassica oleracea*) seeds. *Powder Technol.* **2018**, *358*, 103–109. [CrossRef]
4. Baenas, N.; Gómez-Jodar, I.; Moreno, D.A.; García-Viguera, C.; Periago, P.M. Broccoli and radish sprouts are safe and rich in bioactive phytochemicals. *Postharvest Biol. Technol.* **2017**, *127*, 60–67. [CrossRef]

5. Francis, H.; Debs, E.; Koubaa, M.; Alrayess, Z.; Maroun, R.G.; Louka, N. Sprouts use as functional foods. optimization of germination of wheat (*Triticum aestivum L.*), alfalfa (*Medicago sativa L.*), and radish (*Raphanus sativus L.*) seeds based on their nutritional content evolution. *Foods* **2022**, *11*, 1460. [CrossRef] [PubMed]
6. O'Hare, T.J.; Williams, D.J.; Zhang, B.; Wong, L.S.; Jarrett, S.; Pun, S.; Jorgensen, W.; Imsic, M. Radish sprouts versus broccoli sprouts: A comparison of anti-cancer potential based on glucosinolate breakdown products. *Acta Hortic.* **2009**, *841*, 187–192. [CrossRef]
7. Ahammed, G.J.; Chen, Y.; Liu, C.C.; Yang, Y.X. Light regulation of potassium in plants. *Plant Physiol. Biochem.* **2022**, *170*, 316–324. [CrossRef]
8. Kang, S.; Kim, J.E.; Zhen, S.; Kim, J. Mild-Intensity UV-A radiation applied over a long duration can improve the growth and phenolic contents of sweet basil. *Front. Plant Sci.* **2022**, *13*, 858433. [CrossRef]
9. Moreira-Rodríguez, M.; Nair, V.; Benavides, J.; Cisneros-Zevallos, L.; Jacobo-Velázquez, D. UVA, UVB Light, and methyl jasmonate, alone or combined, redirect the biosynthesis of glucosinolates, phenolics, carotenoids, and chlorophylls in broccoli sprouts. *Int. J. Mol. Sci.* **2017**, *18*, 2330. [CrossRef]
10. Jiang, H.; Li, Y.; He, R.; Tan, J.; Liu, K.; Chen, Y.; Liu, H. Effect of supplemental UV-A intensity on growth and quality of kale under red and blue light. *Int. J. Mol. Sci.* **2022**, *23*, 6819. [CrossRef]
11. He, R.; Zhang, Y.; Song, S.; Su, W.; Liu, H. UV-A and FR irradiation improves growth and nutritional properties of lettuce grown in an artificial light plant factory. *Food Chem.* **2021**, *345*, 128727. [CrossRef] [PubMed]
12. Farag, M.A.; Aboul Naser, A.F.; Zayed, A.; Sharaf El-Dine, M.G. Comparative insights into four major legume sprouts efficacies for diabetes management and its complications: Untargeted versus targeted NMR biochemometrics approach. *Metabolites* **2023**, *13*, 63. [CrossRef] [PubMed]
13. Laila, O.; Murtaza, I.; Muzamil, S.; Imtiyaz Ali, S.; Abid Ali, S.; Ahamad Paray, B.; Gulnaz, A.; Vladulescu, C.; Mansoor, S. Enhancement of nutraceutical and anti-diabetic potential of fenugreek (*Trigonella foenum-graecum*). Sprouts with natural elicitors. *Saudi Pharm. J.* **2023**, *31*, 1–13. [CrossRef] [PubMed]
14. Tsormpatsidis, E.; Henbest, R.G.C.; Davis, F.J.; Battey, N.H.; Hadley, P.; Wagstaffe, A. UV irradiance as a major influence on growth, development and secondary products of commercial importance in Lollo Rosso lettuce 'Revolution' grown under polyethylene films. *Environ. Exp. Bot.* **2008**, *63*, 232–239. [CrossRef]
15. Choi, D.-S.; Nguyen, T.K.L.; Oh, M.-M. Growth and biochemical responses of kale to supplementary irradiation with different peak wavelengths of UV-A light-emitting diodes. *Hortic. Environ. Biotechnol.* **2022**, *63*, 65–76. [CrossRef]
16. Qian, M.; Kalbina, I.; Rosenqvist, E.; Jansen, M.A.K.; Strid, Å. Supplementary UV-A and UV-B radiation differentially regulate morphology in ocimum basilicum. *Photochem. Photobiol. Sci.* **2023**, *22*, 2219–2230. [CrossRef] [PubMed]
17. de Wit, M.; Galvão, V.C.; Fankhauser, C. Light-mediated hormonal regulation of plant growth and development. *Annu. Rev. Plant Biol.* **2016**, *67*, 513–537. [CrossRef]
18. Zhang, X.; Su, N.; Jia, L.; Tian, J.; Li, H.; Huang, L.; Shen, Z.; Cui, J. Transcriptome analysis of radish sprouts hypocotyls reveals the regulatory role of hydrogen-rich water in anthocyanin biosynthesis under UV-A. *BMC Plant Biol.* **2018**, *18*, 227. [CrossRef]
19. Nguyen, L.T.K.; Oh, M.M. Growth and biochemical responses of green and red perilla supplementally subjected to UV-A and deep-blue LED lights. *Photochem. Photobiol.* **2022**, *98*, 1332–1342. [CrossRef]
20. Johnson, G.A.; Day, T.A. Enhancement of photosynthesis in sorghum bicolor by ultraviolet radiation. *Physiol. Plant.* **2010**, *116*, 554–562. [CrossRef]
21. Helsper, J.P.F.G.; Ric de Vos, C.H.; Maas, F.M.; Jonker, H.H.; Van Den Broeck, H.C.; Jordi, W.; Pot, C.S.; Keizer, L.C.P.; Schapendonk, A.H.C.M. Response of selected antioxidants and pigments in tissues of Rosa hybrida and Fuchsia hybrida to supplemental UV-A exposure. *Physiol. Plant.* **2013**, *117*, 171–178. [CrossRef]
22. Mao, P.; Duan, F.; Zheng, Y.; Yang, Q. Blue and UV-A light wavelengths positively affected accumulation profiles of healthy compounds in Pak-choi. *J. Sci. Food Agric.* **2020**, *101*, 1676–1684. [CrossRef] [PubMed]
23. Lemoine, R.; Camera, S.L.; Atanassova, R.; Dédaldéchamp, F.; Allario, T.; Pourtau, N.; Bonnemain, J.-L.; Laloi, M.; Coutos-Thévenot, P.; Maurousset, L.; et al. Source-to-sink transport of sugar and regulation by environmental factors. *Front. Plant Sci.* **2013**, *4*, 272. [CrossRef] [PubMed]
24. Tezuka, T.; Yamaguchi, F.; Ando, Y. Physiological activation in radish plants by UV-A radiation. *J. Photochem. Photobiol. B Biol.* **1994**, *24*, 33–40. [CrossRef]
25. Xu, M.-J.; Dong, J.-F.; Zhu, M.-Y. Effects of germination conditions on ascorbic acid level and yield of soybean sprouts. *J. Sci. Food Agric.* **2005**, *85*, 943–947. [CrossRef]
26. Jia, L.; Tian, J.; Wei, S.; Zhang, X.; Xu, X.; Shen, Z.; Shen, W.; Cui, J. Hydrogen gas mediates ascorbic acid accumulation and antioxidant system enhancement in soybean sprouts under UV-A irradiation. *Sci. Rep.* **2017**, *7*, 16366. [CrossRef]
27. Sun, Y.; Luo, M.; Ge, W.; Zhou, X.; Zhou, Q.; Wei, B.; Cheng, S.; Ji, S. Phenylpropanoid metabolism in relation to peel browning development of cold-stored 'Nanguo' pears. *Plant Sci.* **2022**, *322*, 111363. [CrossRef] [PubMed]
28. Hideg, É.; Jansen, M.A.K.; Strid, K. UV-B exposure, ROS, and stress: Inseparable companions or loosely linked associates? *Trends Plant Sci.* **2013**, *18*, 107–115. [CrossRef]
29. Zhuang, L.; Huang, G.; Li, X.; Xiao, J.; Guo, L. Effect of different LED lights on aliphatic glucosinolates metabolism and biochemical characteristics in broccoli sprouts. *Food Res. Int.* **2022**, *154*, 111015. [CrossRef]

30. He, R.; Li, Y.; Ou, S.; Gao, M.; Zhang, Y.; Song, S.; Liu, H. Regulation of growth and main health-promoting compounds of Chinese kale baby-leaf by UV-A and FR light. *Front. Plant Sci.* **2022**, *12*, 799376. [CrossRef]
31. Xie, C.; Tang, J.; Xiao, J.; Geng, X.; Guo, L. Purple light-emitting diode (LED) lights controls chlorophyll degradation and enhances nutraceutical quality of postharvest broccoli florets. *Sci. Hortic.* **2022**, *294*, 110768. [CrossRef]
32. Bharadwaj, R.P.; Raju, N.G.; Chandrashekharaiah, K.S. Purification and characterization of alpha-amylase inhibitor from the seeds of underutilized legume, mucuna pruriens. *J. Food Biochem.* **2018**, *72*, 2267–2274. [CrossRef]
33. Hossain, U.; Das, A.K.; Ghosh, S.; Sil, P.C. An overview on the role of bioactive α-glucosidase inhibitors in ameliorating diabetic complications. *Food Chem. Toxicol.* **2020**, *145*, 111738. [CrossRef] [PubMed]
34. Guo, L.; Yang, R.; Wang, Z.; Guo, Q.; Gu, Z. Glucoraphanin, sulforaphane and myrosinase activity in germinating broccoli sprouts as affected by growth temperature and plant organs. *J. Funct. Foods* **2014**, *9*, 70–77. [CrossRef]
35. Jia, L.; Wang, T.; Sun, Y.; Zhang, M.; Tian, J.; Chen, H.; Shen, Z.; Abro, H.; Su, N.; Cui, J. Protective effect of selenium-enriched red radish sprouts on carbon tetrachloride-induced liver injury in mice. *J. Food Sci.* **2019**, *84*, 3027–3036. [CrossRef]
36. Su, N.; Liu, Z.; Wang, L.; Liu, Y.; Niu, M.; Chen, X.; Cui, J. Improving the anthocyanin accumulation of hypocotyls in radish sprouts by hemin-induced NO. *BMC Plant Biol.* **2022**, *22*, 224. [CrossRef]
37. Guo, L.; Yang, R.; Wang, Z.; Guo, Q.; Gu, Z. Effect of NaCl stress on health-promoting compounds and antioxidant activity in the sprouts of three broccoli cultivars. *Int. J. Food Sci. Nutr.* **2014**, *65*, 476–481. [CrossRef]
38. Guo, L.; Yang, R.; Gu, Z. Cloning of genes related to aliphatic glucosinolate metabolism and the mechanism of sulforaphane accumulation in broccoli sprouts under jasmonic acid treatment. *J. Sci. Food Agric.* **2016**, *96*, 4329–4336. [CrossRef]
39. Guo, L.; Yang, R.; Wang, Z.; Gu, Z. Effect of freezing methods on sulforaphane formation in broccoli sprouts. *RSC Adv.* **2015**, *5*, 32290–32297. [CrossRef]
40. Chen, C.; Zhang, B.; Huang, Q.; Fu, X.; Liu, R.H. Microwave-assisted extraction of polysaccharides from moringa oleifera Lam. leaves: Characterization and hypoglycemic activity. *Ind. Crops Prod.* **2017**, *100*, 1–11. [CrossRef]
41. Ke, S.; Wei, B.; Qiu, W.; Zhou, T.; Wang, S.; Chen, J.; Chen, J.; Zhang, H.; Jin, W.; Wang, H. Structural characterization and alpha-glucosidase inhibitory and antioxidant activities of fucoidans extracted from saccharina japonica. *Chem. Biodivers.* **2020**, *17*, e2000233. [CrossRef] [PubMed]
42. Ercan, P.; El, S.N. Inhibitory effects of chickpea and Tribulus terrestris on lipase, alpha-amylase and alpha-glucosidase. *Food Chem.* **2016**, *205*, 163–169. [CrossRef] [PubMed]

plants

MDPI

Article

The Cooperation Regulation of Antioxidative System and Hormone Contents on Physiological Responses of *Wedelia trilobata* and *Wedelia chinensis* under Simulated Drought Environment

Ping Huang [1,2,*], Zhiwei Xu [1,2], Weijie He [1,2], Hong Yang [1,2], Bin Li [1,2], Wendian Ding [1,2], Yuze Lei [1,2], Adeel Abbas [1,2,*], Rashida Hameed [1,2], Congyan Wang [1,2], Jianfan Sun [1,2] and Daolin Du [3,*]

1 Institute of Environment and Ecology, School of Environment and Safety Engineering, Jiangsu University, Zhenjiang 212013, China
2 School of Emergency Management, Jiangsu University, Zhenjiang 212013, China
3 Jingjiang College, Jiangsu University, Zhenjiang 212013, China
* Correspondence: huangjiehp@ujs.edu.cn (P.H.); adeel.abbas92@ujs.edu.cn (A.A.); daolindu@163.com (D.D.)

Abstract: Drought-induced metabolic dysregulation significantly enhances the production of reactive oxygen species (ROS), which, in turn, exerts a substantial influence on the oxidation–reduction regulatory status of cells. These ROS, under conditions of drought stress, become highly reactive entities capable of targeting various plant organelles, metabolites, and molecules. Consequently, disruption affects a wide array of metabolic pathways and eventually leads to the demise of the cells. Given this understanding, this study aimed to investigate the effects of different drought stress levels on the growth and development of the invasive weed *Wedelia trilobata* and its co-responding native counterpart *Wedelia chinensis*. Both plants evolved their defense mechanisms to increase their antioxidants and hormone contents to detoxify ROS to avoid oxidative damage. Still, the chlorophyll content fluctuated and increased in a polyethylene-glycol-simulated drought. The proline content also rose in the plants, but *W. chinensis* showed a significant negative correlation between proline and malondialdehyde in different plant parts. Thus, *W. trilobata* and *W. chinensis* exhibited diverse or unlike endogenous hormone regulation patterns under drought conditions. Meanwhile, *W. trilobata* and *W. chinensis* pointedly increased the content of indole acetic acid and gibberellic acid in a different drought stress environment. A positive correlation was found between endogenous hormones in other plant parts, including in the roots and leaves. Both simulated and natural drought conditions exerted a significant influence on both plant species, with *W. trilobata* displaying superior adaptation characterized by enhanced growth, bolstered antioxidant defense mechanisms, and heightened hormonal activities.

Keywords: drought stress effects; metabolite disruption; antioxidant enzymes; adaptation strategies; climate changes

1. Introduction

Various climatological models proffer the prognostication that with the concomitant reduction in precipitation and escalating temperatures, instances of drought will become increasingly recurrent on a global scale [1]. The imposition of drought stress exerts deleterious effects on the trajectory of plant growth and development, constraining access to vital water resources, thereby impeding pivotal physiological processes, such as photosynthesis and respiration, ultimately encumbering the abundance and geographic dispersion of plant populations across the global landscape [2,3]. Moreover, it begets a diminution in cell turgidity, hinders the equilibrium of source–sink dynamics, and disrupts the homeostasis of metabolic processes [4]. The manifestations of drought encompass ostensible morphological and physiological alterations in plants, and their adaptability can transmute them into species possessing drought tolerance attributes, be they native or invasive [5]. Plants evince evolutionary adjustments in response to a plenitude of growth circumstances, and the extent of their tolerance to these vicissitudes exhibits a conspicuous variability [6]. The scrutiny of compromised germination, subterranean structures (i.e., roots), and the quantification of sundry physiological and biochemical constituents serve as instrumental metrics in the assessment of these disparities [7].

Under drought stress conditions, the addition of reactive oxygen species (ROS) triggers oxidative damage, resulting in substantial harm to cellular organelles [8]. This oxidative damage, which is a consequence of excessive ROS accumulation, extends its adverse effects to various cellular components, including the electron transport chain; lipid peroxidation within chloroplasts and mitochondria; and the deactivation of essential enzymatic activity, proteins, and nucleic acids. These detrimental processes ultimately culminate in a notable reduction in photosynthesis and, consequently, a decrease in crop yield [9]. To counteract the detrimental impact of ROS, plants employ a range of antioxidant enzymes, such as Ascorbate peroxidase (APX), catalase (CAT), dehydroascorbate reductase (DHAR), Glutathione reductase (G.R.), glutathione S-transferase (GST), peroxidase (POD), and superoxide dismutase (SOD). These enzymes assume a pivotal role in mitigating the accumulation of reactive oxygen species (ROS) [10,11]. More specifically, these enzymatic systems are instrumental in upholding a delicate equilibrium between the production and signaling of ROS, thereby reinforcing the plant's capacity to endure adverse environmental conditions. Among the array of environmental challenges, drought stress looms as a significant concern, impacting a plethora of plant species [12,13]. In an investigation involving three commonly encountered Lantana variants exposed to constantly changing environment. The functions of heat-resistant antioxidant enzymes, specifically peroxidase (POD), superoxide dismutase (SOD), and catalase (CAT), in the elimination of ROS were explored, which provided insights that shed light on the potential significance of these enzymes in strengthening the plant's resilience against adverse environmental factors [14].

Plants employ various strategies to adapt to fluctuating water stress conditions, including osmotic regulation, antioxidant enzyme activities, and a range of endogenous hormones [15]. These intrinsic plant hormones play essential roles in regulating physiological functions, enabling plants to effectively cope with drought stress [16]. Among these phytohormones, abscisic acid (ABA) and indole acetic acid (IAA) hold special significance as drought-responsive hormonal agents [17]. They exert authoritative control over numerous physiological and biochemical plant processes, ultimately enhancing drought resistance across various plant species. For instance, during drought stress, the concentration of abscisic acid (ABA) increases, leading to stomatal closure, reducing transpiration, and fortifying the plant's resistance to drought [18]. Indole acetic acid (IAA), which is another key endogenous hormone, is prevalent in various plant tissues and organs, particularly those associated with robust growth metabolism in roots and leaves [19]. Its concentration can be adjusted in response to arid environments, although the mechanisms governing IAA regulation during drought are complex and vary between different plant species [20]. In the context of this research, the focus turned to *W. trilobata*, which is a low-lying invasive species known for its remarkable adaptability. Native to Central America, this plant successfully established itself in tropical and subtropical regions through invasions [14]. Introduced in the latter part of the twentieth century to China's Guangdong, Guangxi, and Fujian provinces as an ornamental plant, its invasive nature was quickly identified, resulting in its classification as an invasive species on a global scale [21]. Prior research examining the impact of drought stress on oxidative arrangements and hormonal activities predominantly concentrated on cultivated plants [22]. However, there exists a notable gap in comprehensive research regarding the consequences of drought stress on the regulation of antioxidative systems, hormone levels, and the subsequent physiological responses in *W. trilobata* and *W. chinensis*. Consequently, to facilitate comparative analysis, we designed this study to investigate how different simulated drought conditions influence the regulation of antioxidative systems and hormone levels in *W. trilobata* and *W. chinensis* at the physiological level.

2. Results

Drought stress exerts a considerable impact on the growth and development of plants, often resulting in a noticeable reduction in biomass. Our findings indicate that both *W. trilobata* and *W. chinensis* experienced significant inhibition under PEG-induced simulated drought conditions, as well as natural drought conditions (Figure 1A,B). In comparison with the control group (C.K.), *W. trilobata* exhibited a substantial decrease in the number of leaves, plant height, root length, and lateral roots by 60%, 83.16%, 25.51%, and 68.87%, respectively.

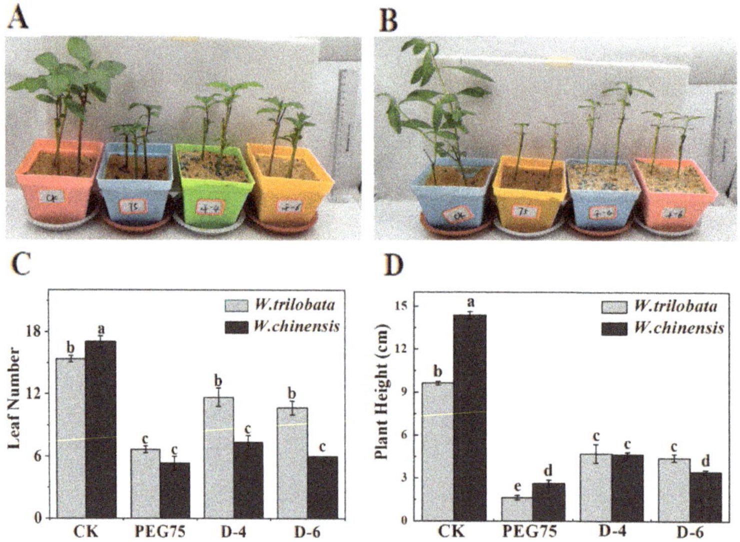

Figure 1. Effects of simulated drought stress on the growth of *W. trilobata* and *W. chinensis* (**A**). Growth phenotype of *W. trilobata*. (**B**) Growth phenotype of *W. chinensis*. (**C**) The effect of simulated drought stress on number of leaves of two plants. (**D**) The effect of simulated drought stress on plant height of two plants (mean ± SE, n = 3). Different lowercase letters in the figure indicate significant differences between different treatments ($p < 0.05$).

W. chinensis showed a decrease of 70.59% in leaf number, 81.85% in plant height, 54.90% in root length, and 53.14% in lateral roots (Figures 1 and 2). This shows that *W. trilobata* and *W. chinensis* produced different results regarding root development under the natural drought treatment: *W. trilobata* and *W. chinensis* had significantly fewer lateral roots in a natural drought environment, but compared with the control group (C.K.), *W. trilobata* increased its root length by 14.51% and 37.13%, and decreased its lateral root number by 42.23% and 60.01% (Figure 2), while *W. chinensis* showed no significant difference in root length (Figure 2). At the same time, compared with the C.K., *W. trilobata* and *W. chinensis* had reduced chlorophyll contents by 36.08% and 24.48% in natural drought. In comparison, the chlorophyll content of *W. chinensis* increased by 7.19% in PEG-simulated drought (Figure 3). The analysis of the aboveground and underground biomass and root–shoot ratio of *W. trilobata* and *W. chinensis* under simulated drought conditions provided insights into their clonal reproductive capacity and ecological adaptability (Figure 3). Both *W. trilobata* and *W. chinensis* exhibited a notable reduction in aboveground biomass when subjected to both PEG-induced simulated and natural drought treatments. However, a significant decrease in underground biomass was observed only in *W. chinensis*. In comparison with the control group (C.K.), both plant species displayed a significant increase in the root–shoot ratio (Figure 3). In comparison with the control group (C.K.), under both PEG-induced simulated drought and natural drought conditions, *W. trilobata* exhibited a obvisously decrease in aboveground biomass. The correlation analysis revealed significant positive correlations between the height of *W. trilobata* and the number of leaves, as well as the chlorophyll content. Additionally, the lateral root number showed significant positive correlations with the number of leaves, plant height, and chlorophyll content, while displaying a meaningful negative correlation with the root–shoot ratio (Table 1, $p < 0.01$). Similarly, the height of *W. chinensis* exhibited significant positive correlations with the number of leaves and root length. The lateral root number also showed significant positive correlations with the number of leaves and plant height, while exhibiting a significant negative correlation with the root–shoot ratio (Table 2).

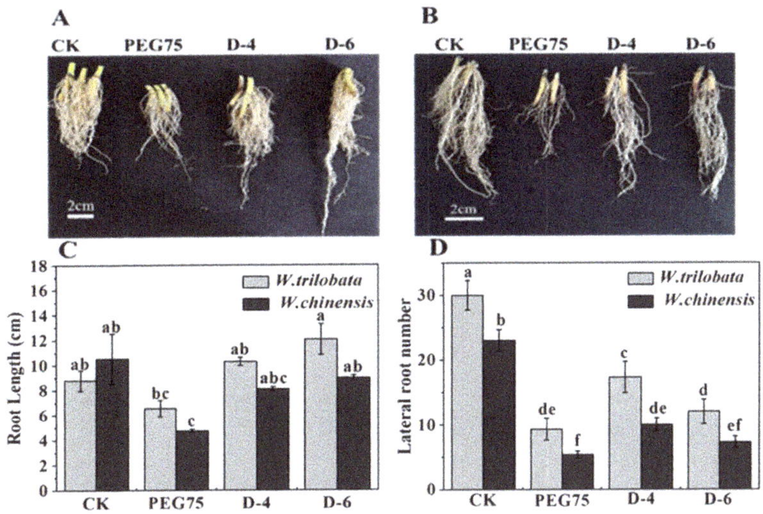

Figure 2. The influence of simulated drought stress on root growth in *W. trilobata* and *W. chinensis* was examined. (**A**) Root growth in *W. trilobata*. (**B**) Root growth in *W. chinensis*. (**C**) Root growth in both *W. trilobata* and *W. chinensis*. (**D**) Number of lateral roots in the two plants (mean ± standard error, n = 3). The lowercase letters in the figure denote significant differences observed between the treatments ($p < 0.05$). CK: the conditions of natural drought simulation, PEG75: Polyethylene glycol, D-4: water applied after four days, D-6: water applied after six days.

Table 1. Correlation analysis of the effects of simulated drought on growth index, antioxidant system, and endogenous hormone ratio of *W. trilobata*.

	LN	PH	RL	CH	RSR	LRN	Pro	MDA	POD	CAT	AI	AG	IG
LN	1	0.922 **	0.275	0.493	−0.562	0.881 **	−0.777 **	−0.851 **	−0.381	−0.484	−0.265	0.235	0.199
PH		1	0.192	0.615 *	−0.509	0.945 **	−0.856 **	−0.872 **	−0.468	−0.463	−0.191	0.303	0.038
RL			1	−0.559	0.035	0.034	0.174	−0.054	−0.298	0.029	−0.504	0.600 *	0.583 *
CH				1	−0.568	0.746 **	−0.875 **	−0.728 **	−0.260	−0.346	0.134	−0.274	−0.403
RSR					1	−0.589 *	0.591 *	0.699 *	0.001	0.124	−0.178	0.110	0.068
LRN						1	−0.944 **	−0.941 **	−0.349	−0.428	−0.128	0.253	0.007
Pro							1	0.868 **	0.390	0.478	−0.012	0.007	0.245
MDA								1	0.221	0.262	0.195	−0.218	−0.174
POD									1	0.629 *	0.519	−0.015	0.138
CAT										1	0.308	0.140	0.200
Al											1	−0.281	−0.584 *
AG												1	0.681 *
IG													1

LN: number of leaves, PH: plant height, RL: root length, CH: chlorophyll content, RSR: root–stem ratio, Pro: proline, MDA: malonaldehyde, POD: peroxidase, CAT: catalase, AI: ABA/IAA value, AG: ABA/GA value, IG: IAA/GA value. The symbol of "*" indicates significant correlations ($p < 0.05$) and "**" signifies exceedingly momentous correlation ($p < 0.01$).

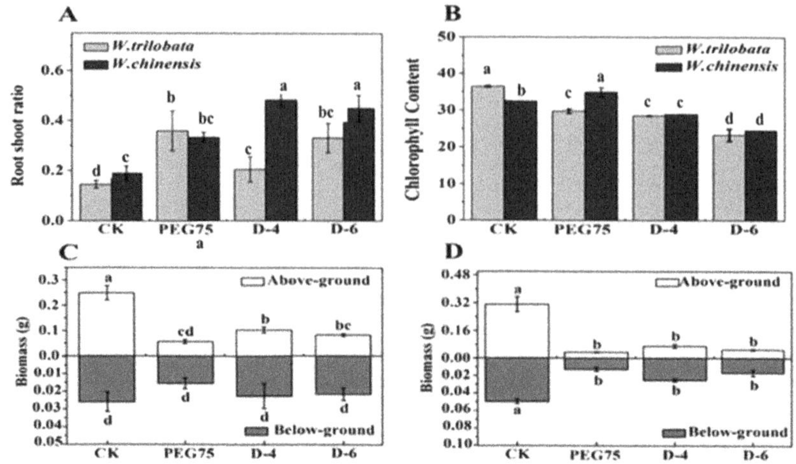

Figure 3. The effects of simulated drought stress on the root–shoot ratio (**A**), chlorophyll content (**B**), and biomass of *W. trilobata* (**C**), and biomass of *W. chinensis* (**D**) (mean ± S.E., n = 3). Different lowercase letters in the figure indicate significant differences between different treatments ($p < 0.05$).

Table 2. Correlation analysis of the effects of simulated drought on growth index, antioxidant system, and endogenous hormone ratio of *W. chinensis*.

	LN	PH	RL	CH	RSR	LRN	Pro	MDA	POD	CAT	AI	AG	IG
LN	1	0.970 **	0.838 **	−0.572	0.001	0.909 *	−0.562	−0.466	−0.235	−0.769 **	−0.322	−0.464	−0.372
PH		1	0.799 **	−0.547	−0.080	0.919 **	−0.606 *	−0.515	−0.188	−0.742	−0.214	−0.525	−0.447
RL			1	−0.772 **	0.145	0.639 *	−0.358	−0.033	−0.278	−0.436	−0.531	−0.060	−0.011
CH				1	−0.193	−0.375	0.380	−0.309	0.530	0.134	0.493	−0.266	−0.183
RSR					1	0.026	0.322	0.430	0.394	0.361	−0.55 **	0.342	0.395
LRN						1	−0.432	−0.640 *	0.077	−0.685 *	−0.175	−0.643 *	−0.519
Pro							1	0.471	0.378	0.549	−0.120	0.508	0.642 *
MDA								1	−0.236	0.677 *	−0.464	0.977 **	0.892 **
POD									1	0.352	−0.010	−0.279	−0.176
CAT										1	−0.079	0.625 *	0.513
Al											1	−0.376	−0.416
AG												1	0.949 **
IG													1

LN: number of leaves, PH: plant height, RL: root length, CH: chlorophyll content, RSR: root–stem ratio, Pro: proline, MDA: malonaldehyde, POD: peroxidase, CAT: catalase, AI: ABA/IAA value, AG: ABA/GA value, IG: IAA/GA value. The symbol of "*" indicates significant correlations ($p < 0.05$) and "**" signifies exceedingly momentous correlation ($p < 0.01$).

Our research findings demonstrate that *W. trilobata* and *W. chinensis* exhibited significant increases in proline content by 47.46% and 29.31%, respectively, during natural drought compared with the control group (C.K.) (Figure 4). Additionally, the result of Figure 4 illustrates that *W. trilobata* exhibited higher levels of MDA (malondialdehyde) content than *W. chinensis* under normal growth conditions. However, when exposed to drought stress, *W. trilobata* experienced a substantial and dramatic increase in MDA content, reaching its peak during the PEG-simulated drought, with a 97.64% increase compared with the control group. The correlation analysis revealed a negative correlation between the proline and malondialdehyde (MDA) contents in both *W. trilobata* and *W. chinensis* with the number of leaves, plant height, and lateral roots under both the PEG-simulated and natural drought treatments. However, *W. chinensis* exhibited a significant negative correlation with the number of leaves, plant height, and lateral roots (Tables 1 and 2).

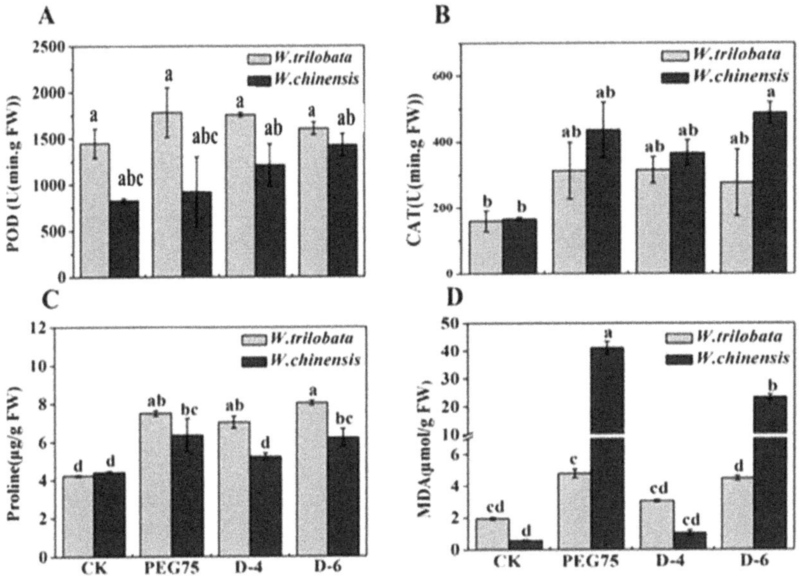

Figure 4. The impacts of simulated drought stress on peroxidase (**A**), catalase (**B**), proline (**C**), and malondialdehyde (**D**) in the leaves of the two plants were examined (mean ± standard error, n = 3). Distinct lowercase letters in the figure represent significant differences observed between the different treatments ($p < 0.05$).

Endogenous Hormones Activities

The research found that *W. trilobata* showed significantly increased ABA contents in its leaves and roots under drought conditions compared with the C.K., and the ABA content in the *W. chinensis* leaves was the highest in natural drought, and the ABA content in the *W. chinensis* roots reached the maximum value during PEG-simulated drought (Figure 5). Meanwhile, *W. trilobata* and *W. chinensis* extensively increased the content of IAA in the stem in a drought environment. At the same time, the changes of IAA in leaves and roots showed different patterns (Table 3). *W. trilobata* significantly increased Gibberellic acid (GA) content in its leaves and stems during PEG-simulated drought, but *W. chinensis* significantly decreased the GA content in its roots and stems during PEG-simulated drought (Figure 6). Furthermore, *W. trilobata* significantly increased the GA content in its roots, and *W. chinensis* significantly decreased the GA content in its roots during natural drought. Comparative analysis of *W. trilobata* and *W. chinensis* hormone combination and ratio exhibited changes in the overground and underground parts. *W. trilobata* significantly reduced the ABA/IAA and ABA/GA ratios in the underground part of the plant but significantly increased the IAA/GA ratio in the overground part of the plant, and *W. chinensis* significantly decreased the ABA/IAA ratio in the underground part of the plant in both drought treatments. The ABA/GA ratio was significantly increased in the aboveground part of *W. chinensis* during PEG-simulated drought, and the IAA/GA ratio was significantly increased in the underground part of *W. chinensis* during natural drought. These results show that *W. trilobata* and *W. chinensis* exhibited different or unlike endogenous hormone regulation patterns under drought conditions. The correlation analysis discovered a significant positive correlation between the ABA/GA (abscisic acid/gibberellic acid) ratio and root length in *W. trilobata*, as well as a significant positive correlation between the IAA/GA (indole acetic acid/gibberellic acid) ratio and root length under drought conditions (Table 1). However, in *W. chinensis*, the ABA/IAA ratio showed a positive correlation with the number of leaves, plant height, and lateral roots, and a negative correlation with the root–shoot ratio and peroxidase activity. The ABA/GA ratio was meaningfully negatively correlated with the root length and positively correlated with the chlorophyll and malondialdehyde contents (Table 2).

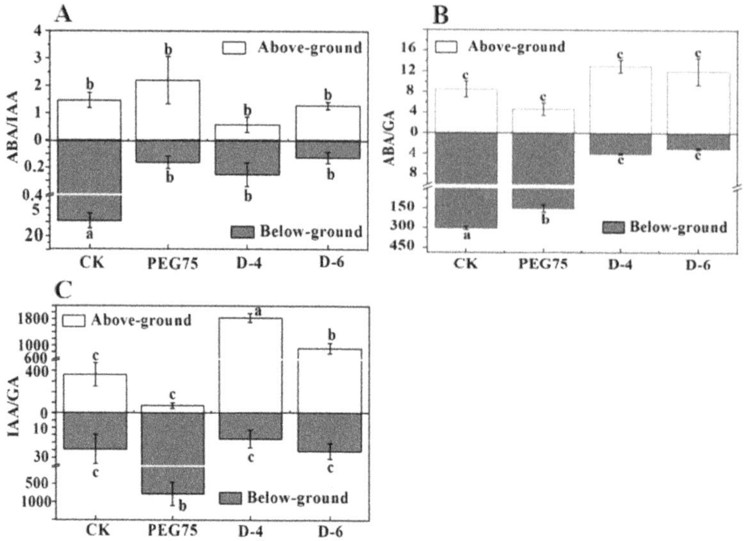

Figure 5. Effects of simulated drought stress on the ABA/IAA (**A**), ABA/GA (**B**), and IAA/GA (**C**) ratios in both the aboveground and underground parts of *W. trilobata*. The data presented represent the mean values with standard errors (mean ± standard error, n = 3). In the figure, lowercase letters are used to indicate significant differences observed between the treatments ($p < 0.05$).

Table 3. The influence of varied drought conditions on hormone contents in *W. trilobata* and *W. chinensis*.

Treatment	ABA Content (ng/g FW)			IAA Content (ng/g Fw0P)			GAA Content (ng/g Fw)		
	Leaf	Stem	Root	Leaf	Stem	Root	Leaf	Stem	Root
W. trilobata									
CK	77.41 ± 12.59 d	24.34 ± 5.90 d	77.94 ± 10.2 c	62.84 ± 7.40 d	2234.83 ± 387.99 b	8.91 ± 1.50 d	35.94 ± 12.88 d	9.26 ± 6.48 b	0.59 ± 0.30 d
PEG75	318.36 ± 85.53 a	64.94 ± 11.25 a	327.05 ± 42.6 b	165.12 ± 29.10 c	1769.06 ± 213.94 c	2105.66 ± 182.3 b	264.77 ± 19.08 a	32.26 ± 10.09 a	3.96 ± 1.67 c
D-4	153.05 ± 34.82 c	46.83 ± 10.09 bc	379.40 ± 19.02 a	349.72 ± 23.0 a	9274.31 ± 129.98 a	1872.16 ± 358.9 bc	73.33 ± 16.02 bc	5.03 ± 0.28 b	114.71 ± 11.82 b
D-6	287.63 ± 29.42 b	54.07 ± 12.91 b	420.10 ± 62.78 a	227.20 ± 16.62 b	7002.72 ± 184.6 b	4138.90 ± 314.8 a	194.65 ± 27.6 b	8.70 ± 2.96 b	156.41 ± 20.60 a
W. Chinensis									
CK	155.52 ± 47.64 b	99.37 ± 2.78 b	174.28 ± 25.9 b	101.58 ± 9.48 d	68.98 ± 2.75 d	2710.41 ± 380.3 c	32.77 ± 14.8 c	55.05 ± 9.76 a	223.18 ± 31.4 a
PEG75	280.98 ± 50.67 a	235.12 ± 15.81 a	345.74 ± 77.5 a	215.80 ± 19.33 c	1223.51 ± 76.6 c	5087.35 ± 286.3 a	43.38 ± 18.01 c	4.40 ± 0.8 bc	92.93 ± 26.49 b
D-4	123.71± 21.89 bc	213.57 ± 33.9 a	114.71 ± 32.50 c	4325.41 ± 331.1 a	6937.79 ± 116.5 a	3521.73 ± 224.01 b	142.81± 75.8 b	10.23 ± 3.99 b	1.38 ± 0.52 d
D-6	302.76± 29.25 a	203.84 ± 52.2 a	337.05 ± 40.72 a	3037.17 ± 358.3 b	5485.87 ± 327.1 b	2129.40 ± 243.7 d	1007.08 ± 146.5 a	52.03 ± 13.01 a	30.31 ± 4.51 c

First four lines describe *W. trilobata* and second four lines describe *W. Chinensis*. Different lowercase letters in each column represent a distinct treatment condition and comparisons within the treatment groups. ABA: abscisic acid, IAA: indole acetic acid, GAA: gibberellic acid, CK: the conditions of natural drought simulation, PEG75: Polyethylene glycol, D-4: water applied after four days, D-6: water applied after six days.

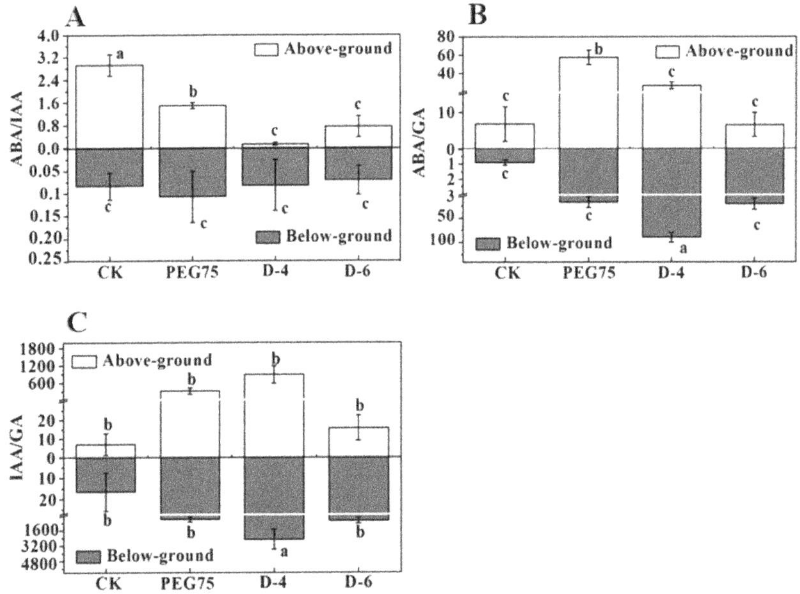

Figure 6. The impacts of simulated drought stress on the aboveground and underground ABA/IAA (**A**), ABA/GA (**B**), and IAA/GA (**C**) ratios in *W. chinensis* were examined (mean ± standard error, n = 3). Lowercase letters in the figure indicate significant differences between the treatments ($p < 0.05$).

3. Discussion

The plant growth index and biomass directly reflect the plant response to drought stress. Different scholars worldwide believe that Polyethylene glycol (PEG) as an osmotic regulator can be used to simulate drought [23]. PEG 6000 simulated drought stress closest to the soil drought level. When plants cope with drought stress, their growth were negatively affected, while their different physiological and photosynthesis factors as well as chlorophyll fluorescence were suppressed obviously [24]. different physiological factors and photosynthesis and chlorophyll fluorescence decreased Insufficient soil water directly leads to slow plant growth and reduced plant height and leaf area, which leads to significantly reduced photosynthesis in plants and a sharp decrease in plant biomass under natural drought stress [25]. The results of this study show that both plants appeared with different phenotypes, growth indexes, antioxidant systems, and the ratio of endogenous hormones under simulated drought conditions (Figures 1, 2, 5 and 6). PEG-manufactured drought and natural drought had different influences on *W. trilobata* and *W. chinensis* in terms of their inhibitory effects on their growth (Figures 1 and 2), which were mainly featured in the following aspects compared with natural drought: PEG-simulated drought had more significant inhibition on leaf number, plant height, root length, lateral root number and chlorophyll content of the two plants (Figures 1–3). PEG-simulated drought can restrain the plant height, root length, and underground biomass of Coix lachryma-Jobi L seedlings [26]. In addition, PEG-simulated drought can affect the seed germination rate, germination index, and relative water content (RWC) of *Coix lachryma* seedlings [27].

In natural drought conditions, as the severity of drought increased, both plants exhibited a more pronounced decrease in the number of leaves, plant height, and chlorophyll content [28]. Conversely, the root–shoot ratio showed a more notable increase. *W. trilobata* demonstrated a significant increase in root length, suggesting its adaptation to drought by enhancing water absorption through an increased root length (Figure 2). In contrast, *W. chinensis* exhibited a significant decrease in lateral root number, with no significant change in root length (Figure 2). This study revealed that *W. trilobata* exhibited greater drought tolerance compared with *W. chinensis* under simulated drought conditions [21]. *W. trilobata* displayed enhanced adaptability and phenotypic plasticity when compared with other plant species. Furthermore, their investigation discovered a potential association between the invasion of *W. trilobata* and the richness of the soil fungal community [29].

Drought stress can disrupt the plant cell membrane system, and numerous studies demonstrated the accumulation of CAT, POD, MDA, and free proline in plants as a response to drought stress [30,31]. This accumulation serves to maintain the dynamic equilibrium of the antioxidant system. CAT plays a protective role in preventing the destruction of cell membranes, while POD aids in the elimination of excessive reactive oxygen species within the plant [32]. MDA serves as the product of lipid peroxidation in plant cells [33]. The accumulation capacity of free proline serves as an indicator of a plant's drought resistance [34]. The substance under investigation interacts and forms cross-links with proteins localized on the cell membrane, thereby exerting a detrimental impact on the structure and functionality of the biofilm [35]. Furthermore, the constituents within this substance possess the capability to serve as indicators, reflecting the magnitude of cellular membrane damage [36]. The findings of this investigation demonstrate that under simulated drought conditions, both *W. trilobata* and *W. chinensis* exhibited a significant expansion in the levels of CAT, proline, and MDA within their leaves, while no significant alteration was observed in the POD content (Figure 4). Furthermore, within *W. chinensis* leaves, the CAT, proline, and MDA contents displayed a positive correlation with the severity of drought during natural drought conditions [37]. These observations suggest that *W. trilobata* and *W. chinensis* alleviate oxidative damage induced by drought stress by augmenting the concentration of free proline and modulating the activity of antioxidant enzymes [38]. Moreover, the rise in MDA content within *W. chinensis* was significantly greater than that in *W. trilobata* under the simulated drought conditions induced by PEG (Figure 4), implying that *W. chinensis* might encounter more pronounced oxidative damage to its cellular components under this particular treatment [39]. Additionally, it is worth noting that endogenous hormones play a remarkably fundamental role in regulating plant growth [40]. Different drought stress levels increase the ABA content (Table 3) in leaves of Populus euphratica, induce stomatal closure to reduce transpiration, and enhance the drought tolerance of *P. euphratica* [41]. ABA content in European rapeseed was positively correlated with the degree of drought and the most obvious increase in ABA content in leaves and root tips [42]. The results of this study show that compared with the control group, *W. trilobata* significantly increased the ABA content in its leaves and roots under drought conditions, while the ABA content in *W. chinensis's* leaves was the highest during natural drought and in its roots during PEG drought [43]. This indicates that the response of different tissue parts of these two plants to the drought environment was different. *W. trilobata* and *W. chinensis* exhibited different endogenous hormones in the two simulated drought conditions, indicating different adaptation mechanisms: The hormone contents in aboveground parts and underground parts of *W. trilobata* (Table 3) followed ABA > IAA > GA during PEG-simulated drought [44]. The hormone content in aboveground parts and underground parts of *W. chinensis* followed IAA > ABA > GA during PEG-simulated drought; the distribution pattern of IAA > ABA > GA in the aboveground and underground of the two plants under natural drought was the same (Table 3) [45]. This result shows that *W. trilobata* and *W. chinensis* responded differently to the two different modes of drought, and these responses were related to the polar transport of endogenous hormones.

Plant hormones are trace organic compounds that regulate plant growth and development and usually play a critical function in plant life activities [46]. Abscisic acid (ABA) is ubiquitous in higher plants and involves many physiological activities, such as flowering time, seed dormancy, and fruit ripening [47]. Under drought stress, abscisic acid accumulates in leaves, stems, and roots to promote stomatal closure and water absorption by roots, thus improving the drought resistance of plants [48]. The chemical essence of Auxin (IAA) is indoleacetic acid, which is widely present in root tips, stem tips, and young leaves [49].

Under drought stress, the content of IAA in plants increases, and after rehydration, the content of IAA in plants shows a declining trend. Gibberellin (GA) is widely found in plants of more than 100 species, and its essence is a diterpenoid compound [50]. Gibberellins are primarily associated with promoting plant growth, including stem elongation, cell division, and seed germination [51]. These findings illuminate the intricate and species-specific responses of these plants to various drought stress conditions, underscoring the complexity of the hormonal regulation and growth patterns of *W. trilobata* and *W. chinensis.*

4. Materials and Methods

4.1. Experimental Materials and Design

For this study, specimens of *W. trilobata* were collected from Haikou, Hainan Province, China (20°02′45.97″ N, 110°11′38.39″ E), while *W. chinensis* samples were obtained from the greenhouse of the College of Environmental and Safety Engineering, Zhenjiang, Jiangsu University, China (32°12.02 N). Potted cultivation was employed, utilizing four distinct treatment conditions, with each treatment being replicated 12 times. The conditions of natural drought simulation were set as C.K. (watered every two days), mild drought (watered every four days), and moderate drought (watered every six days). The simulated drought concentration of PEG 6000 was 75 g/L. The plant growth substrate was a mixture of screened, washed, sterilized, and dried river sand and vermiculite (river sand:Vermiculite = 5:1) with same weight that weighed 300 g. The experiment required the choice of healthy branches and the same thickness of *W. trilobata* and *W. chinensis* regenerated stem pieces; each stem segment comprisedwith two leaves, and the selected stem segments were first soaked in sodium hypochlorite solution (5%) for surface disinfection and sterilization for 10 min, washed in sterile water 5 times, and then soaked in 0.1 times Hoagland nutrient solution with a randomized complete block design. When the new root grew at the end of the stem segment after 2–3 days, the cuttings were cut vertically in a square plastic pot, and three cuttings were cut in each pot. The cuttings were watered and soaked in the greenhouse of Jiangsu University; the temperature was 25 °C, the humidity was 60%, and the light cycle was 16 h/8 h (day/night). The drought treatment was applied after the appearance of new leaves, and a few roots appeared in the stem segment.

4.2. Phenotypic Growth Index

The aerial parts (leaves and stems), as well as the subterranean parts (roots), of both *W. trilobata* and *W. chinensis* were harvested for analysis. Various parameters, including the fresh weight, dry weight, plant height, leaf count, root length, and chlorophyll content, were measured. To prepare the samples, the freshly collected leaves, stems, and roots were carefully washed with clean water and subsequently subjected to an oven-drying process. The samples were initially exposed to a temperature of 105 °C for 10 min, followed by a subsequent drying period at 65 °C for 72 h. The sum of the dry leaf weight, dry stem weight, and root dry weight was the total biomass, where an analytical balance was used to measure them. The plants were taken photos with a digital camera, while the number of leaves, plant height, and root length were analyzed with Image J 1.38e (https://imagej.nih.gov/ij/) (accessed on 15 October 2021). Before harvesting the plants, the chlorophyll contents of the fresh leaves were measured using a portable chlorophyll meter. The measurement time was controlled to be from 9 a.m. to 10 a.m. The repeated measurements with different treatments should be made to ensure that the placement of the flowerpot was fixed as much as possible, as far apart as possible and took the average value for data analysis.

4.3. Biochemical Indices

The assessment of antioxidant defense systems encompassed measurements of the peroxidase (POD) and catalase (CAT) enzyme activities, as well as the determination of the proline and malondialdehyde (MDA) contents. The activity of POD was determined using the guaiacol method, while the activity of CAT was assessed using the UV absorption method. The proline content was determined using the acidic ninhydrin method, whereas the content of MDA was quantified using the thiobarbituric acid method [52].

4.4. Endogenous Hormone

The relative content of abscisic acid (ABA), indole acetic acid (IAA), and gibberellic acid (GA) in the leaves, stems, and roots of *W. trilobata* and *W. chinensis* was quantified using an ELISA kit. The endogenous hormone kit used for this analysis was procured from the Crop Chemical Control Laboratory at China Agricultural University [53].

A total of 0.5 g of plant tissue was stored at −80 °C, 2 mL extraction solution was added, and the mixture was ground under ice and transferred to a 10 mL tube. After rinsing the mortar with 2 mL of solution, it was transferred and refrigerated for 4 h. Following centrifugation at 3500 r/min for 8 min, the supernatant was collected. A total of 1 mL of solution was added to the precipitation, mixed, and refrigerated for 1 h before another centrifugation at 3500 r/min for 8 min. The supernatants were combined, the volume was recorded, and the residue was discarded. The supernatant underwent C-18 column processing with specific steps. The sample was then transferred to a 5 mL tube, concentrated, and dried under vacuum, followed by the addition of 1 mL of sample diluent to make up the volume. Next, for testing, the standard solution was diluted twice the standards were diluted in sequence twofold, with concentrations ranging from 0 to the maximum (50 ng/mL for GA, 100 ng/mL for IAA and ABA). Standards went in the first two columns of a 96-well enzyme plate, with 2 wells for each concentration. Test samples filled the remaining wells. A total of 50 μL of antibody was added to each well, mixed well, and incubated at 37 °C for 0.5 h. After washing the plate four times, 100 μL of secondary antibody was added to each well and incubated again at 37 °C for 0.5 h. The washing process was repeated, the substrate for color development was added, the reaction was terminated with sulfuric acid after color development, and colorimetry was performed using a microplate reader.

4.5. Data Analysis

SPSS (26.0) software was used to perform one-way or multivariate analysis of variance and correlation analysis on each factor. Multiple comparisons were conducted using Duncan's test, with $p < 0.05$ indicating significant levels between treatments. The results of statistical analysis were plotted using origin 2018 (origin lab Co., Northhampton, MA, USA) software.

5. Conclusions

The research outcomes elucidate the physiological reactions of two closely related plant species, namely, *W. trilobata* and *W. chinensis*, to drought-induced stress. Notably, the investigation unveiled that under the influence of drought stress, *W. trilobata* manifested a substantial elevation in ABA levels within both its leaves and roots in comparison with the control cohort. Conversely, *W. chinensis* exhibited the highest ABA concentration in its leaves when subjected to natural drought, while the ABA content in its roots peaked under PEG-simulated drought. Moreover, within a drought milieu, both *W. trilobata* and *W. chinensis* showcased conspicuous increases in indole-3-acetic acid content in their stems. Intriguingly, the response of IAA in the leaves and roots displayed disparate patterns. For instance, *W. trilobata* had significantly augmented gibberellic acid content in its leaves and stems when exposed to PEG-simulated drought. In contrast, *W. chinensis* exhibited a noteworthy reduction in GA content within its roots and stems under identical conditions. The study also disclosed that amid natural drought circumstances, *W. trilobata* manifested a substantial upswing in GA content within its roots, whereas *W. chinensis* exhibited a marked downturn in GA content in its roots. These findings shed light on the intricate and species-specific reactions of these plants to diverse drought stress conditions, emphasizing the complexity of their hormonal regulation and growth patterns in response to environmental challenges.

Author Contributions: P.H. and D.D. conceived and designed the experiments; Z.X., W.H., H.Y. and B.L. conducted the experiments; Z.X., W.H., A.A., W.D. and Y.L. analyzed the data; P.H., A.A., Z.X. and W.H. wrote the manuscript; P.H., A.A. and D.D. were involved in the related discussion; R.H., C.W. and J.S. helped to improve the quality of the manuscript; P.H. provided supervision and funding acquisition. All authors have read and agreed to the published version of the manuscript.

Funding: This work was supported by the National Natural Science Foundation of China (32271587, 31200316, 31971427), Carbon Peak and Carbon Neutrality Technology Innovation Foundation Jiangsu province (BK20220030), the Jiangsu Collaborative Innovation Center of Technology and Material of Water Treatment, Study Abroad Scholarship of Jiangsu University, and Senior Talent Fund of Jiangsu University (11JDG150).

Data Availability Statement: Data are available in the article.

Conflicts of Interest: The authors declare no conflict of interest.

References

1. Kennish, M.J.; Paerl, H.W.; Crosswell, J.R. *Climate Change and Estuaries*; CRC Press: Boca Raton, FL, USA, 2023.
2. Chaudhary, C. *Organic methods of Weed Control. A Monthly Peer Reviewed Magazine for Agriculture and Allied Sciences*; The Agriculture Magazine address: Jaipur, India, 2022; p. 74.
3. Lobanov, V. Nutrient Transfer in Aquaponic Systems–Optimizing Microbial Processes for Greater Circularity and Economic Viability. Ph.D. Thesis, University of Gothenburg, Gothenburg, Sweden, 2023.
4. Mahdid, M.; Simonneau, T. Rapid responses and physiological events of leaf growth in response to water stress induced by poly ethylene glycol in maize (*Zea mays* L.). *Alger. J. Environ. Sci. Technol.* **2023**, *9*, 2947–2955.
5. Sayan, R.C.; Nagabhatla, N. The co-constitution of regional politics and massive infrastructures in the Transaqua water project. *Territ. Politics Gov.* **2022**. [CrossRef]
6. Liang, J.; Liu, B.; Dong, X.; Wang, Y.; Cai, W.; Zhang, N.; Zhang, H. Decoding the Role of Gut Microbiota in Alzheimer's Pathogenesis and Envisioning Future Therapeutic Avenues. *Front. Neurosci.* **2023**, *17*, 1242254. [CrossRef] [PubMed]
7. Zainab, R.; Hasnain, M.; Ali, F.; Dias, D.A.; El-Keblawy, A.; Abideen, Z. Exploring the bioremediation capability of petroleum-contaminated soils for enhanced environmental sustainability and minimization of ecotoxicological concerns. *Environ. Sci. Pollut. Res.* **2023**, *30*, 104933–104957. [CrossRef] [PubMed]
8. Chakraborty, T.; Sarma, A. Toxicity of Nanostructures and Nanodrugs. In *Phytoantioxidants and Nanotherapeutics*; Wiley: Hoboken, NJ, USA, 2022; pp. 267–287.
9. Suzuki, N. Fine tuning of ROS, redox and energy regulatory systems associated with the functions of chloroplasts and mitochondria in plants under heat stress. *Int. J. Mol. Sci.* **2023**, *24*, 1356. [CrossRef] [PubMed]
10. Hu, L.; Wang, X.; Zou, Y.; Wu, D.; Gao, G.; Zhong, Z.; Liu, Y.; Hu, S.; Fan, H.; Zhang, B. Effects of inorganic and organic selenium intervention on resistance of radish to arsenic stress. *Ital. J. Food Sci.* **2022**, *34*, 44–58. [CrossRef]
11. Zhong, M.; Yue, L.; Chen, Q.; Wang, H.; Lei, B.; Yang, X.; Kang, Y. Spermidine carbon dots enhance thermotolerance by modulating photosynthesis and cellular redox homeostasis in tomato. *Environ. Sci. Nano* **2023**, *10*, 595–610. [CrossRef]
12. Phillips, A.L. Interrogating the Multi-Stress Tolerance of Oryza Australiensis Using Novel Genomic and Phenomic Strategies. Ph.D. Thesis, University of Adelaide, Adelaide, Australia, 2022.
13. Zhang, Z. Development of Bioinformatic Solution to Enhance Metabolomics Data Quality and Its Application in Plant Research. Ph.D. Thesis, University of British Columbia, Vancouver, BC, Canada, 2023.
14. Huang, P.; Wang, H.; Xu, Z.; Yang, H.; Du, Y.; Chen, S.; Abbas, A.; Sun, P.; Du, D. The Responses of Invasive *Wedelia trilobata* and Native *Wedelia chinensis* to Levofloxacin Hydrochloride: Implication for Biological Invasion. *Pol. J. Environ. Stud.* **2023**, *32*, 3605–3615. [CrossRef]
15. Abbas, K.; Li, J.; Gong, B.; Lu, Y.; Wu, X.; Lü, G.; Gao, H. Drought Stress Tolerance in Vegetables: The Functional Role of Structural Features, Key Gene Pathways, and Exogenous Hormones. *Int. J. Mol. Sci.* **2023**, *24*, 13876. [CrossRef]
16. Bouremani, N.; Cherif-Silini, H.; Silini, A.; Bouket, A.C.; Luptakova, L.; Alenezi, F.N.; Baranov, O.; Belbahri, L. Plant growth-promoting rhizobacteria (PGPR): A rampart against the adverse effects of drought stress. *Water* **2023**, *15*, 418. [CrossRef]
17. Ali, M.; Pan, Y.; Liu, H.; Cheng, Z. Melatonin interaction with abscisic acid in the regulation of abiotic stress in Solanaceae family plants. *Front. Plant Sci.* **2023**, *14*, 1271137. [CrossRef] [PubMed]
18. Khan, F.; Siddique, A.B.; Shabala, S.; Zhou, M.; Zhao, C. Phosphorus plays key roles in regulating plants' physiological responses to abiotic stresses. *Plants* **2023**, *12*, 2861. [CrossRef]
19. Jiang, J.; Wang, Z.; Chen, Z.; Wu, Y.; Mu, M.; Nie, W.; Zhao, S.; Cui, G.; Yin, X. Identification and Evolutionary Analysis of the Auxin Response Factor (ARF) Family Based on Transcriptome Data from Caucasian Clover and Analysis of Expression Responses to Hormones. *Int. J. Mol. Sci.* **2023**, *24*, 15357. [CrossRef] [PubMed]
20. Kaya, C. Microbial modulation of hormone signaling, proteomic dynamics, and metabolomics in plant drought adaptation. *Food Energy Secur.* **2023**. [CrossRef]

21. Azeem, A.; Mai, W.; Tian, C.; Javed, Q. Dry Weight Prediction of *Wedelia trilobata* and *Wedelia chinensis* by Using Artificial Neural Network and MultipleLinear Regression Models. *Water* **2023**, *15*, 1896. [CrossRef]
22. Yu, X.; Liu, Y.; Cao, P.; Zeng, X.; Xu, B.; Luo, F.; Yang, X.; Wang, X.; Wang, X.; Xiao, X. Morphological Structure and Physiological and Biochemical Responses to Drought Stress of Iris japonica. *Plants* **2023**, *12*, 3729. [CrossRef] [PubMed]
23. Yang, X.; Lu, M.; Wang, Y.; Wang, Y.; Liu, Z.; Chen, S. Response Mechanism of Plants to Drought Stress. *Horticulturae* **2021**, *7*, 50. [CrossRef]
24. Castañeda Presa, V. Medicago Truncatula Response to Water-Deficit Stress: Whole Plant Perspectives. Ph.D. Thesis, Universidad Pública de Navarra, Pamplona, Spain, 2021.
25. Li, H.; Yang, Y.; Wang, H.; Liu, S.; Jia, F.; Su, Y.; Li, S.; He, F.; Feng, C.; Niu, M. The receptor-like kinase ERECTA confers improved water use efficiency and drought tolerance to poplar via modulating stomatal density. *Int. J. Mol. Sci.* **2021**, *22*, 7245. [CrossRef]
26. Li, L.; Yan, X.; Li, J.; Tian, Y. Physiological and FtCHS gene expression responses to PEG-simulated drought and cadmium stresses in tartary buckwheat seedlings. *J. Plant Growth Regul.* **2021**, *41*, 3518–3529. [CrossRef]
27. Binodh, A.; Thankappan, S.; Ravichandran, A.; Mitra, D.; Alagarsamy, S.; Panneerselvam, P.; Senapati, A.; Sami, R.; Al-Mushhin, A.; Aljahani, A. Synergistic Modulation of Seed Metabolites and Enzymatic Antioxidants Tweaks Moisture Stress Tolerance in Non-Cultivated Traditional Rice Genotypes during Germination. *Plants* **2022**, *11*, 775. [CrossRef]
28. Rehman, A.; Weng, J.; Li, P.; Yu, J.; Rahman, S.u.; Khalid, M.; Shah, I.H.; Gulzar, S.; Chang, L.; Niu, Q. Differential Response of Two Contrasting Melon (*Cucumis melo* L.) Genotypes to Drought Stress. *J. Plant Biol.* **2023**, *66*, 519–534. [CrossRef]
29. Sun, J.; Khattak, W.A.; Abbas, A.; Nawaz, M.; Hameed, R.; Javed, Q.; Bo, Y.; Khan, K.A.; Du, D. Ecological adaptability of invasive weeds under environmental pollutants: A review. *Environ. Exp. Bot.* **2023**, *215*, 105492. [CrossRef]
30. Lahijanian, S.; Eskandari, M.; Akhbarfar, G.; Azizi, I.; Afazel, M.; Ghobadi, C. Journal of Agriculture and Food Research. *J. Agric. Food Res.* **2023**, *11*, 100495.
31. Mohamadzadeh, M.; Janmohammadi, M.; Abbasi, A.; Sabaghnia, N.; Ion, V. Physiochemical response of Cicer arietinum to zinc-containing mesoporous silica nanoparticles under water stress. *BioTechnologia J. Biotechnol. Comput. Biol. Bionanotechnol.* **2023**, *104*, 263–273. [CrossRef] [PubMed]
32. Santoyo, G.; Kumar, A.; Aamir, M.; Sivakumar, U. Mitigation of Plant Abiotic Stress by Microorganisms. In *Applicability and Future Directions*; Elsevier: Amsterdam, The Netherlands, 2022.
33. Mas-Bargues, C.; Escriva, C.; Dromant, M.; Borrás, C.; Vina, J. Lipid peroxidation as measured by chromatographic determination of malondialdehyde. Human plasma reference values in health and disease. *Arch. Biochem. Biophys.* **2021**, *709*, 108941. [CrossRef]
34. Liu, X.; Xia, B.; Purente, N.; Chen, B.; Zhou, Y.; He, M. Transgenic Chrysanthemum indicum overexpressing cin-miR396a exhibits altered plant development and reduced salt and drought tolerance. *Plant Physiol. Biochem.* **2021**, *168*, 17–26. [CrossRef] [PubMed]
35. Spesia, M.B.; Durantini, E.N. Evolution of phthalocyanine structures as photodynamic agents for bacteria inactivation. *Chem. Rec.* **2022**, *22*, e202100292. [CrossRef] [PubMed]
36. Wei, G.; Huang, R.; Zhang, G.; Jiang, B.; Zhu, J.; Guo, Y.; Han, G.; Wei, X.; Dai, H. A comprehensive insight into the thermal runaway issues in the view of lithium-ion battery intrinsic safety performance and venting gas explosion hazards. *Appl. Energy* **2023**, *349*, 121651. [CrossRef]
37. El Mamoun, I.; Bouzroud, S.; Zouine, M.; Smouni, A. The Knockdown of AUXIN RESPONSE FACTOR 2 Confers Enhanced Tolerance to Salt and Drought Stresses in Tomato (*Solanum lycopersicum* L.). *Plants* **2023**, *12*, 2804. [CrossRef]
38. Abdelaal, K.; Alsubeie, M.S.; Hafez, Y.; Emeran, A.; Moghanm, F.; Okasha, S.; Omara, R.; Basahi, M.A.; Darwish, D.B.E.; Ibrahim, M.F. Physiological and biochemical changes in vegetable and field crops under drought, salinity and weeds stresses: Control strategies and management. *Agriculture* **2022**, *12*, 2084. [CrossRef]
39. Gupta, S.; Kulkarni, M.G.; White, J.F.; Stirk, W.A.; Papenfus, H.B.; Doležal, K.; Ördög, V.; Norrie, J.; Critchley, A.T.; Van Staden, J. Categories of various plant biostimulants–mode of application and shelf-life. In *Biostimulants for Crops from Seed Germination to Plant Development*; Elsevier: Amsterdam, The Netherlands, 2021; pp. 1–60.
40. Barbier, F.; Fichtner, F.; Beveridge, C. The strigolactone pathway plays a crucial role in integrating metabolic and nutritional signals in plants. *Nat. Plants* **2023**, *9*, 1191–1200. [CrossRef] [PubMed]
41. Rao, S.; Tian, Y.; Zhang, C.; Qin, Y.; Liu, M.; Niu, S.; Li, Y.; Chen, J. The JASMONATE ZIM-domain–OPEN STOMATA1 cascade integrates jasmonic acid and abscisic acid signaling to regulate drought tolerance by mediating stomatal closure in poplar. *J. Exp. Bot.* **2023**, *74*, 443–457. [CrossRef] [PubMed]
42. Batool, M.; El-Badri, A.M.; Hassan, M.U.; Haiyun, Y.; Chunyun, W.; Zhenkun, Y.; Jie, K.; Wang, B.; Zhou, G. Drought stress in Brassica napus: Effects, tolerance mechanisms, and management strategies. *J. Plant Growth Regul.* **2022**, *42*, 21–45. [CrossRef]
43. MUHİE, S.; MEMİŞ, N.; ÖZDAMAR, C.; GÖKDAŞ, Z.; Demir, I. Biostimulant priming for germination and seedling quality of carrot seeds under drought, salt and high temperature stress conditions. *Int. J. Agric. Environ. Food Sci.* **2021**, *5*, 352–359. [CrossRef]
44. Sapakhova, Z.; Raissova, N.; Daurov, D.; Zhapar, K.; Daurova, A.; Zhigailov, A.; Zhambakin, K.; Shamekova, M. Sweet Potato as a Key Crop for Food Security under the Conditions of Global Climate Change: A Review. *Plants* **2023**, *12*, 2516. [CrossRef] [PubMed]
45. Lin, H.Y.; Li, W.H.; Lin, C.F.; Wu, H.R.; Zhao, Y.P. International biological flora: Ginkgo biloba. *J. Ecol.* **2022**, *110*, 951–982. [CrossRef]

46. Wang, L.; Zou, Y.; Kaw, H.Y.; Wang, G.; Sun, H.; Cai, L.; Li, C.; Meng, L.-Y.; Li, D. Recent developments and emerging trends of mass spectrometric methods in plant hormone analysis: A review. *Plant Methods* **2020**, *16*, 54. [CrossRef]
47. Pálfi, P.; Riyazuddin, R.; Bakacsy, L.; Szepesi, Á. 6 FunctionsinAbiotic of PolyaminesTolerance. In *Advancements in Developing Abiotic Stress-Resilient Plants: Basic Mechanisms to Trait Improvements*; CRC Press: Boca Raton, FL, USA, 2022; p. 97.
48. Yan, J.; Li, H.; Li, Y.; Zhang, N.; Zhang, S. Abscisic acid synthesis and root water uptake contribute to exogenous methyl jasmonate-induced improved tomato drought resistance. *Plant Biotechnol. Rep.* **2022**, *16*, 183–193. [CrossRef]
49. El-Banna, M.F.; Farag, N.B.; Massoud, H.Y.; Kasem, M.M. Exogenous IBA stimulated adventitious root formation of Zanthoxylum beecheyanum K. Koch stem cutting: Histo-physiological and phytohormonal investigation. *Plant Physiol. Biochem.* **2023**, *197*, 107639. [CrossRef]
50. Mauli, M.M.; Meneghetti, A.M.; Nóbrega, L.H. Terpenes Behavior in Soil. In *Terpenoids: Recent Advances in Extraction, Biochemistry and Biotechnology*; Bentham Science Publishers: Potomac, MD, USA, 2022; p. 169.
51. Yuan, G.; Lian, Y.; Wang, J.; Yong, T.; Gao, H.; Wu, H.; Yang, T.; Wang, C. AtHSPR functions in gibberellin-mediated primary root growth by interacting with KNAT5 and OFP1 in Arabidopsis. *Plant Cell Rep.* **2023**, *42*, 1629–1649. [CrossRef]
52. Wang, D.; Gao, Y.; Sun, S.; Lu, X.; Li, Q.; Li, L.; Wang, K.; Liu, J. Effects of salt stress on the antioxidant activity and malondialdehyde, solution protein, proline, and chlorophyll contents of three Malus species. *Life* **2022**, *12*, 1929. [CrossRef] [PubMed]
53. Gao, L.; Cai, M.; Zeng, L.; Zhang, Q.; Zhu, H.; Gu, X.; Peng, C. Adaptation of the Invasive Plant (*Sphagneticola trilobata* L. Pruski) to a High Cadmium Environment by Hybridizing With Native Relatives. *Front. Plant Sci.* **2022**, *13*, 905577. [CrossRef] [PubMed]

plants

MDPI

Article

Unraveling the Role of AtSRT2 in Energy Metabolism, Stress Responses, and Gene Expression during Osmotic Stress in *Arabidopsis thaliana*

Alberto Obrecht [1,2] and Manuel Paneque [2,*]

1 Doctoral Program in Biotechnology, Universidad de Santiago de Chile, Av. Lib. Bdo. O'Higgins 3363, Estación Central, Santiago 9170022, Chile; alberto.obrecht@usach.cl

2 Department of Environmental Sciences and Natural Resources, Faculty of Agricultural Sciences, University of Chile, Santa Rosa 11.315, La Pintana, Santiago 8820808, Chile

* Correspondence: mpaneque@uchile.cl

Abstract: Sirtuins participate in chromatin remodeling and gene expression regulation during stress responses. They are the only deacetylases that couple the cellular NAD^+-dependent energy metabolism with transcriptional regulation. They catalyze the production of nicotinamide, inhibiting sirtuin 2 (SIR2) activity in vivo. The SIR2 homolog, AtSRT2, deacetylates non-histone proteins associated with mitochondrial energy metabolism. To date, AtSRT2 mechanisms during stress responses in *Arabidopsis thaliana* remain unclear. The transduction of mitochondrial metabolic signals links the energy status to transcriptional regulation, growth, and stress responses. These signals induce changes by regulating nuclear gene expression. The present study aimed to determine the role of SRT2 and its product nicotinamide in the development of *A. thaliana* and the expression of osmotic stress-response genes. Leaf development was greater in *srt2+* plants than in the wild type, indicating that SET2 plays a role in energy metabolism. Treatment with polyethylene glycol activated and inhibited gene expression in *srt2-* and *srt2+* lines, respectively. Therefore, we concluded that SRT2-stimulated plant growth and repressed signaling are associated with osmotic stress.

Keywords: sirtuin; Sir2; osmotic stress; *Arabidopsis*; NAD+, mannitol; PEG

1. Introduction

The silent information regulator 2 (SIR2) family of proteins (sirtuins) are NAD^+-dependent deacetylases that lack structural or sequence homology with other deacetylases [1]. They can detect the energy status of the cell depending on the concentration of NAD^+ and deacetylated proteins [2,3]. Their activity depends on the $NAD^+/NADH$ ratio and increases under conditions of nutritional or energy deficiency in mammals and yeasts [4,5].

Sirtuin-mediated deacetylation is coupled with NAD^+ hydrolysis and the production of nicotinamide, which alters the NAD^+ levels and triggers cellular changes [6–9]. Sirtuins were initially described as nuclear proteins with histone deacetylase activity that causes chromatin compaction and gene silencing. They have also been linked to increased longevity in yeast [10–13]. Recent studies have demonstrated that sirtuins are located in different organelles and deacetylate several non-histone proteins [14,15].

Sirtuins have been identified in bacteria (*Bacillus subtilis*) [16], archaea (*Sulfolobus solfataricus* L. major) [17], yeasts (*Saccharomyces cerevisiae*) [18], nematodes (*Caenorhabditis elegans*) [19], fruit flies (*Drosophila melanogaster*) [20], humans (*Homo sapiens*) [21], and plants (*Oryza sativa* and *Arabidopsis thaliana*) [22,23]. Protein deacetylation promotes changes in the enzymatic activity or function of target proteins [24,25]. Two SIR2 homologs, namely, sirtuin 1 (SRT1) and sirtuin 2 (SRT2), have been identified in *O. sativa* [22,26], *Vitis vinifera* [27,28],

Solanum lycopersicum [29,30], and *A. thaliana* [23,31,32]. The functions of these enzymes have been studied in some organisms but not in plants [33].

To date, SRT1 has primarily been found in the nuclei of the studied plant species, whereas SRT2 is ubiquitous in different organelles [34]. SRT2 is localized in the mitochondria of *O. sativa* (OsSRT29) [22], mitochondria and chloroplasts of *V. vinifera* (VvSRT2) [28], and the nuclei and cytoplasm of *S. lycopersicum* (SiSRT2) [30]. SRT2 expression in grapevine is likely related to photosynthetic activity. Levels of VvSRT2 in the leaves correspond to the progression through developmental stages, where increased expression of VvSRT2 correlates with increased photosynthesis [28]. The rates of photosynthesis and VvSRT2 expression were found to be high during the inflorescence stage and gradually decreased over time [35]. However, the expression of VvSRT2 remained high in the leaves and was expressed in the flowers and fruits, but it was not expressed in the roots [36].

Little is known about the sirtuins in *A. thaliana* (ArSRT2) and other plant species [23,31,32]. AtSRT2 is localized in the nucleus [32] or mitochondria [23]. It negatively regulates biotic stress responses by silencing genes involved in salicylic acid signaling [32] and deacetylating non-histone proteins [23]. In addition, it physically interacts with various proteins, including ATP synthase and ATP/ADP transporters [23].

In AtSRT2 T-DNA insertion mutants, many proteins that interact with AtSRT2 are hyperacetylated [23]. Compared to that in wild-type (WT) plants, acetylation on the gamma and delta/epsilon subunits of ATP synthase was increased 1.9- and 2.1-fold, respectively; on subunit 8, it was increased 2.3-fold; and on subunit ATP17, it was increased 10.2-fold [23]. ATP17 is specific to plant ATP synthases. In addition, acetylation of ATP/ADP carrier proteins (AAC1 and AAC3) was increased 1.6-fold in AtSRT2 mutants compared to that in WT plants [23]. Changes in protein acetylation, linked to energy metabolism, did not result in any significant phenotypic changes. However, respiratory coupling to ATP synthesis was decreased, possibly through suppression of ATP synthase activity or an increase in mitochondrial decoupling in *srt2*-mutant plants [23].

SRT2-mediated deacetylation increases ATP synthase activity [23]. However, deacetylation of ATP/ADP transporters decreases its activity, as deacetylation increases ADP entry and decreases the ATP/ADP ratio in *srt2* mutants [23]. Disruption of SRT2 increases the concentration of amino acids, including glutamine and glycine, and decreases the concentration of important metabolites, including sugars (fructose, glucose, erythritol, and myoinositol), amino acids (serine, proline, arginine, threonine, tyrosine, and alanine), and organic acids (shikimic, ascorbic, pyruvic, fumaric, and gamma-aminobutyric acids) [23]. Furthermore, SRT2 regulates the energy metabolism in cells, inducing changes in metabolites that may serve as signaling molecules. This activates or represses the downstream pathways, allowing plants to adapt to their current energy status.

Although AtSRT2 is associated with the regulation of energy metabolism [23], its role in the stress response of *A. thaliana* has not been fully elucidated. Mitochondrial metabolic signals can induce changes in other organelles [37], including the nucleus, by regulating nuclear gene expression. These signals are transduced by proteins such as the target of rapamycin (TOR) and SNF1-related kinases (SnRK1), which couple the energy and nutritional status of the cell to transcriptional regulation, growth, and stress responses [38,39].

The objective of the present study was to evaluate the role of SRT2 during the development and osmotic stress response in *A. thaliana* through phenotypic analyses. In *A. thaliana*, SRT2 is located in the mitochondria [23]. Therefore, in this study, we observed the phenotypic changes in plant development and the transcription of stress-response genes related to SRT2-dependent changes in mitochondrial metabolism. Moreover, we investigated the effect of *srt2* overexpression and silencing on plant development and stress-related gene expression.

2. Results and Discussion

2.1. Effect of SRT2 on the Development of A. thaliana

Phenotypic changes caused by SRT2 deletion or overexpression during the development of *A. thaliana* were evaluated by examining the morphology of the seeds, roots, leaves, and siliques.

Gene expression in the *srt2+*, *srt2-*, and WT lines was analyzed using reverse-transcription polymerase chain reaction (RT-PCR). The deletion line (*srt2-*) exhibited minimal SRT2 expression, whereas the overexpression line (*srt2+*) showed higher expression levels than the WT line (Figure 1A). The *srt2-* plants had noticeably shorter and thinner leaves compared to those of the *srt2+* and WT plants (Figure 1B). However, there were no observed abnormalities in the development of seeds (Figure 1C), seedlings (Figure 1D), roots (Figure 1E), or silique morphology (Figure 1F) among the three genetic lines.

Figure 1. *Srt2* affects the leaf size in Arabidopsis thaliana, as shown by a qualitative, phenotypic analysis of *A. thaliana* genetic lines, *srt2+*, *srt2-*, and wild type (WT). (**A**) Semi-quantitative RT-PCR of the *srt2-* deletion line and *srt2+* overexpression line. (**B**) Plants grown in vermiculite–peat soils for 23 long days. (**C**) Seed morphology and size. (**D**) Development of 3-day-old seedlings cultivated in MS medium. (**E**) Root development in 16-day-old seedlings cultivated in vertical plates in MS medium. (**F**) Silique morphology and size.

The *srt2+* plants had larger leaves and were more vigorous than *srt2-* and WT plants (Figure 1B). These differences were greater in plants grown during longer days. In contrast, the two mutant lines developed normally, with no significant differences in seed characteristics, seedling morphology, root development, or silique morphology. The length of the siliques, number of seeds per silique, number of siliques per plant, and seed width, length, and area were measured (Table 1). Compared to the WT plants, the *srt2+* and *srt2-* plants had shorter siliques, fewer seeds per silique, and fewer siliques per plant. These differences were discrete and did not decrease the germination percentage or seedling viability.

Table 1. *Srt2* affects the silique length and number of seeds per silique in *Arabidopsis thaliana*. The parameters of interest were analyzed in the *srt2+*, *srt2-*, and WT lines; these included the length of the mature siliques (60 siliques per row); number of seeds per silique (60 siliques per row); number of siliques per plant (24 plants per row); and seed width, length, and area as an average of four seeds. All plants evaluated in this study were cultivated in soil.

Genotype	WT	*srt2-*	*srt2+*
Feature	Mean ± SD	Mean ± SD	Mean ± SD
Silique length (mm)	13.93 ± 0.16	13.53 ± 0.12 *	13.01 ± 0.12 *
Seed number per silique	48.67 ± 0.95	44.49 ± 0.82 *	46.63 ± 0.89
Siliques number per plant	90.21 ± 4.21	82.41 ± 4.92	87.042 ± 4.03
Seed width (mm)	0.40 ± 0.03	0.45 ± 0.02	0.44 ± 0.02
Seed length (mm)	0.72 ± 0.03	0.75 ± 0.01	0.73 ± 0.03
Seed area (mm^2)	0.23 ± 0.04	0.26 ± 0.03	0.25 ± 0.02

*: Significant differences in silique length and number of seeds per silique compared to those of the WT control, as calculated using the Mann–Whitney U test ($p < 0.05$).

We observed that neither SRT2 deletion nor its overexpression led to major changes in the morphology of seeds and siliques during the development of *A. thaliana*. The siliques were shorter in the mutant plants than in the WT plants (Table 1). The reduction in silique length was greater in *srt2+* plants than in *srt2-* and WT plants. In addition, the number of seeds per silique was reduced in *srt2+* and *srt2-* plants (Table 1), and the reduction was greater in *srt2-* than in *srt2+*. The number of siliques per plant did not significantly differ among the three genetic lines (Table 1).

In comparison with those of the WT plants, the length of the siliques and the number of seeds per silique were slightly lower in the mutant plants. However, we could not infer whether SRT2 directly affected the growth of siliques or seed formation. Possible changes in energy metabolism caused by SRT2 deletion or overexpression were more pronounced in vegetative tissues, such as leaves, than in reproductive tissues.

In comparison with that in WT plants, the silique length in the mutant plants was decreased, and the reduction was more pronounced in *srt2+* plants than in *srt2-* plants (Table 1). In addition, the number of seeds per silique was lower in the mutant plants than in the WT plants, and the reduction in seed number was greater in the *srt2-* plants than in *srt2+* plants. No significant differences were observed in the number of siliques per plant among the three lines. Previous reports have indicated that the inhibition of SRT2 enhances auxin production [40], and although auxins are crucial phytohormones for plant growth optimization [41], excessive levels can inhibit growth once optimal concentrations are exceeded [42]. Therefore, it is hypothesized that SRT inhibition could lead to unregulated auxin production or transport, subsequently inhibiting growth. Conversely, if SRT2 is implicated in auxin production or transport [10,40], its overexpression might reduce or inhibit auxin production or transport, thereby hindering growth if auxin levels fall below optimal concentrations.

Leaf growth, as measured by width and length, was higher in the *srt2+* plants than in the WT plants on the last day of measurement, corresponding to maximum leaf development (Figure 2A,C,D). In contrast, leaf growth was lower in the *srt2-* line than in the WT line (Figure 2A,B).

Leaf growth was higher in the *srt2+* plants cultivated for 23 long days (16 h days) than in the WT plants (Figure 2A). Leaf development in plants cultivated under a short photoperiod (8 h days) was similar in the three genetic lines. However, *srt2+* plants still had wider leaves than those of the WT plants on day 36 (Figure 2C). The most evident differences were observed on the final day of measurement (Figure 2A–C). Longer *srt2+* leaves were slightly thinner (Figure 2C), but still slightly shorter (Figure 2D), than those of the WT plants before maximum leaf development. This observation was reversed starting on day 29.

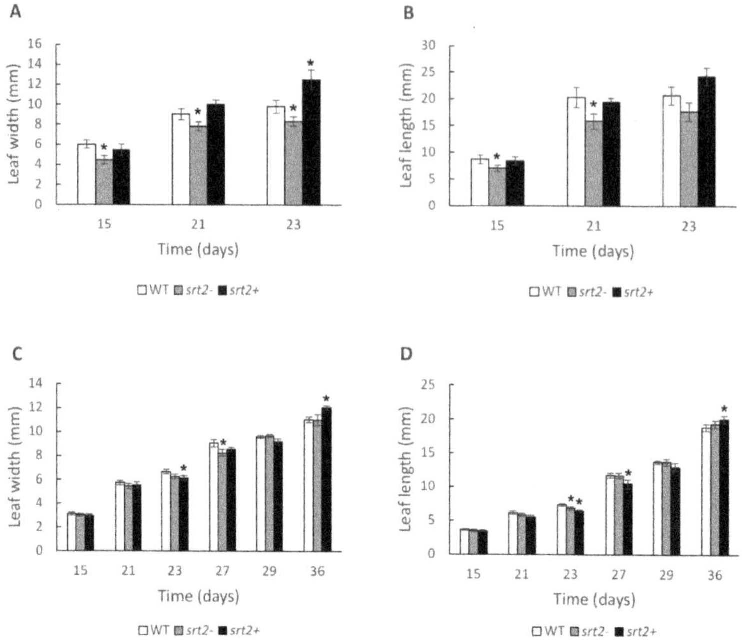

Figure 2. *Srt2* affects the leaf length and width in *Arabidopsis thaliana* grown in long- and short-day photoperiods. The leaves of *srt2*+, *srt2*-, and WT genetic lines of *A. thaliana* grown in vermiculite-peat soil (3:1) were measured on days 15, 21, and 23 for the long-day photoperiod conditions and on days 15, 21, 23, 27, 29, and 36 for the short-day photoperiod conditions. (**A**) Width of the leaves in the long-day photoperiod. (**B**) Length of the leaves during the long-day photoperiod. (**C**) Width of the leaves during the short-day photoperiod. (**D**) Length of the leaves during the short-day photoperiod (n = 9; two leaves per plant and nine plants per row). The asterisks indicate significant differences compared with the WT control, as calculated using the Mann–Whitney U test ($p < 0.05$).

Plant growth was the highest in the *srt2*+ plants, particularly during the long-day photoperiod (Figure 1A). This suggests that SRT2 may be responsible for increasing energy-utilization efficiency. Sirtuins in yeasts and mammals affect several processes depending on their subcellular locations [34,43]. SRT2 is exclusively located in the mitochondria and binds to the mitochondrial proteins involved in energy metabolism, including ATP synthase and ATP/ADP transporters [23].

SRT2 modulates protein activity via lysine deacetylation. In the present study, the coupling of mitochondrial respiration and ATP synthesis was reduced in the *srt2*- plants compared to that in the *srt2*+ plants. Moreover, SRT2 deletion is known to decrease glucose levels and change the concentrations of other metabolites, including amino acids, organic acids, and sugars [23]. Therefore, SRT2 can promote changes in the plant's energy status, which can, in turn, trigger changes in protein synthesis/degradation and gene expression/repression via the activation or repression of the TOR pathway and its antagonist SnRK1. TOR and SnRK1 kinases can sense the energy status of the cell and modulate protein transcription and synthesis to adapt to a new energy state, resulting in either growth or autophagy [38,44].

The observed phenotypic differences among the three genetic lines can be attributed to differences in mitochondrial coupling or ATP synthase activity, both of which are affected by SRT2 activity. SRT2 overexpression may have increased several parameters, including plant photosynthetic efficiency, the ATP/ADP ratio, and glucose levels. Increased glucose levels can activate the TOR signaling pathway, which is mediated by Glc-TOR [44]. Together with an increase in energy efficiency, the activation of TOR can explain the increase in plant vigor and leaf growth in the *srt2+* line. Therefore, TOR may be an effector of SRT2 activity in the mitochondria.

2.2. Root Length under Stress Conditions

Treatment with 50 mM of mannitol and 2.5% polyethylene glycol (PEG) increased the root length compared to that of the control (grown in only MS medium) at 16 days post-germination (Figure 3). Under mannitol treatment, both *srt2-* and *srt2+* lines exhibited shorter roots compared to the WT, whereas only the *srt2+* line had shorter roots compared to WT when treated with PEG (Figure 3).

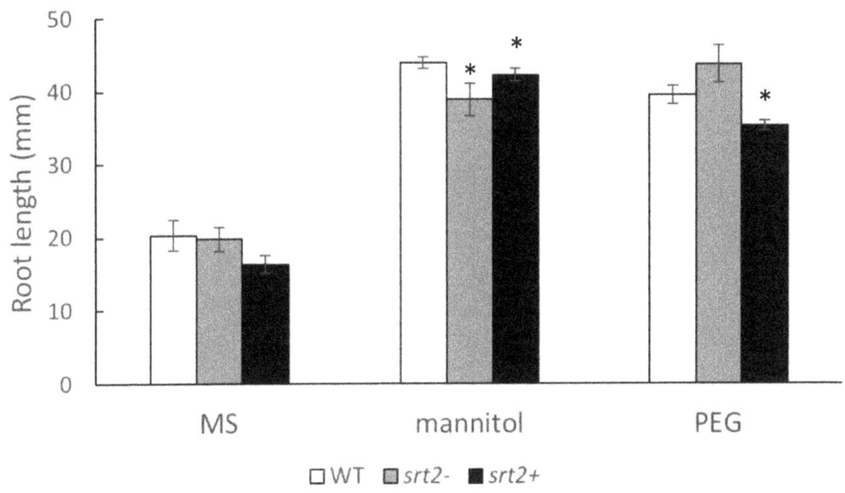

Figure 3. Osmotic stress affects the root length of the three genetic lines of *Arabidopsis thaliana* plants grown under osmotic stress. The root length of *A. thaliana* seedlings grown on vertical plates was measured after 16 days of treatment with the control MS medium, 50 mM of mannitol, or 2.5% polyethylene glycol (PEG). The asterisks indicate significant differences compared to the WT control, as calculated using the Mann–Whitney U test ($p < 0.05$).

All three lines responded similarly to osmotic stress, both for mannitol and PEG, by extending their roots into the culture medium. Differences in root growth among the three plant lines were discrete. According to König et al. (2014) [23], SRT2 does not directly influence plant phenotype but plays a role in metabolite transport. Consequently, its mechanisms of action may not manifest in noticeable morphological changes.

2.3. Membrane Lipoperoxidation

The plants were grown in multi-well plates and subjected to osmotic stress by treatment with 300 mM of mannitol and 30% PEG. This methodology allowed multiple assays to be performed simultaneously under the same experimental conditions. Ten seeds were allowed to germinate per well. After 10 days of germination, stress-induced production of malondialdehyde (MDA) was quantified. The MDA levels in the *srt2+* plants were higher than those in the WT plants. The rate of lipoperoxidation was higher in the *srt2-* plants grown in MS medium only and after 3 and 7 days of incubation with mannitol than in WT plants (Figure 4A,B).

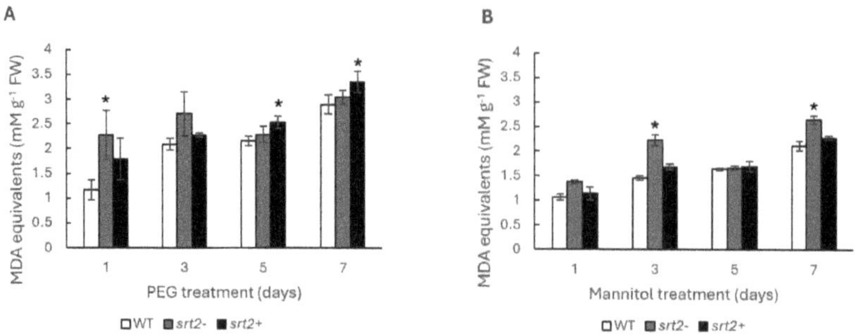

Figure 4. Membrane lipoperoxidation in plants subjected to two types of osmotic stress. The malondialdehyde (MDA) level per gram of fresh weight was determined after 1, 3, 5, and 7 days of treatment with (**A**) 30% polyethylene glycol (PEG) or (**B**) 300 mM of mannitol in the *srt2+*, *srt2-*, and WT lines of *Arabidopsis thaliana* cultivated in a liquid medium for 10 days. The asterisks indicate significant differences compared with the WT control, as calculated using the Mann–Whitney U test ($p < 0.05$) (n = 3).

The increase in MDA levels was a function of the duration of stressor exposure in all three genetic lines. No significant differences in MDA levels were observed among the three genetic lines after the different treatments. The putative increase in energy efficiency in the *srt2+* line did not increase the neutralization of membrane lipoperoxides induced by the treatments. This may be because the experiments were conducted on seedlings during the first week of growth, and in this early stage plant, the morphology among three genetic lines was similar. However, phenotypic differences in response to osmotic stress are possible in adult plants, which may be the result of an increase in the stress response in *srt2+* plants.

2.4. SRT2 Is Critical for Activating Osmotic Stress Response Genes

Treatment with 300 mM of mannitol increased the expression of the marker genes *kin2* and *rd22* in the WT plants. An increase in *kin2* and *rd22* expression was expected, as these genes are markers of osmotic stress. After the treatment with mannitol for 4 h, the expression levels of *kin2* and *rd22* were significantly lower in the mutant plants than in the WT plants (Figure 5).

After 4 h of incubation with 30% PEG, *kin2* and *rd22* expression increased in *srt2-* and WT plants but decreased in *srt2+* plants. The expression of these genes in response to osmotic stress was significantly different among the groups evaluated and was more pronounced after 4 h of incubation with 30% PEG. Therefore, SRT2 indirectly modulates the expression of these genes, and this activity is part of the stress-signaling pathway in *A. thaliana*. The post-treatment expression profiles of both genes were similar (Figure 5).

This study aimed to assess the role of SRT2 in the activation or repression of stress marker genes in response to osmotic stress caused by these two stressors. After incubation with PEG for 4 h, the expression of these genes was suppressed in the *srt2+* line but activated in the *srt2-* line. In contrast, mannitol treatment suppressed their expression in both mutant lines. These results indicated that gene repression was independent of SRT2 expression.

Therefore, SRT2 modulated gene expression by indirectly silencing *kin2* and *rd22*. This activity is part of the stress signaling pathway in response to PEG treatment in *A. thaliana*. However, mannitol did not have the same effect on gene expression. This may be because mannitol can penetrate the apoplast and is taken up by cells [45], attenuating the effective osmotic gradient between the medium and the symplast. Another reason may be that the hydric potential generated by mannitol (−0.74 MPa) is much lower than that generated by PEG (−10.27 MPa).

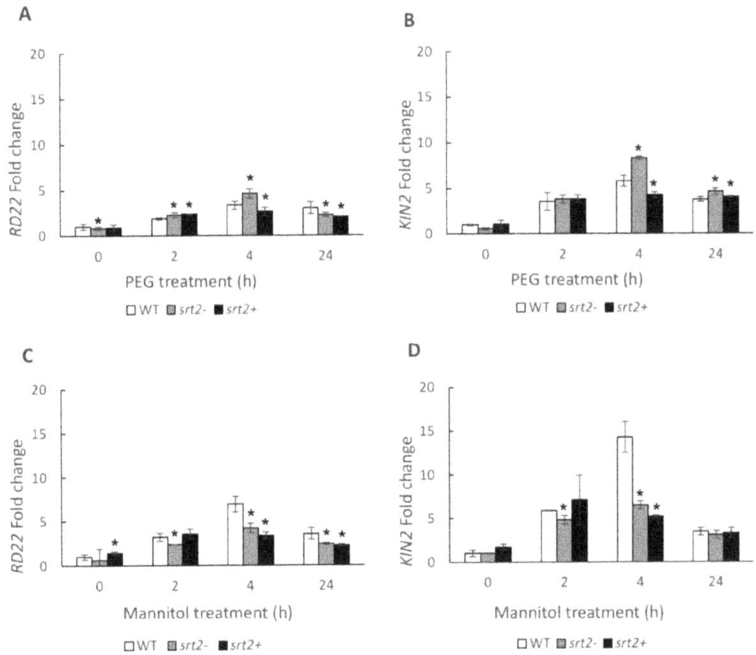

Figure 5. Short-term changes in the expression of osmotic stress marker genes (*kin2* and *rd22*) in plants subjected to two types of osmotic stress. Changes in their expression in the *srt2+*, *srt2-*, and WT control plants of *Arabidopsis thaliana* grown in a liquid medium for 10 days were determined after 0, 2, 4, and 24 h of treatment. The plants were grown in (**A**,**B**) MS medium + 30% polyethylene glycol (PEG), or (**C**,**D**) MS medium + 300 mM of mannitol. *Actin2* was used as a housekeeping gene. The results are relative to the control WT expression at t = 0 h. The asterisks indicate significant differences compared with the WT control, as calculated using the Mann–Whitney U test ($p < 0.05$).

2.5. Nicotinamide's Contribution to Stress Signaling

SRT2-mediated signaling during the osmotic stress response may be nicotinamide-dependent. The differences in the expression of osmotic stress marker genes induced by nicotinamide were measured. In the presence of 2.5 mM of nicotinamide, the expression of *rd22* was slightly increased in all three genetic lines, and the expression was the highest after 4 h of incubation (Figure 6). After 2 and 4 h of treatment, *kin2* was activated in all three genetic lines. However, its expression in the *srt2+* and *srt2-* plants declined rapidly beyond 4 h after the treatment.

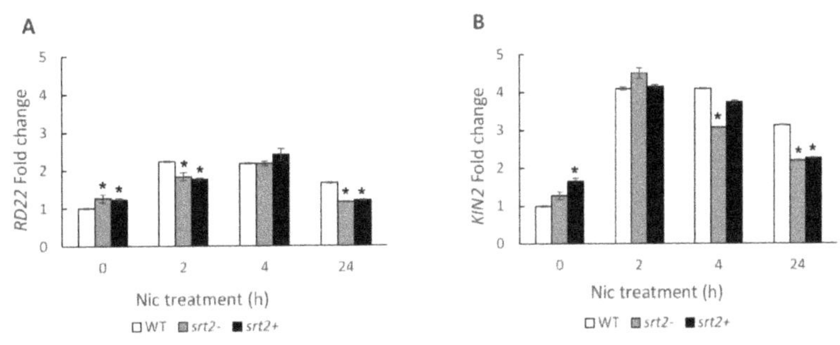

Figure 6. Short-term changes in the expression of osmotic stress marker genes (*kin2* and *rd22*) in plants treated with 2.5 mM of nicotinamide. The seedlings of *srt2+*, *srt2-*, and WT control lines of

Arabidopsis thaliana were grown in a liquid medium for 10 days and treated with 2.5 mM of nicotinamide. The fold changes in the osmotic stress marker genes, (**A**) *kin2* and (**B**) *rd22*, were determined after 0, 2, 4, and 24 h of treatment with nicotinamide. Actin2 was used as a housekeeping gene. The results are presented relative to the control WT expression at t = 0 h. The asterisks indicate significant differences compared with the WT control using the Mann–Whitney U test ($p < 0.05$) (n = 3).

Nicotinamide did not induce SRT2-mediated osmotic stress signaling (Figure 6). This finding indicated that it did not directly participate in the activation of SRT2-mediated stress signaling in *A. thaliana*. The SRT2-dependent, nicotinamide-independent defense response model proposed for *A. thaliana* is depicted in Figure 7.

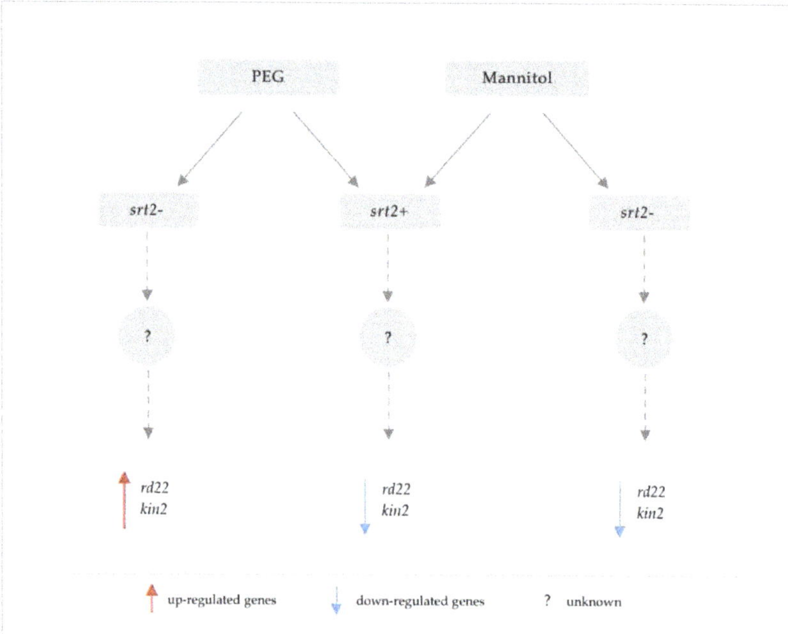

Figure 7. Proposed mechanism of SRT2 activity in the sirtuin-dependent, nicotinamide-independent defense response. Diagrammatic representation of osmotic stress-dependent genes with sensitivity to mannitol and polyethylene glycol (PEG) in the *srt2* mutants compared with those in the wild type (WT).

Mitochondrial SRT2 produces signals associated with the regulation of stress marker gene expression. These signals, including changes in mitochondrial coupling, energy efficiency, and increased glucose levels, can be sensed and transmitted to the nucleus to modulate gene expression and increase the plant's ability to adapt to a new state of energy or stress. SnRKs and TOR promote the coupling of these retrograde signals, as these molecules are synthesized in the mitochondria and affect the nuclear processes, such as transcription.

In the mitochondria, SRT2 binds to and deacetylates critical proteins related to cellular respiration, including ATP/ADP transporters, AAC1, and AAC3, as well as ATP synthase. Proteins in the *srt2+* plants were hyperacetylated, which led to decreased ATP synthase activity, mitochondrial coupling, photosynthetic rate, and concentration of sugars such as glucose (Figure 8).

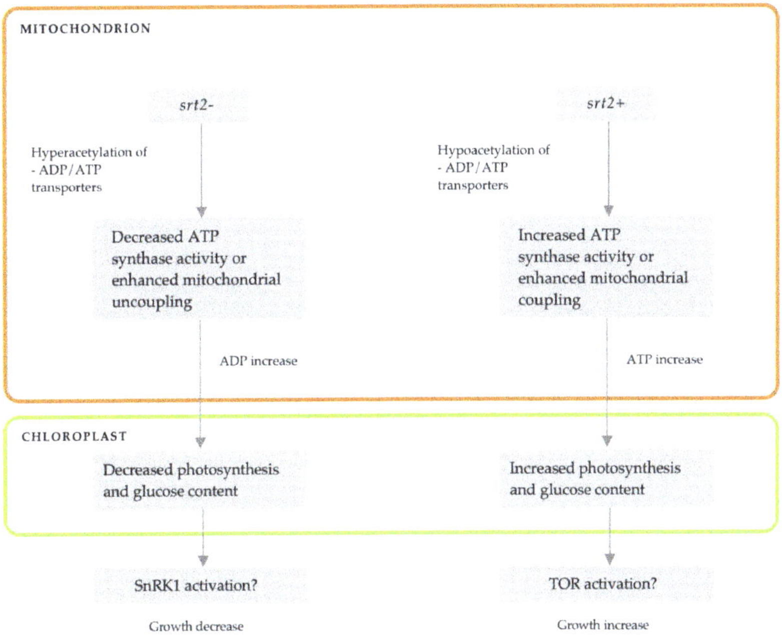

Figure 8. Possible mechanism underlying the variations in leaf growth. Disruption of SRT2 causes hyperacetylation of critical mitochondrial proteins, resulting in a decrease in ATP synthase activity and an increase in the concentration of photosynthetic products such as glucose. This model proposes that SRT2 overexpression has the opposite effect, reducing mitochondrial protein acetylation and increasing ATP synthase activity and glucose levels. The putative increase in glucose levels may be responsible for the increased leaf growth in *srt2*+ plants, which is mediated by TOR.

Low energy levels can activate SnRK1 and trigger autophagy. This may have been the cause of the reduced leaf development in *srt2*- plants. In contrast, these mitochondrial proteins may be hypoacetylated in *srt2*+ plants, resulting in increased ATP synthase activity or mitochondrial coupling. This could result in a higher energy state in which SnRK1 is inhibited and TOR is activated, explaining the increased leaf growth (Figure 8).

TOR is involved in osmotic stress signaling [46] and promotes root growth by modulating gene expression. Glc-TOR signaling also involves several chromatin modulators, transcription factors, signaling regulators, and growth and stress response proteins [44]. Therefore, the activation of TOR affects nuclear gene expression and may be partially responsible for the changes in gene expression observed in this study.

3. Materials and Methods

3.1. Plant Material, Growth Conditions, and Stress Treatment

A. thaliana ecotype Columbia (Col) was used as the WT control. The *srt2* mutants (Salk_149295) were obtained from ABRC (http://www.arabidopsis.org/abrc (accessed on 15 august 2003), and homozygous mutant lines were identified according to the Salk protocol (http://signal.salk.edu/tdnaprimers.2.html (accessed on 28 July 2005). Plants homozygous for T-DNA insertion were confirmed by PCR amplification using primers corresponding to sequences flanking the T-DNA insertion site and gene-specific primers. The primers used to identify the homozygous plants are listed in Table 2. Transgenic plants overexpressing AtSRT2 (OE) were generated in the Col-0 background. The AtSRT2-CDS3 sequence was amplified from the cDNA of WT (Col-0) plants using high-fidelity DNA polymerase KOD-plus (Toyobo, Osaka, Japan). The forward and reverse primer sequences are listed in Table 3. The amplified DNA was inserted into the pENTR/D-TOPO cloning vector (Invitrogen, Taastrup, Denmark), followed by site-directed recombination to generate the binary vectors pGWB2 and pGWB5 for SRT2-OE and 35S:SRT2-GFP, respectively [47]. The binary pGWB5-AtSRT2 plasmid was transformed into *Agrobacterium tumefaciens* strain GV3101 (pMP90). Transformation of *A. thaliana* was performed using the floral dip method [48]. To screen the transformants, seeds were grown on plates containing Murashige and Skoog (MS) medium supplemented with 40 μg/mL of hygromycin B (Roche Diagnostics, Mannheim, Germany). The resistant plants were then transferred to the soil for further analysis. The seeds were surface-sterilized and sown on the MS agar plates containing full-strength MS salts, 0.8% (*w*/*v*) agar, and 1% (*w*/*v*) sucrose. They were stratified in the dark at 4 °C for four days and then either transferred to a growth chamber with a 16 h/8 h light (350 μmol $m^{-2}s^{-1}$)/dark cycle at 23 °C or directly sown in the soil after stratification under the same conditions.

Table 2. Primers used for the identification of homozygous plants.

Oligonucleotide	Sequence
SALK_N649295-PCR1-D	5′-CGC AGA GAG AGA ACA AAA TCG-3′
SALK_N649295-PCR1-R	5′-TTC CAC ATT CTG TGC TAA CCC-3′
SALK_N649295-PCR2-D	5′-TGG TTC ACG TAG TGG GCC ATC G-3′
SALK_N649295-PCR2-D	5′-TTC CAC ATT CTG TGC TAA CCC-3′
SALK_N58811-PCR1-D	5′-AGA GTT CAT AAA AAC AAT GAA TCA AG-3′
SALK_N58811-PCR1-R	5′-AAC TCA TTG CAT TTG CAT AGG-3′
SALK_N58811-PCR2-D	5′-AAC TCA TTG CAT TTG CAT AGG-3′
SALK_N58811-PCR2-R	5′-TGG TTC ACG TAG TGG GCC ATC G-3′
SALK_N631994-PCR1-D	5′-TTT AAG GCA TTT TCA AGG CTG-3′
SALK_N631994-PCR1-R	5′-GGA GTT TAC TCG ATC AAG CCG-3′
SALK_N631994-PCR1-D	5′-GGA GTT TAC TCG ATC AAG CCG-3′
SALK_N631994-PCR1-R	5′-TGG TTC ACG TAG TGG GCC ATC G-3′

Table 3. Primers used to amplify the CDS3 of AtSRT2.

Oligonucleotide	Sequence
AtSRT2-CDS3-LP	5′-AGAAGCCGCACTCTAAGCAC-3′
AtSRT2-CDS3-RP	5′-AGTTACGCTGGATGGAGGAG-3′

A. thaliana seeds were surface-sterilized and sown in 24-well tissue culture clusters at 20 seeds per well (Costar Corp., Cambridge, MA, USA). Each well contained 1 mL of sterile MS medium (MS salts; ICN Hubber, Barcelona, Spain) supplemented with 0.5% sucrose. The seeds were grown with shaking (150 rpm) in a culture room under a 16 h day (26 °C)/8 h night (22 °C) diurnal cycle. Fresh medium (500 μL) was added to each well eight days after sowing, and the experiments were conducted after two days. The medium remaining in the wells was removed prior to initiating the treatment, and 1 mL of fresh medium was added to each well. The test compounds were diluted directly in the wells to their final concentrations from the stock solutions, which were prepared by separately mixing 300 mM of D-mannitol and polyethylene glycol (PEG; molecular weight of 6000 kDa) from Sigma-Aldrich (St. Louis, MO, USA), with water to obtain a 30% (*w*/*w*) aqueous solution. To study the germination of plants grown on the agar plates, the surface-sterilized seeds were planted in 1/2 MS medium supplemented with 0.8% phytagel, cold-treated for two days, and grown under the same conditions as the soil-grown plants. This medium allows solid culture, long root production, and easy root cleaning [49,50]. The plants were sown onto a mesh and cultured for 16 days before careful extraction and washing with sterile water. All experiments were performed independently at least three times, yielding highly reproducible results.

3.2. Greenhouse Growth Conditions

The seeds were stratified in the dark at 4 °C for 4 days and then either transferred to a growth chamber with 16 h/8 h light or 8 h/16 h light (350 µmol $m^{-2}s^{-1}$)/dark cycle at 23 °C or directly sown in the soil after stratification under the same conditions. A nutritive solution (pH 6.5) comprising the following macro and micronutrients was applied to the plants: 4 mM of KNO_3, 4 mM of $Ca(NO_3)_2$, 1.5 mM of $MgSO_4$, 0.75 mM of KH_2PO_4, 0.035% (p/v) EDTA-$FeSO_4$, 70 µM of H_3BO_3, 14 µM of $MnCl_2$, 0.5 µM of $CuSO_4$, 1 µM of $ZnSO_4$, 0.2 µM of Na_2MoO_4, 10 µM of NaCl, and 0.01 µM of $CoCl_2$.

3.3. Measurement of Leaf Properties

The dimensions of the plant leaves grown on long and short days were measured. Leaf length and width were measured using the AxioVision Rel. Software (version 4.6). The number of siliques growing on the main stem, their length, and number of seeds per silique were measured for each plant. The length of the mature siliques of plants grown in the soil was determined by capturing the images of the siliques and analyzing them later using EZ-Rhizo software (version 1.0.0.0). Three technical replicates were designed for each genetic line, and the median and standard error were calculated for each.

3.4. Germination Assay

Germination experiments were performed using fully ripened seeds of *A. thaliana*. The seeds were surface-sterilized and plated in groups of 50–100 in each plastic Petri dish containing MS medium without hormones or vitamins [51] supplemented with 2% (*w*/*v*) sucrose and 0.7% (*w*/*v*) agar and adjusted to pH 7.0. The seeds were incubated for 24 h at 4 °C and subsequently grown in a growth chamber at 24 °C under continuous white light (8.4 µmol $s^{-1}m^{-2}$) for 50 h. Ruptures of the testa and endosperm were scored by observation using a binocular microscope [52]. Four replicates were used for each treatment group. The seeds were photographed using a Canon G10 digital camera (Bensheim, Germany) attached to a stereomicroscope (Stemi 2000 C; Carl Zeiss, Oberkochen, Germany). AxioVision Rel. software (version 4.6) and Adobe Photoshop CS3 were used to analyze the images.

3.5. Lipid Peroxidation

Malondialdehyde (MDA) assay was conducted to estimate the level of lipid peroxidation in the leaf tissues following the method described by Hodges et al. [53] with minor modifications. Briefly, the leaf discs (0.1 g) were ground, suspended in 2 mL of ice-cold 1% *w*/*v* trichloroacetic acid (TCA), and centrifuged at 10,000× *g* for 5 min. Subsequently, 0.25 mL of the supernatant was mixed with 1 mL of 0.5% *w*/*v* thiobarbituric acid (TBA), incubated at 100 °C for 30 min, and immediately chilled. Differences in absorbance at 532 and 600 nm were measured using a Shimadzu UV-1601 spectrophotometer (Tokyo, Japan). An extinction coefficient of 155 mM^{-1} cm^{-1} was used for MDA determination.

3.6. RNA Extraction and First Strand cDNA Synthesis

The sampled plants were frozen in liquid nitrogen and ground to a fine powder using a mortar and pestle; 100 mg of this material was used for RNA isolation. Total RNA was extracted using the TRIzol reagent (TIANGEN, Beijing, China) in accordance with the manufacturer's instructions. The isolated RNA was treated with RNase-free DNase (Takara Bio, Kusatsu, Japan) to eliminate genomic DNA contamination. The purity and concentration of the RNA samples were measured with a Q5000 micro-volume UV-Vis spectrophotometer (Quawell Technology, San Jose, CA, USA). The integrity of the samples was examined by agarose gel electrophoresis. RNA samples with 260/280 nm absorbencies between 1.9 and 2.1 were used for subsequent experiments. First-strand cDNA synthesis was performed using a cDNA synthesis kit (Sangon, Shanghai, China) in accordance with the manufacturer's instructions, with a total volume of 20 µL containing 2 µg of total RNA. RNA extraction and cDNA synthesis were performed in triplicate for each sample. The cDNA solution was diluted 10 times with nuclease-free water, and the aliquots were stored at −20 °C until further use in RT-qPCR [54,55].

3.7. *qRT-PCR Analysis*

qRT-PCR analysis was performed in 96-well plates on an Mx3005 multiplex system (Stratagene, La Jolla, CA, USA) using EvaGreen-based PCR assay. Each reaction mixture had a final volume of 20 μL containing 1.0 μL of the template, 10.0 μL of 2× SuperReal Premix Plus (SYBR Green, Tiangen Biotech, Beijing, China), 0.4 μL of 50× ROX Reference Dye, 0.6 μL of each primer (10 μmol/μL), and 6.2 μL of double-distilled water. The reaction steps were as follows: pre-denaturation at 94 °C for 40 s, followed by 38 cycles of denaturation at 94 °C for 10 s and annealing at 57.2 °C for 30 s, and a final extension step at 72 °C for 10 min [54]. Three technical replicates were used for each cDNA sample.

3.8. *Molecular Markers of Abiotic Stress Response*

The expressions of molecular markers of abiotic stress response, including *kin2* (At5g15970) [56–61], *rd22* (At5g25610) [59,62–66], and the constitutive expression of *ACT2* (At3g18780), which is commonly used as a reference gene for qRT-PCR, were studied. Table 4 shows the primer sequences used, amplicon length in base pairs (bp), and melting temperature (Tm).

Table 4. Primers for reference genes and *srt2* for RT-qPCR.

Gene	Gene Code	Primer Sequence (Forward/Reverse)	Amplicon Length (bp)	Tm °C
Act2	At3g18780	5′-TGCCAATCTACGAGGGTTTC-3′	203	57.3
		5′-TGAGGTTTCCATCTCCTGCT-3′		57.3
Kin2-LP	At5g15970	5′-ACCAACAAGAATGCCTTCCA-3′	146	55.2
		5′-ACTGCCGCATCCGATATACT-3′		57.3
RD22-LP	At5g25610	5′-GGGCTGTTTCCACTGAGGT-3′	160	58.8
		5′-GTCGTCATCATCGCCTTGT-3′		56.7
SRT2-LP	At5g09230	5′-AGTTACGCTGGATGGAGGAG-3′	177	59.4
		5′-AGAAGCCGCACTCTAAGCAC-3′		59.4

3.9. *Statistical Analyses*

All the graphs were prepared in Microsoft Excel. Statistical analyses were carried out using GRAPHPAD PRISM 8 software. The Mann–Whitney U test was used to obtain significant differences compared with the WT ($p < 0.05$ was considered statistically significant).

4. Conclusions

In *Arabidopsis thaliana*, the sirtuin SRT2 is a mitochondrial deacetylase that binds to and deacetylates proteins involved in the electron transport chain. The results of this study indicated that SRT2 was not a major participant in plant development but regulated energy efficiency by modulating its cellular energy status and/or leaf growth under normal growth conditions. This was demonstrated in the *srt2+* overexpressing line, which exhibited greater leaf growth than that of the WT control. Moreover, our results indicated that SRT2 may be involved in signaling under conditions of osmotic stress generated by PEG or mannitol, as the deletion or overexpression of SRT2 induced changes in the expression of stress marker genes. However, SRT2 did not affect plant resistance to osmotic stress in the first developmental stages, as evidenced by no significant intergroup differences in MDA levels, initial seedling growth, or root growth.

Author Contributions: Conceptualization, A.O. and M.P.; formal analysis: A.O. and M.P.; investigation: A.O.; methodology, A.O. and M.P.; writing of (original draft): A.O.; writing (review, and editing): M.P. All authors have read and agreed to the published version of the manuscript.

Funding: This work was supported by the FONDECYT under grant 11070234.

Data Availability Statement: Data are contained within the article.

Conflicts of Interest: The authors declare no conflicts of interest.

References

1. Grubisha, O.; Smith, B.C.; Denu, J.M. Small molecule regulation of Sir2 protein deacetylases. *FEBS J.* **2005**, *272*, 4607–4616. [CrossRef]
2. Zhang, Q.; Piston, D.W.; Goodman, R.H. Regulation of corepressor function by nuclear NADH. *Science* **2002**, *295*, 1895–1897. [CrossRef] [PubMed]
3. Pétriacq, P.; de Bont, L.; Tcherkez, G.; Gakière, B. NAD: Not just a pawn on the board of plant-pathogen interactions. *Plant Signal. Behav.* **2013**, *8*, e22477. [CrossRef]
4. Guarente, L. Sirtuins, aging, and metabolism. *Cold Spring Harb. Symp. Quant. Biol.* **2011**, *76*, 81–90. [CrossRef] [PubMed]
5. Zhong, L.; Mostoslavsky, R. Fine tuning our cellular factories: Sirtuins in mitochondrial biology. *Cell Metab.* **2011**, *13*, 621–626. [CrossRef] [PubMed]
6. Canto, C.; Auwerx, J. NAD+ as a signaling molecule modulating metabolism. *Cold Spring Harb. Symp. Quant. Biol.* **2011**, *76*, 291–298. [CrossRef] [PubMed]
7. Verdin, E. NAD(+) in aging, metabolism, and neurodegeneration. *Science* **2015**, *350*, 1208–1213. [CrossRef] [PubMed]
8. Berglund, T.; Ohlsson, A.B.; Rydström, J.; Jordan, B.R.; Strid, Å. Effect of nicotinamide on gene expression and glutathione levels in tissue cultures of Pisum sativum. *J. Plant Physiol.* **1993**, *142*, 676–684. [CrossRef]
9. Berglund, T.; Wallström, A.; Nguyen, T.V.; Laurell, C.; Ohlsson, A.B. Nicotinamide; antioxidative and DNA hypomethylation effects in plant cells. *Plant Physiol. Biochem.* **2017**, *118*, 551–560. [CrossRef]
10. Grozinger, C.M.; Chao, E.D.; Blackwell, H.E.; Moazed, D.; Schreiber, S.L. Identification of a class of small molecule inhibitors of the sirtuin family of NAD-dependent deacetylases by phenotypic screening. *J. Biol. Chem.* **2001**, *276*, 38837–38843. [CrossRef]
11. Tanny, J.C.; Moazed, D. Coupling of histone deacetylation to NAD breakdown by the yeast silencing protein Sir2: Evidence for acetyl transfer from substrate to an NAD breakdown product. *Proc. Natl. Acad. Sci. USA* **2001**, *98*, 415–420. [CrossRef]
12. Berger, F.; Ramirez-Hernandez, M.H.; Ziegler, M. The new life of a centenarian: Signalling functions of NAD(P). *Trends Biochem. Sci.* **2004**, *29*, 111–118. [CrossRef]
13. Adams, J.D.; Klaidman, L.K., Jr. Sirtuins, Nicotinamide and Aging, A Critical Review. *Lett. Drug Des. Discov.* **2007**, *4*, 44–48. [CrossRef]
14. Martínez-Redondo, P.; Vaquero, A. The Diversity of Histone Versus Nonhistone Sirtuin Substrates. *Genes Cancer* **2013**, *4*, 148–163. [CrossRef] [PubMed]
15. Szucko, I. Sirtuins: Not only animal proteins. *Acta Physiol. Plant* **2016**, *38*, 237. [CrossRef]
16. Gardner, J.G.; Escalante-Semerena, J.C. In Bacillus subtilis, the sirtuin protein deacetylase, encoded by the srtN gene (formerly yhdZ), and functions encoded by the acuABC genes control the activity of acetyl coenzyme A synthetase. *J. Bacteriol.* **2009**, *191*, 1749–1755. [CrossRef] [PubMed]
17. Bell, S.D.; Botting, C.H.; Wardleworth, B.N.; Jackson, S.P.; White, M.F. The interaction of Alba, a conserved archaeal chromatin protein, with Sir2 and its regulation by acetylation. *Science* **2002**, *296*, 148–151. [CrossRef]
18. Lu, S.P.; Lin, S.J. Regulation of yeast sirtuins by NAD+ metabolism and calorie restriction. *Biochim. Biophys. Acta Proteins Proteom.* **2010**, *1804*, 1567–1575. [CrossRef] [PubMed]
19. Wirth, M.; Paap, F.; Fischle, W.; Wenzel, D.; Agafonov, D.E.; Samatov, T.R.; Wisniewski, J.R.; Jedrusik-Bode, M. HIS-24 linker histone and SIR-2.1 deacetylase induce H3K27me3 in the Caenorhabditis elegans germ line. *Mol. Cell Biol.* **2009**, *29*, 3700–3709. [CrossRef]
20. Whitaker, R.; Faulkner, S.; Miyokawa, R.; Burhenn, L.; Henriksen, M.; Wood, J.G.; Helfand, S.L. Increased expression of Drosophila sir2 extends life span in a dose dependent manner. *Aging* **2013**, *5*, 682–691. [CrossRef]
21. Frye, R.A. Characterization of Five Human cDNAs with Homology to the Yeast SIR2 Gene: Sir2-like Proteins (Sirtuins) Metabolize NAD and May Have Protein ADP-Ribosyltransferase Activity. *Biochem. Biophys. Res. Commun.* **1999**, *260*, 273–279. [CrossRef] [PubMed]
22. Zhong, X.; Zhang, H.; Zhao, Y.; Sun, Q.; Hu, Y.; Peng, H.; Zhou, D.X. The rice NAD(+)-dependent histone deacetylase OsSRT1 targets preferentially to stress- and metabolism-related genes and transposable elements. *PLoS ONE* **2013**, *8*, e66807. [CrossRef] [PubMed]
23. König, A.C.; Hartl, M.; Pham, P.A.; Laxa, M.; Boersema, P.J.; Orwat, A.; Kalitventseva, I.; Plöchinger, M.; Braun, H.P.; Leister, D.; et al. The Arabidopsis class II sirtuin is a lysine deacetylase and interacts with mitochondrial energy metabolism. *Plant Physiol.* **2014**, *164*, 1401–1414. [CrossRef] [PubMed]

24. Finkemeier, I.; Laxa, M.; Miguet, L.; Howden, A.J.; Sweetlove, L.J. Proteins of diverse function and subcellular location are lysine acetylated in Arabidopsis. *Plant Physiol.* **2011**, *155*, 1779–1790. [CrossRef] [PubMed]
25. Wu, X.; Oh, M.H.; Schwarz, E.M.; Larue, C.T.; Sivaguru, M.; Imai, B.S.; Yau, P.M.; Ort, D.R.; Huber, S.C. Lysine acetylation is a widespread protein modification for diverse proteins in Arabidopsis. *Plant Physiol.* **2011**, *155*, 1769–1778. [CrossRef] [PubMed]
26. Zhang, H.; Lu, Y.; Zhao, Y.; Zhou, D.X. OsSRT1 is involved in rice seed development through regulation of starch metabolism gene expression. *Plant Sci.* **2016**, *248*, 28–36. [CrossRef] [PubMed]
27. Busconi, M.; Reggi, S.; Fogher, C.; Bavaresco, L. Evidence of a sirtuin gene family in grapevine (*Vitis vinifera* L.). *Plant Physiol Biochem.* **2009**, *7*, 650–652. [CrossRef] [PubMed]
28. Cucurachi, M.; Busconi, M.; Morreale, G.; Zanetti, A.; Bavaresco, L.; Fogher, C. Characterization and differential expression analysis of complete coding sequences of *Vitis vinifera* L. sirtuin genes. *Plant Physiol. Biochem.* **2012**, *54*, 123–132. [CrossRef]
29. Aiese Cigliano, R.; Sanseverino, W.; Cremona, G.; Ercolano, M.R.; Conicella, C.; Consiglio, F.M. Genome-wide analysis of histone modifiers in tomato: Gaining an insight into their developmental roles. *BMC Genom.* **2013**, *14*, 5. [CrossRef]
30. Zhao, L.; Lu, J.; Zhang, J.; Wu, P.Y.; Yang, S.; Wu, K. Identification and characterization of histone deacetylases in tomato (*Solanum lycopersicum*). *Front. Plant Sci.* **2015**, *5*, 760. [CrossRef]
31. Huang, L.; Sun, Q.; Qin, F.; Li, C.; Zhao, Y.; Zhou, D.X. Down-Regulation of a Silent Information regulator2-Related Histone Deacetylase Gene, OsSRT1, induces DNA fragmentation and cell death in rice. *Plant Physiol.* **2007**, *144*, 1508–1519. [CrossRef]
32. Wang, C.; Gao, F.; Wu, J.; Dai, J.; Wei, C.; Li, Y. Arabidopsis putative deacetylase AtSRT2 regulates basal defense by suppressing PAD4, EDS5, and SID2 expression. *Plant Cell Physiol.* **2010**, *51*, 1291–1299. [CrossRef]
33. Shen, Y.; Issakidis-Bourguet, E.; Zhou, D. Perspectives on the interactions between metabolism, redox, and epigenetics in plants. *J. Exp. Bot.* **2016**, *67*, 5291–5300. [CrossRef] [PubMed]
34. Kupis, W.; Pałyga, J.; Tomal, E.; Niewiadomska, E. The role of sirtuins in cellular homeostasis. *J. Physiol. Biochem.* **2016**, *72*, 371–380. [CrossRef] [PubMed]
35. Lebon, G.; Brun, O.; Magné, C.; Clément, C. Photosynthesis of the grapevine (*Vitis vinifera*) inflorescence. *Tree Physiol.* **2005**, *25*, 633–639. [CrossRef] [PubMed]
36. Aquea, F.; Timmermann, T.; Arce-Johnson, P. Analysis of histone acetyltransferase and deacetylase families of *Vitis vinifera*. *Plant Physiol. Biochem.* **2010**, *48*, 194–199. [CrossRef] [PubMed]
37. Hartl, M.; Finkemeier, I. Plant mitochondrial retrograde signaling: Post-translational modifications enter the stage. *Front. Plant Sci.* **2012**, *253*, 253. [CrossRef]
38. Soto-Burgos, J.; Bassham, D.C. SnRK1 activates autophagy via the TOR signaling pathway in *Arabidopsis thaliana*. *PLoS ONE* **2017**, *12*, e0182591. [CrossRef] [PubMed]
39. Schepetilnikov, M.; Ryabova, L.A. Recent Discoveries on the Role of TOR (Target of Rapamycin) Signaling in Translation in Plants. *Plant Physiol.* **2018**, *176*, 1095–1105. [CrossRef] [PubMed]
40. Singh, S.; Singh, A.; Yadav, S.; Gautam, V.; Singh, A.; Sarkar, A.K. Sirtinol, a Sir2 protein inhibitor, affects stem cell maintenance and root development in *Arabidopsis thaliana* by modulating auxin-cytokinin signaling components. *Sci. Rep.* **2017**, *14*, 42450. [CrossRef]
41. Zhao, Y. Auxin biosynthesis and its role in plant development. *Annu. Rev. Plant Biol.* **2010**, *61*, 49–64. [CrossRef] [PubMed]
42. Edelmann, H.G. Plant root development: Is the classical theory for auxin-regulated root growth false? *Protoplasma* **2022**, *259*, 823–832. [CrossRef] [PubMed]
43. Ma, X.; Lv, S.; Zhang, C.; Yang, C. Histone deacetylases and their functions in plants. *Plant Cell Rep.* **2013**, *32*, 465–478. [CrossRef] [PubMed]
44. Xiong, Y.; Sheen, J. The role of target of rapamycin signaling networks in plant growth and metabolism. *Plant Physiol.* **2014**, *164*, 499–512. [CrossRef] [PubMed]
45. Hohl, M.; Schopfer, P. Water relations of growing maize coleoptiles. Comparison between mannitol and polyethylene glycol 6000 as external osmotica for adjusting turgor pressure. *Plant Physiol.* **1991**, *95*, 716–722. [CrossRef] [PubMed]
46. Deprost, D.; Yao, L.; Sormani, R.; Moreau, M.; Leterreux, G.; Nicolaï, M.; Bedu, M.; Robaglia, C.; Meyer, C. The *Arabidopsis* TOR kinase links plant growth, yield, stress resistance and mRNA translation. *EMBO Rep.* **2007**, *9*, 864–870. [CrossRef] [PubMed]
47. Nakagawa, T.; Kurose, T.; Hino, T.; Tanaka, K.; Kawamukai, M.; Niwa, Y.; Toyooka, K.; Matsuoka, K.; Jinbo, T.; Kimura, T. Development of series of gateway binary vectors, pGWBs, for realizing efficient construction of fusion genes for plant transformation. *J. Biosci. Bioeng.* **2007**, *104*, 34–41. [CrossRef]
48. Clough, S.J.; Bent, A.F. Floral dip: A simplified method for Agrobacterium-mediated transformation of *Arabidopsis thaliana*. *Plant J.* **1998**, *16*, 735–743. [CrossRef]
49. Armengaud, P.; Zambaux, K.; Hills, A.; Sulpice, R.; Pattison, R.J.; Blatt, M.R.; Amtmann, A. EZ-Rhizo: Integrated software for the fast and accurate measurement of root system architecture. *Plant J.* **2009**, *57*, 945–956. [CrossRef]
50. Julkowska, M.M.; Hoefsloot, H.C.; Mol, S.; Feron, R.; de Boer, G.J.; Haring, M.A.; Testerink, C. Capturing Arabidopsis root architecture dynamics with ROOT-FIT reveals diversity in responses to salinity. *Plant Physiol.* **2014**, *166*, 1387–1402. [CrossRef]
51. Murashige, T.; Skoog, F. A revised medium for rapid growth and bioassays with tobacco cultures. *Physiol. Plant.* **1962**, *15*, 473–497. [CrossRef]
52. Gul, B.; Weber, D.J. Effect of dormancy relieving compounds on the seed germination of non-dormant Allenrolfea occidentalis under salinity stress. *Ann. Bot.* **1998**, *82*, 555–560. [CrossRef]

53. Hodges, D.M.; DeLong, J.M.; Forney, C.F.; Prange, R.K. Improving the thiobarbituric acid-reactive-substances assay for estimating lipid peroxidation in plant tissues containing anthocyanin and other interfering compounds. *Planta* **1999**, *207*, 604–611. [CrossRef]
54. Shi, H.T.; Li, R.J.; Cai, W.; Liu, W.; Wang, C.L.; Lu, Y.T. Increasing nitric oxide content in *Arabidopsis thaliana* by expressing rat neuronal nitric oxide synthase resulted in enhanced stress tolerance. *Plant Cell Physiol.* **2011**, *53*, 344–357. [CrossRef] [PubMed]
55. Sun, X.; Luo, X.; Sun, M.; Chen, C.; Ding, X.; Wang, X.; Yang, S.; Yu, Q.; Jia, B.; Ji, W.; et al. A *Glycine Soja* 14-3-3 protein GsGF14o participates in stomatal and root hair development and drought tolerance in *Arabidopsis thaliana. Plant Cell Physiol.* **2014**, *55*, 99–118. [CrossRef]
56. Levin, D.E.; Hammond, C.I.; Ralston, R.O.; Bishop, J.M. Two yeast genes that encode unusual protein kinases. *Proc. Natl. Acad. Sci. USA* **1987**, *84*, 6035–6039. [CrossRef]
57. Kurkela, S.; Borg-Franck, M. Structure and expression of kin2, one of two cold- and ABA-induced genes of *Arabidopsis thaliana. Plant Mol. Biol.* **1992**, *19*, 689–692. [CrossRef]
58. Wang, H.; Datla, R.; Georges, F.; Loewen, M.; Cutler, A.J. Promoters from kin1 and cor6.6, two homologous *Arabidopsis thaliana* genes: Transcriptional regulation and gene expression induced by low temperature, ABA, osmoticum and dehydration. *Plant Mol. Biol.* **1995**, *28*, 605–617. [CrossRef] [PubMed]
59. Huang, D.; Wu, W.; Abrams, S.R.; Cutler, A.J. The relationship of drought-related gene expression in *Arabidopsis thaliana* to hormonal and environmental factors. *J. Exp. Bot.* **2008**, *11*, 2991–3007. [CrossRef]
60. Lata, C.; Prasad, M. Role of DREBs in regulation of abiotic stress responses in plants. *J. Exp. Bot.* **2011**, *14*, 4731–4748. [CrossRef]
61. Jia, W.; Zhang, L.; Wu, D.; Liu, S.; Gong, X.; Cui, Z.; Cui, N.; Cao, H.; Rao, L.; Wang, C. Sucrose transporter AtSUC9 mediated by a low sucrose level is involved in Arabidopsis abiotic stress resistance by regulating sucrose distribution and ABA accumulation. *Plant Cell Physiol.* **2015**, *56*, 1574–1587. [CrossRef] [PubMed]
62. Yamaguchi-Shinozaki, K.; Shinozaki, K. The plant hormone abscisic acid mediates the drought-induced expression but not the seed-specific expression of rd22, a gene responsive to dehydration stress in *Arabidopsis thaliana. Mol. Gen. Genet* **1993**, *238*, 17–25. [CrossRef] [PubMed]
63. Ding, X.; Hou, X.; Xie, K.; Xiong, L. Genome-wide identification of BURP domain-containing genes in rice reveals a gene family with diverse structures and responses to abiotic stresses. *Planta* **2009**, *230*, 149–163. [CrossRef] [PubMed]
64. Abe, H.; Yamaguchi-Shinozaki, K.; Urao, T.; Iwasaki, T.; Hosokawa, D.; Shinozaki, K. Role of Arabidopsis MYC and MYB homologs in drought- and abscisic acid-regulated gene expression. *Plant Cell* **1997**, *9*, 1859–1868. [CrossRef]
65. Matus, J.T.; Aquea, F.; Espinoza, C.; Vega, A.; Cavallini, E.; Santo, S.D.; Cañón, P.; de la Guardia, A.R.H.; Serrano, J.; Tornielli, G.B.; et al. Inspection of the grapevine BURP superfamily highlights an expansion of RD22 genes with distinctive expression features in berry development and ABA-mediated stress responses. *PLoS ONE* **2014**, *9*, e110372. [CrossRef]
66. Chen, T.; Li, W.; Hu, X.; Guo, J.; Liu, A.; Zhang, B. A cotton MYB transcription factor, GbMYB5, is positively involved in plant adaptive response to drought stress. *Plant Cell Physiol.* **2015**, *56*, 917–929. [CrossRef]

plants

MDPI

Communication

The Invasive Plant *Amaranthus spinosus* L. Exhibits a Stronger Resistance to Drought than the Native Plant *A. tricolor* L. under Co-Cultivation Conditions When Treated with Light Drought

Congyan Wang [1,2,3,*], Yingsheng Liu [1], Chuang Li [1], Yue Li [1] and Daolin Du [4,*]

1 School of Environment and Safety Engineering, Jiangsu University, Zhenjiang 212013, China; 3202202017@stmail.ujs.edu.cn (Y.L.); 2222209081@stmail.ujs.edu.cn (C.L.); 2222209050@stmail.ujs.edu.cn (Y.L.)
2 Jiangsu Collaborative Innovation Center of Technology and Material of Water Treatment, Suzhou University of Science and Technology, Suzhou 215009, China
3 Key Laboratory of Forest Plant Ecology, Ministry of Education, Northeast Forestry University, Harbin 150040, China
4 Jingjiang College, Jiangsu University, Zhenjiang 212013, China
* Correspondence: liuyuexue623@ujs.edu.cn or liuyuexue623@163.com (C.W.); ddl@ujs.edu.cn or daolin-du@163.com (D.D.)

Abstract: Drought may facilitate the invasion process of invasive plants, mainly because invasive plants can obtain a stronger growth competitiveness than native plants under drought. It is therefore imperative to illuminate the mechanisms underlying the successful invasion of invasive plants under drought, with a particular focus on the differences in the resistance of invasive and native plants to drought. This study aimed to elucidate the differences in the resistance between the invasive plant *Amaranthus spinosus* L. and the native plant *A. tricolor* L. to drought under a gradient of drought. The resistance of co-cultivated *A. spinosus* to drought was significantly higher than that of co-cultivated *A. tricolor* under light drought. Hence, *A. spinosus* may obtain a stronger competitive advantage than *A. spinosus* under co-cultivation conditions when treated with light drought. The resistance of the two plants to drought may be predominantly influenced by their height and biomass. This present study also defines a method for evaluating the stress resistance of a given plant species to stress by calculating the stress resistance index. This present study offers a robust theoretical foundation for determining the stress resistance of a given plant species and the environmental management of *A. spinosus* under drought.

Keywords: biomass; competitive advantage; external stress; invasive plants; stress resistance index

1. Introduction

All organisms are situated within a particular ecological environment. It is important to note, however, that the ecological environment in which organisms reside is not always homeostatic. During the growth process, the survival of organisms may be threatened. In particular, the ecosystems in which organisms reside are often subject to a certain degree of external stress, including, but not limited to, drought, fires, and high temperatures [1–4]. Such conditions may exert a certain influence on the growth and metabolic processes of the organisms in question [5–8]. It is evident that organisms possess the capacity to resist external stresses to a certain extent, thereby enabling them to survive within the ecological environment and ultimately prevail in the struggle for survival. In particular, the stronger resistance of organisms to external stresses may confer upon them a competitive advantage, particularly in the expansion of their ecological niche within a given habitat, specifically when subjected to stronger selective pressure mediated by those external stresses. More importantly, a stronger resistance to external stresses can promote the maintenance of a relatively stable state, thereby facilitating adaptation to surrounding habitats.

It is of greater consequence that organisms, and in particular plant species, are indispensable for the functioning of ecosystems [9–12]. It is of paramount importance to recognize that the resistance of each plant species to external stresses may fluctuate due to the variations in the species identity and/or the severity of external stresses. It is unfortunate that, at present, there is no method for quantifying the resistance of a given plant species to external stresses.

Currently, one of the most significant global environmental concerns is the advanced increase in climate change, largely due to the ongoing rise in global temperatures and the gradual onset of annual rainfall variability [13–16]. Presently, the area of the world's land surface covered by arid and semi-arid districts is ~3.83×10^7 km^2, representing ~40% of the total land surface area of the planet [17,18]. Furthermore, it is anticipated that drought events will become more frequent and severe in the future, particularly in arid and semi-arid regions. The extent of these regions is also expected to continue to expand due to the increasing frequency and intensity of climate change and anthropogenic activities [14–17]. It is also important to note that drought can facilitate the invasion process of invasive plants (IPSs), as IPSs can often outcompete native plants in the context of drought [5,6,8,19]. However, IPSs can exert a considerable influence on native ecosystems. In particular, IPS can disrupt the structure and ecological function of native ecosystems, which can result in the loss of native biodiversity [7,20–22]. It is therefore imperative to elucidate the mechanisms underlying the successful invasion of IPSs in the context of drought, with a particular focus on the differences in the resistance of IPSs and native plants to drought. Nevertheless, the advancement of research concerning the discrepancy in the resistance between invasive and native plants to external stresses, particularly in relation to a spectrum of drought conditions, is currently limited.

The aim of this present study was to propose a methodology for quantifying the resistance of a given plant species to external stresses. A case study was conducted to ascertain the discrepancies in the resistance between the invasive plant *Amaranthus spinosus* L. and the native plant *A. tricolor* L. to drought under a gradient of drought (including the control, a light drought, and a heavy drought). The two plants are both herbaceous species of *Amaranthus*, belonging to the Amaranthaceae family. At the family level, Amaranthaceae was found to have the fourth highest species number of IPSs (25 species), following the families of Asteraceae (49 species), Fabaceae (35 species), and Poaceae (28 species) in East China [23]. At the generic level, the species number of IPSs included in *Amaranthus* (with 16 species, representing ~5.35% of the total species number of IPSs in East China) is significantly higher than that included in other genera in East China (Yan et al., 2021). Furthermore, *A. spinosus* originated in North America, and the species number of IPSs in China that originated in North America is significantly higher than that of species originating in other countries and/or regions [23–26]. It is of particular importance to note that *A. spinosus* is currently registered as one of the most notorious IPSs in China, largely due to the significant threats it poses to the structure and ecological function of native ecosystems. Furthermore, *A. spinosus* and *A. tricolor* can co-occur in the same community, including wastelands, farmland edges, and areas near major roads in southern Jiangsu, China. The areas in China where the two plants have been distributed are also among the areas most severely affected by drought [15–17].

The objective of this present study was to propose a question: Does the invasive plant *A. spinosus* exhibit a greater resistance to drought than the native plant *A. tricolor* under a gradient of drought? In addition, this present study also aims to define a methodology for quantifying the stress resistance of a given plant species to external stresses by calculating the stress resistance index.

2. Materials and Methods

2.1. Experimental Design

The seeds of *A. spinosus* were collected in October 2019 from Zhenjiang, Jiangsu, China (32.11° N, 119.53° E). The selected ecosystems were classified as wastelands. *Amaranthus spinosus* was the sole species of IPS present in the sampled communities. It is plausible that the sampled *A. spinosus* individuals may have dispersed naturally within the sampled communities. Zhenjiang is located within a north subtropical monsoon humid climate zone. The annual mean temperature is ~17.0 °C, with a monthly mean temperature reaching a maximum of ~30.5 °C in August and a minimum of ~3.1 °C in December. The annual precipitation is ~919.5 mm, with the monthly mean precipitation reaching a maximum of ~208.7 mm in July and a minimum of ~9.2 mm in May. The annual sunshine duration is ~2014.0 h, with a monthly mean sunshine time reaching a maximum of ~218.1 h in July and a minimum of ~101.5 h in November [27]. The seeds of *A. tricolor* cv. *xinbai* were procured from Qingxianchunfeng Vegetable Cultivars Breeding Base, Cangzhou, China, at a local vegetable market.

A pot competitive co-culture experiment was conducted to assess the growth of the two plants. This experiment was conducted in June 2020. Seeds of the two plants were sown in garden pots (manufacturer: Huazhihuijing Co. Ltd., Suqian, Jiangsu, China; diameter: ~30 cm; height: ~17.2 cm). Six uniform, robust seedlings of *A. spinosus* and/or *A. tricolor* were planted per garden pot. The pasture yellow soil (manufacturer: Zhongfangnongmu Co. Ltd., Taizhou, Jiangsu, China; pH value: ~6.5; organic content: ≥40%; ~5 kg per garden pot) was selected as the growth substratum. The rationale for utilizing pasture yellow soil as the growth substratum was to eliminate the possibility of IPSs becoming established in soil obtained from a natural field. Three distinct forms of plant cultivation were employed: (1) six *A. spinosus* seedlings were planted per garden pot in the monoculture; (2) three *A. spinosus* seedlings and three *A. tricolor* seedlings were planted per garden pot in the co-culture; and (3) six *A. tricolor* seedlings were planted per garden pot in the monoculture. All garden pots were treated with a gradient of artificially simulated drought through the addition of varying quantities of water; specifically, this included (1) the control (100% of the normal monthly mean precipitation of June of Zhenjiang, which equates to ≈430 mL/d per garden pot); (2) a light drought (60% of the normal monthly mean precipitation of June of Zhenjiang, which equates to ≈258 mL/d per garden pot); and (3) a heavy drought (20% of the normal monthly mean precipitation of June of Zhenjiang, which equates to ≈86 mL/d per garden pot). In this present study, the gradient of artificially simulated drought was identified in accordance with the Chinese industry standards (No.: GB/T 20481; name: Classification of Meteorological Drought of China) [28]. This present study employed a range of plant cultivation combinations (i.e., monocultural *A. spinosus*, co-cultivated *A. spinosus* and *A. tricolor*, and monocultural *A. tricolor*) and a gradient of drought combinations (i.e., the control, a light drought, and a heavy drought). Three replicates were conducted for each treatment. The experimental design of the present study is illustrated in Figure 1.

Seedlings of the two plants were cultivated in an artificial greenhouse at Jiangsu University, Zhenjiang (32.2061° N, 119.5128° E), under conditions of natural sunlight.

Upon completion of the pot competitive co-culture experiment, all individuals of the two plants were used for analysis. This included the determination of the values of the plants' functional traits, the biomass stability index, the levels of osmotic adjustment and the activities of antioxidant enzymes, the stress resistance index of the two plants to drought, the relative competition intensity and relative dominance indices of *A. spinosus*, and the stress intensity index of drought on the growth of the two plants.

In particular, the analyzing method and the values of the plants' functional traits (including height, ground diameter, fresh and dry weights, water content, leaf length and width, specific leaf area, leaf fresh and dry weight weights, leaf water content, and leaf chlorophyll and nitrogen contents), the biomass stability index, the levels of osmotic adjustment (including malondialdehyde and proline contents) and the activities of antioxidant enzymes (including catalase, peroxidase, and superoxide dismutase activities) of the two plants, the relative competition intensity and the relative dominance indices of *A. spinosus*, and the stress intensity index of drought on the growth of the two plants have been previously described in a related study [6].

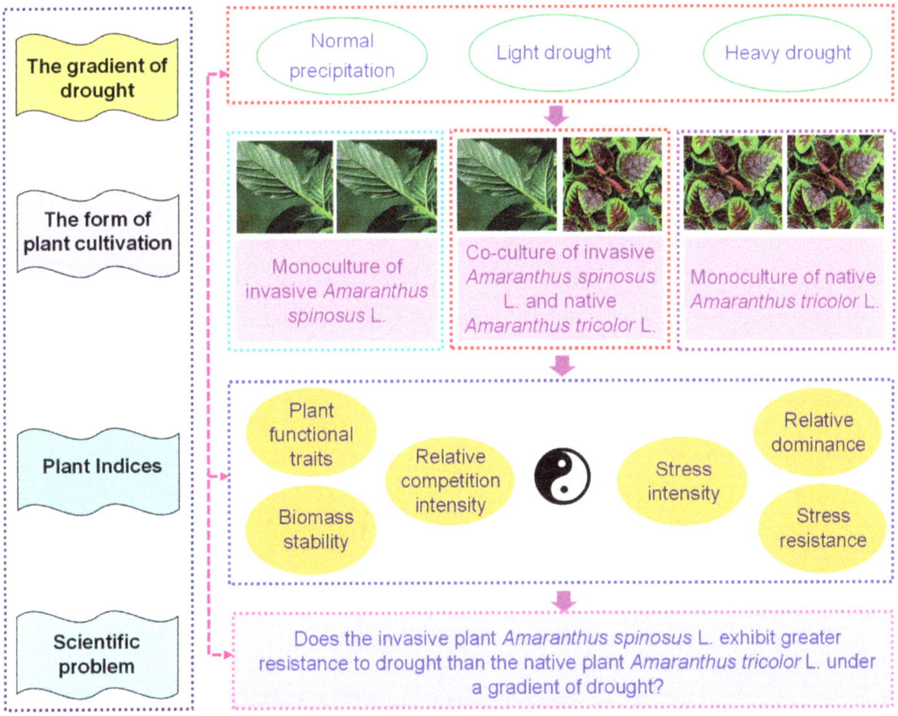

Figure 1. The chart of the experiment design of this present study.

2.2. Method for Determining the Stress Resistance Index

To assess the degree of resistance exhibited by a given plant species to stress (which in this study is also referred to as drought), the stress resistance index (*SRI*) was calculated according to the following formula:

$$SRI = 1 - \frac{B_{ns} - B_s}{B_{ns}}$$

where B_s and B_{ns} represent the total biomass of the plant species i under the condition with stress and the condition without stress, respectively. In particular, plant species with a higher value of *SRI* demonstrate a greater degree of stress resistance compared to those with a lower value of *SRI*.

2.3. Statistical Analysis

Shapiro–Wilk's test and Bartlett's test were conducted to determine the extent of departure from the normality and the homogeneity of the examined variances, respectively. A multiple comparison with the Tukey's test was employed to assess the statistical differences in the values of the stress resistance index of the two plants across different treatments. Correlation patterns between the values of the plants' functional traits, the levels of osmotic adjustment and the activities of antioxidant enzymes, and the stress resistance index of the two plants were assessed by using correlation analysis and principal component analysis (PCA), respectively. $p \leq 0.05$ was considered to signify a statistically significant difference. Statistical analyses were conducted using IBM SPSS Statistics 26.0 (IBM, Inc., Armonk, NY, USA).

3. Results and Discussion

The values of the stress resistance index of co-cultivated *A. spinosus* treated with a light drought were found to be significantly higher than that of monocultural *A. spinosus* treated with a light drought ($p < 0.05$; Figure 2). However, no significant differences were observed in the values of the stress resistance index between monocultural *A. spinosus* treated with a heavy drought and co-cultivated *A. spinosus* treated with a heavy drought ($p > 0.05$; Figure 2). Thus, the resistance of co-cultivated *A. spinosus* to drought was found to be significantly higher than that of monocultural *A. spinosus* under a condition with a light drought. This may be attributed to the higher biomass stability index of co-cultivated *A. spinosus* and the lower stress intensity index of drought for the growth of co-cultivated *A. spinosus* compared to monocultural *A. spinosus* when treated with a light drought [6]. It can therefore be concluded that *A. spinosus* displays a more pronounced competitive advantage under the co-cultivation conditions compared to the monoculture condition under a condition with a light drought. Nevertheless, no significant differences were observed in the stress resistance index between monocultural *A. tricolor* and co-cultivated *A. tricolor* regardless of the gradient of drought ($p > 0.05$; Figure 2). Thus, the form of plant cultivation did not have a significant impact on the resistance of *A. tricolor* to drought.

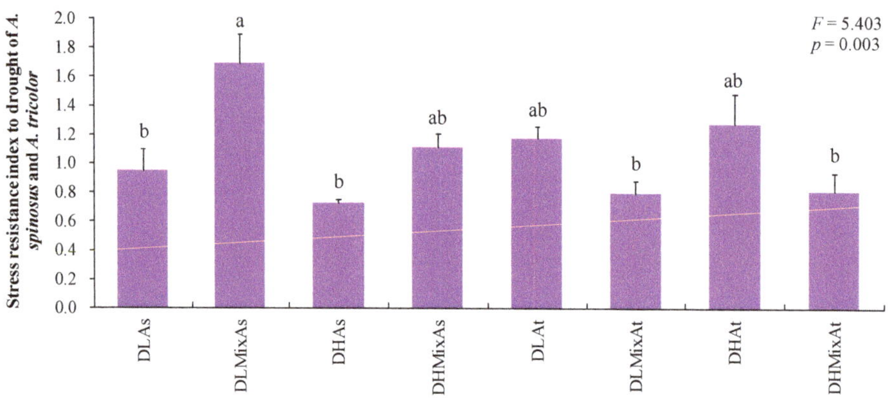

Figure 2. Differences in the values of the stress resistance index of *A. spinosus* and *A. tricolor* among different treatments. Bars (mean and standard error, $n = 3$) with different lowercase letters represent statistically significant differences at 0.05 probability. Abbreviations: DLAs, monocultural *A. spinosus* treated with a light drought; DLMixAs, co-cultivated *A. spinosus* treated with a light drought; DHAs, monocultural *A. spinosus* treated with a heavy drought; DHMixAs, co-cultivated *A. spinosus* treated with a heavy drought; DLAt, monocultural *A. tricolor* treated with a light drought; DLMixAt, co-cultivated *A. tricolor* treated with a light drought; DHAt, monocultural *A. tricolor* treated with a heavy drought; DHMixAt, co-cultivated *A. tricolor* treated with a heavy drought.

More importantly, the stress resistance index of co-cultivated *A. spinosus* treated with a light drought was found to be significantly higher than that of co-cultivated *A. tricolor* treated with a light drought ($p < 0.05$; Figure 2). However, no significant differences were observed in the stress resistance index between co-cultivated *A. spinosus* treated with a heavy drought and co-cultivated *A. tricolor* treated with a heavy drought ($p > 0.05$; Figure 2). It was thus demonstrated that the resistance of co-cultivated *A. spinosus* to drought was significantly higher than that of co-cultivated *A. tricolor* under the condition of light drought. This phenomenon may be attributed to the higher biomass stability index of co-cultivated *A. spinosus* and the lower stress intensity index of drought for the growth of co-cultivated *A. spinosus* compared to co-cultivated *A. tricolor* when treated with a light drought [6]. An additional potential explanation for this phenomenon is that co-cultivated *A. spinosus* exhibits a greater height and leaf chlorophyll and nitrogen contents compared to co-cultivated *A. tricolor* when treated with a light drought [6]. The greater height and leaf chlorophyll and nitrogen contents of co-cultivated *A. spinosus* may confer a competitive advantage in light acquisition and its absorption and utilization efficiency compared to co-cultivated *A. tricolor* under the condition of light drought. In general, light is a vital component of the growth process for plants [29–32]. It was thus demonstrated that co-cultivated *A. spinosus* displayed markedly greater resistance to drought in comparison to co-cultivated *A. tricolor* under the condition of light drought. It can thus be proposed that *A. spinosus* may exhibit a more pronounced competitive advantage than *A. tricolor* under co-cultivations condition when treated with a light drought.

The results of the correlation analysis ($p < 0.01$; Table 1) and principal component analysis (Figure 3) indicate that the stress resistance index of the two plants may be predominantly influenced by their height and biomass. This can be attributed to the fact that height and biomass represent a key factor in determining the competitive capacity for light acquisition and the growth competitiveness of plants [33–36]. However, the competitive capacity for light acquisition and the growth competitiveness of plants are essential for plant fitness [37–40]. Consequently, the competitive capacity for light acquisition and their growth competitiveness represent pivotal elements influencing their stress tolerance.

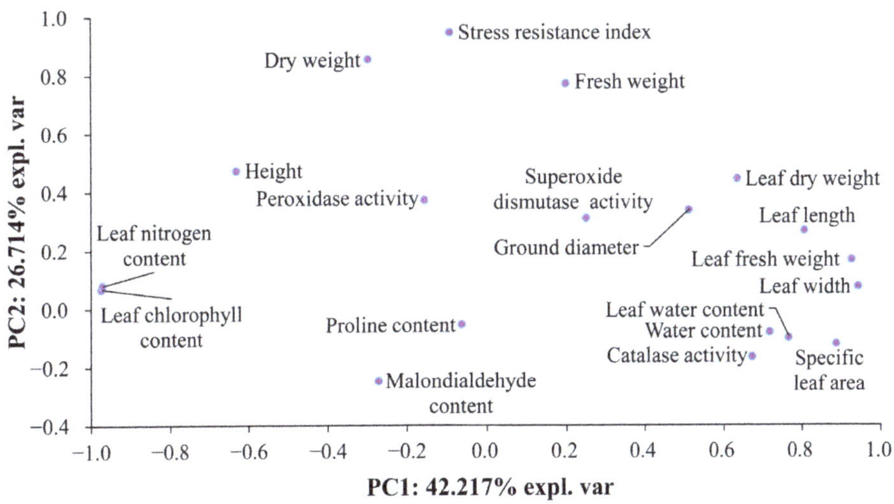

Figure 3. PCA of the correlation patterns of the values of the plants' functional traits, the levels of osmotic adjustment and the activities of antioxidant enzymes, and the stress resistance index of *A. spinosus* and *A. tricolor*. The *X*-axis and the *Y*-axis account for 42.217% and 26.714% of the total variation, respectively.

It is therefore of the utmost importance to halt or even stop the invasive process of *A. spinosus*, particularly under co-cultivation conditions when treated with drought, especially in the case of a light drought. The findings of this study also provide a substantial practical basis for the environmental management of *A. tricolor*, including the implementation of effective early-warning prevention and control measures, particularly in the context of a gradient of drought. It is of particular importance to modify the arid conditions in areas invaded by *A. tricolor*. This is to diminish the competitive advantage of *A. tricolor*, particularly in regard to its stress resistance, especially under the condition of a light drought.

In conclusion, this present study proposes a methodology for evaluating the stress resistance of a given plant species to stress by calculating the stress resistance index. This present study provides a robust theoretical basis for determining the stress resistance of a given plant species to stress. Moreover, this study offers a robust theoretical foundation for the environmental management of *A. tricolor* under drought.

Table 1. Correlations (r) between the stress resistance index of *A. spinosus* and *A. tricolor* and the values of the plants' functional traits, and the levels of osmotic adjustment and the activities of antioxidant enzymes in *A. spinosus* and *A. tricolor* (n = 3).

		Stress Resistance Index
Height	r	0.534 **
	p	**0.007**
Ground diameter	r	0.347
	p	0.097
Fresh weight	r	0.769 ***
	p	**<0.0001**
Dry weight	r	0.892 ***
	p	**<0.0001**
Water content	r	−0.146
	p	0.495
Leaf length	r	0.266
	p	0.210
Leaf width	r	0.019
	p	0.929
Specific leaf area	r	−0.201
	p	0.347
Leaf fresh weight	r	0.029
	p	0.893
Leaf dry weight	r	0.210
	p	0.324
Leaf water content	r	−0.076
	p	0.726
Leaf chlorophyll content	r	0.121
	p	0.573
Leaf nitrogen content	r	0.134
	p	0.533
Malondialdehyde content	r	−0.309
	p	0.142
Proline content	r	−0.123
	p	0.567
Catalase activity	r	−0.160
	p	0.454
Peroxidase activity	r	0.410 *
	p	**0.047**
Superoxide dismutase activity	r	0.315
	p	0.133

*, ** and *** indicate statistically significant differences at the 0.05, 0.01, and 0.001 probability levels, respectively. $p \leq 0.05$ is shown in bold.

Author Contributions: C.W.: Conceptualization; Methodology; Formal analysis; Funding acquisition; Project administration; Supervision; Writing—original draft; Y.L. (Yingsheng Liu): Methodology; Writing—review and editing; C.L.: Methodology; Writing—review and editing; Y.L. (Yue Li): Methodology; Writing—review and editing; D.D.: Funding acquisition; Project administration; Writing—review and editing. All authors have read and agreed to the published version of the manuscript.

Funding: This study was funded by Special Research Project of School of Emergency Management, Jiangsu University (KY-C-01), Open Science Research Fund of Key Laboratory of Forest Plant Ecology, Ministry of Education (Northeast Forestry University), China (K2020B02), National Natural Science Foundation of China (32071521), Carbon Peak and Carbon Neutrality Technology Innovation Foundation of Jiangsu Province (BK20220030), and Jiangsu Collaborative Innovation Center of Technology and Material of Water Treatment (no grant number).

Data Availability Statement: No new data were created or analyzed in this study. Data sharing is not applicable to this article.

Acknowledgments: We are extremely grateful to the anonymous reviewers for their most insightful and constructive comments and valuable editorial efforts, which have enabled us to improve the manuscript significantly.

Conflicts of Interest: The authors declare no conflicts of interest.

Abbreviations

IPS	Invasive plants
SRI	Stress resistance index
PCA	Principal component analysis
DLAs	Monocultural *A. spinosus* treated with a light drought
DLMixAs	Co-cultivated *A. spinosus* treated with a light drought
DHAs	Monocultural *A. spinosus* treated with a heavy drought
DHMixAs	Co-cultivated *A. spinosus* treated with a heavy drought
DLAt	Monocultural *A. tricolor* treated with a light drought
DLMixAt	Co-cultivated *A. tricolor* treated with a light drought
DHAt	Monocultural *A. tricolor* treated with a heavy drought
DHMixAt	Co-cultivated *A. tricolor* treated with a heavy drought

References

1. Rudolph, J.; Gornish, E.S.; Barberán, A. Plant–plant and plant–soil interactions under drought and the presence of invasive buffelgrass (*Cenchrus ciliaris*). *Biol. Invasions* **2024**, *26*, 1281–1293. [CrossRef]
2. Esperon-Rodriguez, M.; Power, S.A.; Tjoelker, M.G.; Marchin, R.M.; Rymer, P.D. Contrasting heat tolerance of urban trees to extreme temperatures during heatwaves. *Urban For. Urban Green.* **2021**, *66*, 127387. [CrossRef]
3. Zhong, S.S.; Xu, Z.L.; Cheng, H.Y.; Wang, Y.Y.; Yu, Y.L.; Du, D.L.; Wang, C.Y. Does drought stress intensify the allelopathy of invasive woody species *Rhus typhina* L.? *Trees* **2023**, *37*, 811–819. [CrossRef]
4. Werner, C.M.; Harrison, S.P.; Safford, H.D.; Bohlman, G.N.; Serata, R. Extreme pre-fire drought decreases shrub regeneration on fertile soils. *Ecol. Appl.* **2021**, *32*, e02464. [CrossRef] [PubMed]
5. Wang, S.; Wei, M.; Cheng, H.Y.; Wu, B.D.; Du, D.L.; Wang, C.Y. Indigenous plant species and invasive alien species tend to diverge functionally under heavy metal pollution and drought stress. *Ecotoxicol. Environ. Saf.* **2020**, *205*, 111160. [CrossRef] [PubMed]
6. Yu, Y.L.; Cheng, H.Y.; Wang, S.; Wei, M.; Wang, C.Y.; Du, D.L. Drought may be beneficial to the competitive advantage of *Amaranthus spinosus*. *J. Plant Ecol.* **2022**, *15*, 494–508. [CrossRef]
7. Ettinger, C.L.; Laforgia, M.L. Invasive plant species interact with drought to shift key functions and families in the native rhizosphere. *Plant Soil* **2024**, *494*, 567–588. [CrossRef]
8. NeSmith, J.E.; Alba, C.; Flory, S.L. Experimental drought and plant invasion additively suppress primary pine species of southeastern US forests. *For. Ecol. Manag.* **2018**, *411*, 158–165. [CrossRef]
9. Khanalizadeh, A.; Eshaghi Rad, J.; Amiri, G.Z.; Zare, H.; Schall, P.; Lexer, M.J. The relationship between plant diversity and aboveground biomass in managed and unmanaged temperate forests. *Eur. J. For. Res.* **2023**, *142*, 1167–1175. [CrossRef]
10. Wuest, S.E.; Schulz, L.; Rana, S.; Frommelt, J.; Ehmig, M.; Pires, N.D.; Grossniklaus, U.; Hardtke, C.S.; Hammes, U.Z.; Schmid, B.; et al. Single-gene resolution of diversity-driven overyielding in plant genotype mixtures. *Nat. Commun.* **2023**, *14*, 3379. [CrossRef]
11. Liang, M.; Baiser, B.; Hallett, L.M.; Hautier, Y.; Jiang, L.; Loreau, M.; Record, S.; Sokol, E.R.; Zarnetske, P.L.; Wang, S. Consistent stabilizing effects of plant diversity across spatial scales and climatic gradients. *Nat. Ecol. Evol.* **2022**, *6*, 1669–1675. [CrossRef] [PubMed]
12. Shumi, G.; Rodrigues, P.; Hanspach, J.; Härdtle, W.; Hylander, K.; Senbeta, F.; Fischer, J.; Schultner, J. Woody plant species diversity as a predictor of ecosystem services in a social–ecological system of southwestern Ethiopia. *Landsc. Ecol.* **2021**, *36*, 373–391. [CrossRef]
13. Easterling, D.R.; Meehl, G.A.; Parmesan, C.; Changnon, S.A.; Karl, T.R.; Mearns, L.O. Climate extremes: Observations, modeling, and impacts. *Science* **2000**, *289*, 2068–2074. [CrossRef] [PubMed]
14. IPCC. Climate Change 2013: The Physical Science Basis. In *Contribution of Working Group I to the Fifth Assessment Report of the Intergovernmental Panel on Climate Change*; Edenhofer, O., Pichs-Madruga, R., Sokona, Y., Farahani, E., Kadner, S., Seyboth, K., Adler, A., Baum, I., Brunner, S., Eickemeier, P., et al., Eds.; Cambridge University Press: Cambridge, UK, 2013.
15. Liu, K.; Jiang, D.B. Projected changes in the dry/wet climate of China under the RCP4.5 scenario. *Chin. J. Atmos. Sci.* **2015**, *39*, 489–502.
16. Jiang, J.; Jiang, D.B.; Lin, Y.H. Changes and projection of dry/wet areas over China. *J. Atmos. Sci.* **2017**, *41*, 43–56.
17. Reynolds, J.F.; Smith, D.M.S.; Lambin, E.F.; Turner, B.L.I.; Walker, B. Global desertification: Building a science for dryland development. *Science* **2007**, *316*, 847–851. [CrossRef] [PubMed]

18. Pointing, S.B.; Belnap, J. Microbial colonization and controls in dryland systems. *Nat. Rev. Microbiol.* **2012**, *10*, 551–562. [CrossRef]
19. Leal, R.P.; Silveira, M.J.; Petsch, D.K.; Mormul, R.P.; Thomaz, S.M. The success of an invasive Poaceae explained by drought resilience but not by higher competitive ability. *Environ. Exp. Bot.* **2022**, *194*, 104717. [CrossRef]
20. Beshai, R.A.; Truong, D.A.; Henry, A.K.; Sorte, C.J.B. Biotic resistance or invasional meltdown? Diversity reduces invasibility but not exotic dominance in southern *California epibenthic* communities. *Biol. Invasions* **2023**, *25*, 533–549. [CrossRef]
21. Sharma, A.; Kaur, A.; Kohli, R.K.; Singh, H.P.; Batish, D.R. *Bidens pilosa* (Asteraceae) invasion reshapes the pattern of plant communities and edaphic properties across the north-western Himalayan landscape. *Ecol. Inform.* **2023**, *77*, 102281. [CrossRef]
22. Savage, C.; Savage, K.; Keller, K.R. Effect of *Carpobrotus edulis* invasion history on plant communities. *West. N. Am. Nat.* **2023**, *83*, 484–497. [CrossRef]
23. Yan, J.; Yan, X.L.; Li, H.R.; Du, C.; Ma, J.S. Composition, time of introduction and spatial-temporal distribution of naturalized plants in East China. *Biodivers. Sci.* **2021**, *29*, 428–438. [CrossRef]
24. Yan, X.L.; Liu, Q.R.; Shou, H.Y.; Zeng, X.F.; Zhang, Y.; Chen, L.; Liu, Y.; Ma, H.Y.; Qi, S.Y.; Ma, J.S. The categorization and analysis on the geographic distribution patterns of Chinese alien invasive plants. *Biodivers. Sci.* **2014**, *22*, 667–676.
25. Wang, C.Y.; Liu, J.; Xiao, H.G.; Zhou, J.W.; Du, D.L. Floristic characteristics of alien invasive seed plant species in China. *An. Acad. Bras. Cienc.* **2016**, *88*, 1791–1797. [CrossRef]
26. Mandák, B.; Zákravský, P.; Dostál, P.; Plačková, I. Population genetic structure of the noxious weed *Amaranthus retroflexus* in Central Europe. *Flora* **2011**, *206*, 697–703. [CrossRef]
27. National Bureau of Statistics of China. *Zhenjiang Statistical Yearbook 2023*; China Statistics Press: Beijing, China, 2023.
28. *GB/T 20481-2017*; Grades of Meteorological Drought. National Standard of China. State Administration for Market Regulation of China and Standardization Administration of China: Beijing, China, 2017.
29. Balezentiene, L.; Marozas, V.; Miksa, O. Comparison of the carbon and water fluxes of some aggressive invasive species in baltic grassland and shrub habitats. *Atmos* **2021**, *12*, 12080969.
30. Shen, S.C.; Xu, G.F.; Li, D.Y.; Yang, S.S.; Jin, G.M.; Liu, S.F.; Clements, D.R.; Chen, A.D.; Rao, J.; Wen, L.L.; et al. Potential use of *Helianthus tuberosus* to suppress the invasive alien plant *Ageratina adenophora* under different shade levels. *Bmc Ecol. Evol.* **2021**, *21*, 85. [CrossRef] [PubMed]
31. Jiang, M.; Xuan, S.L.; Muneer, M.A.; Sun, B.; Shi, C.L.; Liu, F.S.; Wu, R.Y. Response of dry matter partition and yield components to waterlogging and sunlight shortage in different growth stages of wheat. *Nat. Hazards* **2022**, *110*, 1133–1152. [CrossRef]
32. Sharma, P.K.; Kumar, S. Soil Physical Productivity and Plant Growth. In *Soil Physical Environment and Plant Growth: Evaluation and Management*; Sharma, P.K., Kumar, S., Eds.; Springer International Publishing: Cham, Switzerland, 2023; pp. 1–32.
33. O'Leary, B.; Burd, M.; Venn, S.E.; Gleadow, R. Integrating the Passenger-Driver hypothesis and plant community functional traits to the restoration of lands degraded by invasive trees. *For. Ecol. Manag.* **2018**, *408*, 112–120. [CrossRef]
34. Aan, A.; Hallik, L.; Kull, O. Photon flux partitioning among species along a productivity gradient of an herbaceous plant community. *J. Ecol.* **2006**, *94*, 1143–1155. [CrossRef]
35. Gross, N.; Suding, K.N.; Lavorel, S.; Roumet, C. Complementarity as a mechanism of coexistence between functional groups of grasses. *J. Ecol.* **2007**, *95*, 1296–1305. [CrossRef]
36. Falster, D.S.; Westoby, M. Plant height and evolutionary games. *Trends Ecol. Evol.* **2003**, *18*, 337–343. [CrossRef]
37. Okubo, S.; Tomatsu, A.; Parikesit; Muhamad, D.; Harashina, K.; Takeuchi, K. Leaf functional traits and functional diversity of multistoried agroforests in West Java, Indonesia. *Agric. Ecosyst. Environ.* **2012**, *149*, 91–99. [CrossRef]
38. Funk, J.L. Differences in plasticity between invasive and native plants from a low resource environment. *J. Ecol.* **2008**, *96*, 1162–1173. [CrossRef]
39. Rijkers, T.; Pons, T.L.; Bongers, F. The effect of tree height and light availability on photosynthetic leaf traits of four neotropical species differing in shade tolerance. *Funct. Ecol.* **2000**, *14*, 77–86. [CrossRef]
40. Pyšek, P.; Richardson, D.M. Traits Associated with Invasiveness in Alien Plants: Where Do we Stand? In *Biological Invasions*; Nentwig, W., Ed.; Springer: Berlin/Heidelberg, Germany, 2007; pp. 97–125.

plants

MDPI

Review

Advances in Plant GABA Research: Biological Functions, Synthesis Mechanisms and Regulatory Pathways

Yixuan Hu [1,†], Xin Huang [1,†], Qinglai Xiao [1], Xuan Wu [1], Qi Tian [1], Wenyi Ma [1], Noman Shoaib [2], Yajie Liu [1], Hui Zhao [1], Zongyun Feng [1,*] and Guowu Yu [1,*]

1 State Key Laboratory of Crop Gene Exploration and Utilization in Southwest China, College of Agronomy, Sichuan Agricultural University, Chengdu 611130, China; 202200303@stu.sicau.edu.cn (Y.H.); 202200322@stu.sicau.edu.cn (X.H.); 202200313@stu.sicau.edu.cn (Q.X.); 202200312@stu.sicau.edu.cn (X.W.); 202200328@stu.sicau.edu.cn (Q.T.); 202200349@stu.sicau.edu.cn (W.M.); zhaohui_hbu@126.com (H.Z.)

2 Chengdu Institute of Biology, Chinese Academy of Sciences, Chengdu 610041, China; noman@cib.ac.cn

* Correspondence: 10036@sicau.edu.cn (Z.F.); 13862@sicau.edu.cn (G.Y.)

† These authors contributed equally to this work.

Abstract: The γ-aminobutyric acid (GABA) is a widely distributed neurotransmitter in living organisms, known for its inhibitory role in animals. GABA exerts calming effects on the mind, lowers blood pressure in animals, and enhances stress resistance during the growth and development of plants. Enhancing GABA content in plants has become a focal point of current research. In plants, GABA is synthesized through two metabolic pathways, the GABA shunt and the polyamine degradation pathway, with the GABA shunt being the primary route. Extensive studies have investigated the regulatory mechanisms governing GABA synthesis. At the genetic level, GABA production and degradation can be modulated by gene overexpression, signaling molecule-induced expression, transcription factor regulation, and RNA interference. Additionally, at the level of transporter proteins, increased activity of GABA transporters and proline transporters enhances the transport of glutamate and GABA. The activity of glutamate decarboxylase, a key enzyme in GABA synthesis, along with various external factors, also influences GABA synthesis. This paper summarizes the biological functions, metabolic pathways, and regulatory mechanisms of GABA, providing a theoretical foundation for further research on GABA in plants.

Keywords: γ-aminobutyric acid; GABA shunt; polyamine degradation pathway; glutamic acid decarboxylase; abiotic stress

1. Introduction

1.1. Physical and Chemical Properties

γ-aminobutyric acid (GABA) is a four-carbon non-protein amino acid with the molecular formula $NH_2(CH_2)_3COOH$, which was first discovered in potato tubers over 70 years ago [1,2]. GABA is a white, flaky, or needle-like crystal that is slightly soluble in hot ethanol, insoluble in most common organic solvents, and highly soluble in water. It is deliquescent and dissociates in aqueous solutions, typically existing as a zwitterion. GABA has a molecular weight of 103.12, an isoelectric point (pI) of 7.19, and a decomposition temperature of 202 °C. It is widely distributed in plants and animals, occurring in the seeds, rhizomes, and tissue fluids of legumes and herbs, while in animals, it is almost exclusively found in nervous tissue.

1.2. Biological Functions of GABA in Plants

GABA can act as a signaling molecule directly involved in the regulation of biotic and abiotic stresses. GABA triggers a defense response in plants when they experience biotic stress, particularly from insect attacks. Research has shown a positive correlation between increased GABA levels and enhanced resistance to such stress. For instance, plants with higher GABA content have demonstrated greater resistance than their wild-type counterparts [3]. Moreover, GABA inhibits neuronal transmission in insects, thereby serving as a defensive mechanism [4]. Additionally, GABA accumulation has been found to mitigate the toxic effects of certain compounds produced during insect feeding [5]. These findings indicate that GABA is crucial in reducing plant damage caused by biotic stress. GABA is a signaling molecule in plants able to enhance plant adaptation to abiotic stresses. A number of studies have shown that GABA can enhance plant resistance to hypoxia [6], salinity, drought [7], high temperature [8], and heavy metal stress [9]. In addition, specific concentrations of GABA can regulate the antioxidant system of plant growth, thus reducing oxidative damage in plants [10], GABA and proline in barley and tobacco leaves can increase their salt tolerance [11], GABA may be as a non-toxic permeability clear free radicals (ROS) produced under salt stress [12], which is consistent with the previous conclusion that GABA can inhibit the production of hydrogen peroxide under salt stress [13], GABA treatment, the germination rate of wheat and maize under salt stress is significantly increased [14]. In addition, experiments on morning glory pepper seeds under salt stress revealed that treatment with 6.0 μmol/L GABA improved both germination ability and salt tolerance [15]. In another study on rice seedlings, GABA treatment resulted in increasing plant height and leaf area, enhancing root strength and salt tolerance [16]. Additionally, different concentrations of GABA have been shown to alleviate the negative effects of drought stress on maize [17]. Research on durum wheat has further indicated that under drought conditions, GABA shunt metabolism is activated, which helps maintain carbon and nitrogen balance, regulate amino acid metabolism, and support plant growth [18].

In addition to its direct involvement as a signaling molecule in the regulation of the physiological state of plants, GABA can also interact with various plant hormones such as ABA, GA, auxin, CTK, and ethylene to help plants cope with abiotic stress [19]. For example, under salt stress, GABA application to *Caragana intermedia* and Poplar can promote the production of ABA and hydrogen peroxide while also regulating the production of ACS, ACO and ethylene [13]. Under drought stress, GABA can not only affect ALMT 9 channels but also promote the synthesis of PAs and inhibit the catabolism of PAs, which can greatly increase the content of different types of PAs in response to drought [20]. Conversely, addition of ABA and auxin under adversity could also increase GABA production [21]. In the face of stress and stress-related response and metabolism, The crosstalk between auxin and other phytohormones can all cause signaling interference from GABA, Overall, GABA and multiple phytohormones are mutually influencing the interacting [22], GABA increases the ethylene content by increasing the expression of genes such as ACC synthetase and ACC oxidase (ACO), Whereas, when the plant CK is deficient, The amount of LEA genes and glyoxylate reductase (GLYR) also decreases (GLYR is the enzyme involved in GABA catabolism) [23].

GABA also functions as a signaling molecule in plants, where it plays a key role in regulating stomatal movement. It has been observed that GABA inhibits ion channels in stomatal guard cells, thereby controlling stomatal opening, water loss, and drought tolerance [24]. In studies involving tomato plants, GABA was found to reduce stomatal conductance and diameter, promoting stomatal closure [25]. In *Arabidopsis*, guard cells produce GABA when reduced water, and GABA suppresses stomatal opening, a negative regulation of [26]. In addition to *Arabidopsis* and other plants also have this regulation, the reason is mainly with water, GABA increased, this signal through anion channel ALMT 9 transduction, and anion channel ALMT9 is the process of the main route [27], to reduce the stomatal opening, also reduce the transpiration water loss, improve the water use efficiency (WUE), enhance the drought resistance. Interestingly, GABA helped plants resist stress under both single and compound stress, but the relative GAD expression of compound stress was not consistent with stomatal movement compared to single stress. Balfagon et al., exposed to high light intensity (HL), heat stress (HS) and its combination (HL + HS), found increased transpiration and stomatal conductivity, which may be related to reduced water loss, lower leaf temperature and the limited photosynthesis rate of *Arabidopsis* itself [28]. Further research suggests that GABA may also be involved in intercellular signaling within stomata [29]. Additionally, GABA signaling is thought to regulate ion fluxes across the protective membranes of stomata, although this requires further investigation [30]. These findings underline the importance of GABA as a signaling molecule in stomatal regulation [31].

Furthermore, GABA has been shown to influence seed germination, enhancing germination rates under certain concentrations. Experiments with licorice seeds treated with varying concentrations of GABA demonstrated that a specific GABA concentration increased the content of seed respiratory metabolites and promoted the tricarboxylic acid cycle, both of which are beneficial for seed germination [32]. Similarly, treatment of soybean seeds with GABA led to an increase in germination potential and vigor index, likely due to a reduction in abscisic acid (ABA) content, an increase in ethylene levels, and the promotion of seed germination regulation by endogenous hormones [33]. Another study found that wheat seeds germinated more quickly under certain treatments, likely due to enhanced GABA shunt metabolism [34], indicating that the careful application of GABA can effectively promote seed germination.

2. Transporters of GABA

A variety of GABA transporters have been found on plant cell membranes and organelles. These transporters mainly include aluminum-activated malate transporters (ALMTs), GABA transporters (GATs), cationic amino acid transporters (CATs), AAP3, ProTs and bidirectional amino acid transporters (BATs). Transporters located in the cell membrane are ALMTs, GATs, AAP3 and ProTs, in addition, ALMTs appear to be recognized as a specific GABA receptor on the surface of plant cell membranes [35]. Transporters located in the organelle membrane are CATs and BATs. GABA is controlled by shuttling across cell membranes and organelles, entering and exiting cells for metabolism [36–38]. The following is an introduction to the functions and mechanisms of these transporters.

2.1. Transporters on Cell Membranes

2.1.1. ALMTs

Aluminum-activated malate transporters (ALMTs) are bidirectional transmembrane anion transporters [39], 12 ALMT family genes have been found in plants [40], and ALMT1 has been found in rice, wheat, rape, *Arabidopsis* and other plants. In 2018, Ramesh found that ALMT1 on plant cell membranes can efficiently transport GABA [39], and in addition, ALMTs respond to various signals, are activated by anions, and negatively regulate GABA [35].The specific transport mechanism of ALMTs is as follows: Previous studies on GABA and malate have shown that anions can activate ALMT1 [41], and H^+-ATPase can input and export amino acids to cells through the protons produced by the plasma membrane, resulting in a potential difference between inside and outside the membrane. When the intracellular pH is low, aluminum ions promote efflux of GABA through ALMT1. When the extracellular are acidified, GABA was transfered into inside of cell through ALMT1. These studies have shown that pH can affect the potential difference between inside and outside the membrane, and then affect the direction of ALMT1's transport of GABA [42–44]. At present, ALMT family genes have been cloned and identified successively, and their protein sequences and transport mechanisms have been analyzed, and ALMT1 is still the most studied. It is generally confirmed that when GABA content increases, the activity of ALMT decreases, that is, GABA negatively regulates the activity of ALMT [42]. Yu Long's study proved that GABA inhibits the transport of anions in wheat by changing the active structure of ALMT1, and this conformational change is similar to the conformational transformation of aspartate aminotransferase (AAT) [45]. At present, there are few studies on regulating the activity of ALMTs to regulate the transport of GABA, and most of them remain in the exploration of the transport mechanism. However, the interaction between

GABA and ALMT can be used as a signal molecule within the implant to regulate the transmembrane transport of GABA.

2.1.2. GATs

GATs is a class of membrane transmembrane transporter protein, GAT gene belongs to AAAP gene family, and four GAT homologous genes (GAT1, GAT2, GAT3, GAT4) have been found in plants [36,46]. At present, GAT1, which is located on the cell membrane, can transport GABA across the membrane and flow between the apoplast and the cytoplasm. Both GAT1 and ALTM1 are transmembrane transporters of GABA, but it is worth noting that Al^{3+} blocks the influx of GABA from apoplast to the cytoplasm via ALMT1, but it has no effect on GAT1 [42]. Up to now, GAT1 has been cloned in potato, *Arabidopsis*, rice and other crops, and related studies have been carried out in *Arabidopsis*. AtGAT1 in *Arabidopsis thaliana* transports GABA across membranes through proton coupling, which is mainly driven by the proton electrochemical potential gradient and the high affinity of AtGAT1 for GABA [46].

2.1.3. AAP3 and ProTs

At present, two GABA transporters (AAP3 and ProT2) on the cell membrane have been identified by heterologous recombination in yeast, but the affinity of these two proteins to GABA is relatively low [47–49]. Regarding the transport function of OsProTs family proteins in rice, it has been reported that proline transporters not only have a transport function for Pro, but also have a transport function for GABA. OsProT2 is heterologous expressed in Xenopus oocytes. It is found that OsProT2 is a proton cotransporter and has specific transport function for L-Pro, but whether it transports GABA remains to be studied [50,51]. In *Saccharomyces cerevisiae* mutants with amino acid transport defects, OsProT1 and OsProT3 can specifically absorb Pro and GABA, and OsProT3 has a stronger transport capacity for Pro than OsProT1, while OsProT2 has no transport properties for various amino acids. This may be due to the inability of OsProT2 to localize to the plasma membrane in yeast [52]. In summary, ProTs and AAP3 can transport GABA, but the affinity is not strong, and its related causes and regulatory factors need to be further studied.

2.2. *Transporters on Organelle Membranes*

2.2.1. CAT9

CATs is a cationic transporter located on the vacuole membrane and belongs to the APC gene family [53–55]. At present, nine CAT genes have been found in plants, among which the transporter encoded by CAT9 is mainly responsible for the bidirectional transport of GABA between vacuole and cytoplasm. At present, CAT9 has been found in tomato, potato, *Arabidopsis*, rice and other crops [53,55], and the transport dynamic mechanism of CAT9 has been verified in tomato. SlCAT9 transport in tomato mainly has two driving forces: one is the proton driving force of chloroplast proton pump, and the other is the chemical potential difference caused by the concentration gradient of substrate [56]. Interestingly, in addition to GABA, SlCAT9 also transports protein amino acids such as glutamic acid and aspartic acid, but the transport conditions are relatively strict. Glutamic acid is the substrate for GABA synthesis, and aspartic acid can also be converted into glutamic acid, so it is speculated that SlCAT9 co-regulates GABA transformation with these amino acids [57]. Studying the co-transport mechanism and regulatory factors of CAT9 will be a promising direction, which can effectively regulate GABA transport. Relevant studies have shown that some external factors may affect the expression of CATs, such as supplemental nitrogen sources and stress. For example, when tea leaves are fed with different nitrogen sources or under stress conditions, the expression of CsCAT9 will be inhibited or induced, thus inhibiting or promoting the transport of GABA. The content of glutamate will increase under nitrogen treatment or stress conditions, which will also promote the synthesis of GABA [58]. AtCAT9 in *Arabidopsis thaliana* can directly or indirectly balance intracellular amino acid concentration in the absence of leaf nitrogen [59]. In addition, the expression level of CATs is also different in different parts, and the expression level of AtCAT9 in *Arabidopsis thaliana* is relatively high in root, stem, leaf and flower tissues, indicating that GABA transport is relatively active in these parts [59].

2.2.2. BAT1

BAT is a bidirectional transmembrane protein on mitochondrial membrane. Seven homologous genes have been identified in plants, belonging to the APC gene family [60,61], in which the transporter encoded by BAT1 can transport amino acids. AtGABP in *Arabidopsis thaliana* is a splice variant of AtBAT1 that can transport GABA [62]. Unlike AAP3 and ProT2, AtGABP cannot transport proline, and the co-expression of AtGABP gene is highly correlated with the expression of *SSADH* gene. It can be speculated that AtGABP is associated with some reactions of GABA metabolism, and may jointly participate in the regulation of GABA shingling and TCA cycle [62]. It is worth noting that AtBAT1 in *Arabidopsis* can transport a variety of amino acids such as arginine and lysine, but has no transport activity for GABA [37]. At present, there is relatively little research on *BAT* genes, and the transport mechanism of BAT transporter and its splicing variant GABP will be a hot direction.

3. Metabolism of GABA in Plants

There are two primary pathways for GABA synthesis in plants: the GABA shunt and the polyamine degradation pathway. The GABA shunt is the predominant pathway for GABA synthesis in most plants, while the polyamine degradation pathway remains less explored. GABA synthesis has been extensively studied in plants such as tomatoes [63], rice [64], watermelon [65], quinoa [66], wheat [67], and corn [68] but research on barley remains limited.

3.1. GABA Shunt

The GABA shunt, also known as the GABA bypass, is the major pathway for GABA synthesis in most higher plants [69]. In the cytoplasm, where there is a high concentration of H^+ ions, glutamic acid (Glu) undergoes an irreversible reaction catalyzed by the enzyme cytosolic glutamate decarboxylase (GAD), leading to the production of GABA.

There are two main pathways for glutamate synthesis in plants (Figure 1). The first pathway involves glucose undergoing glycolysis, Under the catalysis of pyruvate dehydrogenase system, pyruvate generates acetyl-Coenzyme A in the mitochondrial matrix, and then enters the tricarboxylic acid cycle (Kreb's cycle, also known as TCA). The citrate in TCA passes under the catalysis of a series of enzymes to generate the intermediate product α-ketoglutaric acid. This α-ketoglutaric acid can then be catalyzed by glutamate dehydrogenase (GDH). Glutamate dehydrogenase (GDH) can be divided into three types according to their coenzymes: NADH-dependent, NADPH-dependent and NADH/NADPH-dependent [70], NADH-dependent GDH is usually involved in glutamate catabolism, while NADPH-dependent GDH requires nitrogen assimilation, and dual-dependent GDH is usually in mammals [71]. The second pathway involves the plant's response to an excess of free ammonia within the cell, which can be toxic. In this case, glutamine synthetase (GS) catalyzes the reaction of free ammonia with glutamate to form glutamine. This glutamine is then catalyzed by glutamate synthetase (GOGAT) in the presence of α-ketoglutarate, regenerating two molecules of glutamate. Glutamate synthase (GOGAT) in plants can also be divided into two types of coenzyme: Fd-GOGAT and NADH-GOGAT, the former mainly exists in green tissues such as chloroplast and plastid, while the latter mainly exists in non-green tissues such as nodules and stems, participating in nitrogen fixation, and the specific site of the reaction is in the mitochondria [72].The GS/GOGAT cycle is the primary pathway for the assimilation of ammonia and the synthesis of amino acids in plants [73].

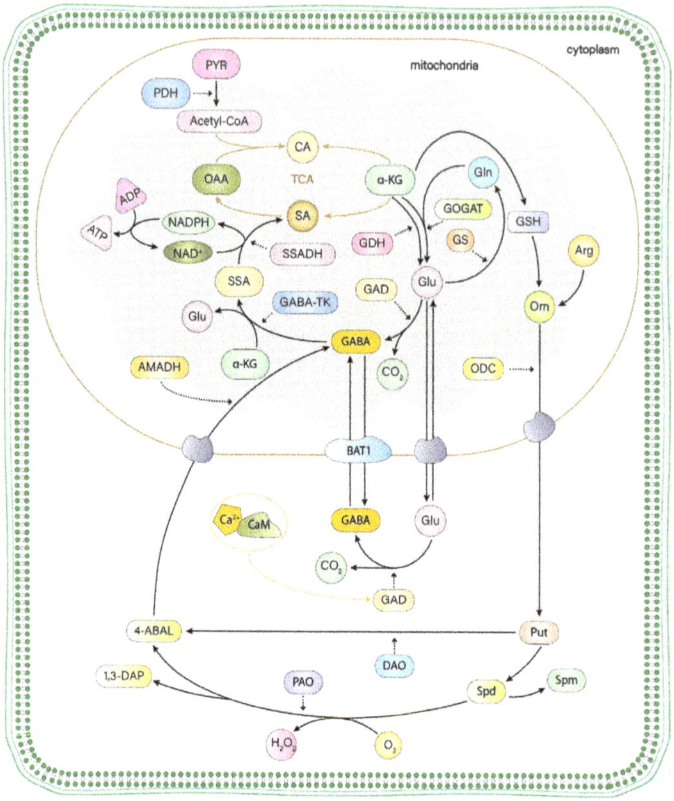

Figure 1. Metabolism of GABA in plants, PDH: Pyruvate Dehydrogenase, Acetyl-CoA: Acetyl Coenzyme A, CA: Citric Acid, OAA: Oxaloacetic Acid, α-KG: α-Ketoglutaric Acid, SA: Succinic Acid, SSA: Succinate Semialdehyde, NADPH: Nicotinamide Adenine Dinucleotide Phosphate, NAD^+: Nicotinamide Adenine Dinucleotide, SSADH: Succinate Semialdehyde Dehydrogenase, Glu: Glutamic Acid, Gln: Glutamine, GS: Glutamine Synthase, GOGAT: Glutamate Synthetase, GDH: Glutamate Dehydrogenase, GSH: Glutathione, Orn: Ornithine, Arg: Arginine, ODC: Ornithine Decarboxylase, AMADH: Amino Acid Dehydrogenase, Put: Putrescine, Spm: Spermine, Spd: Spermidine, PAO: Polyamineoxidase, DAO: Diamine Oxidase, CaM: Calmodulin, GAD: Glutamate Decarboxylase, GABA: γ-aminobutyric Acid, BAT1: GABA Bidirectional Transmembrane Protein, GABA-TK: GABA-transaminase, 4-ABAL: 4-aminobutanal, 1,3-DAP: 1,3-diaminopropane.

After GABA formation by glutamate under the glutamate decarboxylase, the generated GABA is transferred from the cytosol to the mitochondria through the BAT1. GABA and α-ketoglutarate are catalyzed to glutamate (Glu) and succinate semialdehyde (SSA) by GABA transaminase (GABA-TK). Succinate semialdehyde is catalyzed to succinate by succinate semialdehyde dehydrogenase (SSADH), and succinate eventually entering the TCA cycle to be degraded (Tricarboxylic acid, TCA) [21].

3.2. Polyamine Degradation Pathway

The polyamine degradation pathway, also known as the putrescine degradation pathway, is primarily responsible for GABA synthesis in legumes, with less activity observed in grasses [74]. Additionally, it has been found that this pathway contributes to GABA synthesis in plants under stress conditions [75]. Polyamines (PAs) in plants include putrescine (Put), spermine (Spm), and spermidine (Spd). Putrescine is mainly generated through the decarboxylation of ornithine, although it can also result from spermidine decarboxylation. Once formed, putrescine is further catalyzed by diamine oxidase (DAO) to produce 4-aminobutanal, which is then transformed into GABA by aminobutyraldehyde dehydrogenase (AMADH). The synthesized GABA can enter the tricarboxylic acid cycle (TCA cycle) through the catalytic actions of GABA transaminase (GABA-T) and succinic semialdehyde dehydrogenase (SSADH) [76]. Putrescine can also be converted into spermidine and spermine through reactions catalyzed by spermidine synthase and spermine synthase respectively. Spermidine, in turn, can be catalyzed by polyamine oxidase (PAO) to produce 4-aminobutyraldehyde, which further produces GABA [74].

4. GABA Regulation

Based on the synthesis and metabolic pathways of GABA, its regulation can be categorized into three main aspects: gene-level regulation (Figure 2A), regulation of key enzyme activities (Figure 2A), and regulation by external factors (Figure 2B).

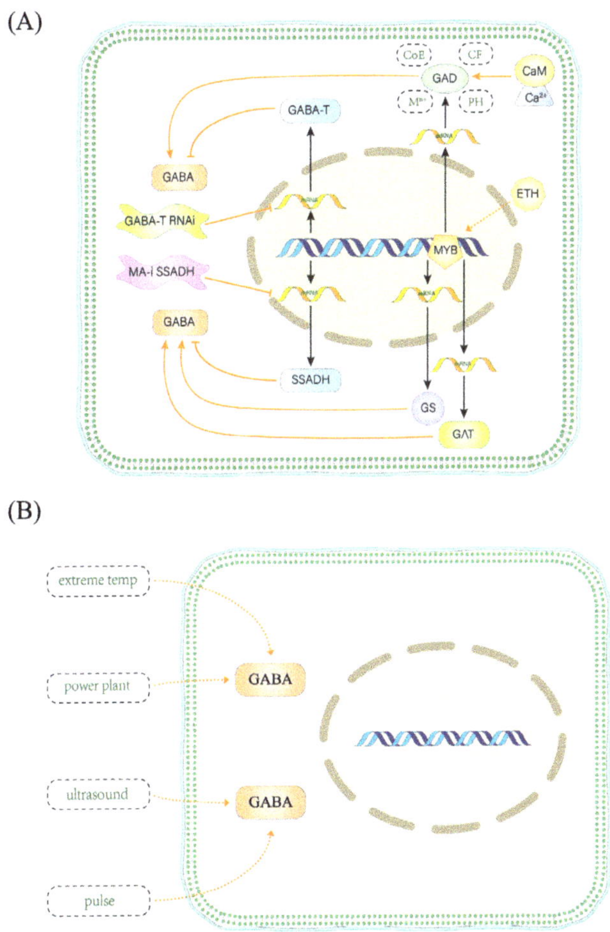

Figure 2. (**A**) Regulation of GABA at the gene and protein levels; (**B**) Regulation of GABA by physical factors. GABA: γ-aminobutyric acid, GAT: Transferase of glutaminase, GATs: Transferase of glutaminase family, ProTs: Proline transferase family, Temp: Temperature, GABA-T: gamma-aminobutyric acid transferase, Pro: proline GS: Transferase of glutaminase, SSADH: succinate semi-aldehyde dehydrogenase,

GABA-T IFN: GABA-T interferon, MA-iSSADH: SSADH gene RNA interference Strain, M^{n+}: Metal ion, CaM: calmodulin, Ca^{2+}: Calcium ion, GAD: glutamate decarboxylase, PH: pH, CoE: Coenzyme CF: Cogroup, ETH: ethylene, MYB: A kind of transcription factor in plants.

4.1. Gene Level Regulation

GAD (glutamate decarboxylase) genes have been identified in a variety of plants (Table 1). Alterations in the structure of some genes, as well as the regulation of transcription factors, influence GABA production. In *Arabidopsis*, there are five GAD family genes: *GAD1*, *GAD2*, *GAD3*, *GAD4*, and *GAD5* [77]. These genes exhibit differential expression across various parts of the plant. For instance, *GAD1* is highly expressed in roots, *GAD2* shows very high expression in seedlings after two weeks and in carpels, buds, wreaths, roots, and stems, while *GAD4* is predominantly expressed in pollen tubes. *GAD3* and *GAD5* have lower expression levels in the same tissues [78–80]. A common feature of these *GAD* family genes is the presence of CAAT-box, TATA-box, and MYC elements in their promoters and they could be binded by transcription factors to regulate the expression of *GAD* genes [81]. Myelocytomatosis oncogenes (MYC) transcription factors (TFs) belong to the bHLH (basic helix-loop-helix) family, which play a central role in plant growth, development, adaptation to biotic and abiotic stress, as well as secondary metabolism [82–85]. At present, They have been mainly studied in rice, wheat and Arabidopsis [86].

Table 1. Number of GAD genes in different plants.

Assortment	Numbers of GADs	Authors of the Reference	Years of Publication
Arabidopsis	5	Shelp, B.J., et al.	2017 [77]
Citrullus lanatus	3	Li, M., et al.	2023 [65]
Oryza sativa Janica	5	Zhou, L.	2015 [64]
Solanum lycopersicum	5	Sun, X., et al.	2022 [63]
Chenopodium quinoa wild	8	Li, L.	2022 [66]
Zea mays L.	5	Wang, Y.	2023 [68]

In addition, MYB transcription factors were involved in regulating various stresses in plant life cycle, and OsMYB55 belongs to one of the MYB transcription factors, which can promote amino acid metabolism during plant growth under high temperature to improve rice yield [87]. In rice, the transcription factor OsMYB55 binds to the promoter regions of target genes, directly activating the expression of several genes, including glutamine synthetase (Os*GS1*, Os*GS2*), glutamine amidotransferase (*GAT1*), and glutamate decarboxylase 3 (*GAD3*) [87]. Overexpression of *OsMYB55* in rice lines resulted in increased GABA content in leaves, especially under high-temperature conditions, with a more obvious effect in transgenic plants compared to wild-type plants. Additionally, total amino acid content increased following the overexpression of *OsMYB55*. Transcriptome analysis identified Os*GS2*, *GAT1*, and *GAD3* as potential targets of *OsMYB55*. This transcription factor has been shown to bind to the promoters of these genes to activate them [87].

In bananas, exogenous ethylene was found to significantly induce the expression of Ma*GAD1*, with regulation primarily occurring at the transcriptional level. Research in *Arabidopsis* revealed that Ma*GAD1* enhances the exogenous ethylene sensing capacity of the upstream components of the ethylene signaling pathway, particularly through the regulation of *ACS4*, a member of the ACC synthase family involved in ethylene biosynthesis [88]. This regulation impacts the endogenous ethylene biosynthesis pathway in *Arabidopsis*. In poplar, members of the *GAD* gene family may be induced by anaerobiosis, low temperature, and drought, as their regulatory elements correspond to these conditions. Poplar also contains two GABA-T family genes with elements related to light response, anaerobic induction, and abscisic acid. Additionally, *PopGABA-T1* includes elements associated with meristem organization and cyclic regulation. Poplar has two SSADH gene families, both of which include elements related to light and gibberellin responses, and *PopSSADH1* also has elements involved in anaerobic induction, defense, and stress response [89]. By regulating these elements and the local environment of the plant, we can regulate GABA metabolism and promote the synthesis of GABA to increase its content.

Inhibiting the expression of *SSADH* and *GABA-T* genes is an effective approach to increase GABA content in plants by reducing GABA degradation (Figure 2A). For instance, RNA interference technology has been employed to construct an *SSADH* gene RNA interference strain (MA-i SSADH), leading to a significant decrease in *SSADH* gene transcript levels compared to the prototrophic strain. In rice, an RNA interference vector targeting the GABA transaminase 1 (*OsGABA-T1*) gene was developed and successfully transformed into the japonica rice variety "Ning Japonica 1" using Agrobacterium-mediated transformation. This method effectively suppressed the expression of *OsGABA-T1* and its related gene *OsGABA-T2* [90]. In conclusion, the expression of *GABA-T* and *SSADH* was inhibited by RNA interference, thus inhibiting GABA degradation and increasing its content.

4.2. *Effect of GAD Enzyme Activity on GABA Synthesis*

GABA production in plants and microorganisms primarily occurs through the GABA shunt, a pathway catalyzed by GAD. It is the key enzyme in this pathway, and its activity can be enhanced by optimizing various factors, including temperature, pH, Ca^{2+}/CaM, the coenzyme pyridoxal phosphate (PLP), and metal ions (Figure 2A).

4.2.1. Temperature Regulation of GAD

The optimal temperature for GAD activity varies significantly across different species, generally ranging from 30 °C to 50 °C in plants and microorganisms [91]. For example, the optimal temperature for GAD in bananas during ripening is 37 °C, with GAD activity decreasing by 23% at 25 °C and by 39% at 55 °C [92]. In potatoes, the optimal temperature for GAD activity is also 37 °C [93]. In lactic acid bacteria, 52 °C is identified as the optimal temperature for GAD activity [94]. Meanwhile, in pumpkin, the optimal reaction temperature for GAD is 30–35 °C; however, the enzyme is heat-sensitive and becomes inactive after short exposure to temperatures above 50 °C [95]. These findings indicate that the optimal conditions for GAD activity vary widely among species, suggesting that adjusting the reaction temperature can significantly improve GAD activity.

4.2.2. The pH Regulation of GAD

The optimal pH for GAD activity varies depending on its source. Typically, GAD from microbial origins operates best within a pH range of 3.8 to 5.5 [96]. For example, the optimal pH for GAD in *E. coli* is between 3.8 and 4.5 [97], while in *Lactobacillus* it is between 4.0 and 5.0 [98], and in *Bacillus megaterium*, it is 4.5 to 5.0 [99]. For plant-derived GAD, the optimal pH is generally weakly acidic, mainly between 5.5 and 6.0. For instance, the optimal pH in soybeans is 5.9. Although soybean GAD can retain some activity between pH 5.0 and 8.0, its activity decreases sharply at pH levels above 8.0, with an 83% reduction at pH 9.0 [100]. Similarly, the optimal pH for GAD in corn embryos is 5.7, with a dramatic decline in activity at pH levels below 4.0 or above 8.0, showing an 80% decrease at pH 9.0 [101]. In rice embryos, the optimal pH for GAD is between 5.5 and 5.8 [102]. These studies demonstrate that pH significantly influences GAD activity, and adjusting the pH of the enzyme reaction can enhance GAD activity, thereby increasing GABA content in plants.

4.2.3. Pyridoxal Phosphate of GAD

The PLP is a key coenzyme for GAD, enhancing GABA synthesis by promoting the decarboxylation of glutamate. Research has shown that GAD from barley embryo extracts, with a Km value of 22 mmol/L for L-glutamate, is activated 3.5-fold in the presence of pyridoxal phosphate. Further studies revealed that GAD in barley embryos exists in two molecular weight forms: a low molecular weight form, which is relatively inactive, and a high molecular weight form, which becomes more active upon storage. The presence of

2-mercaptoethanol, a thiol reagent, shifts the distribution of enzyme activity toward the low molecular weight form. However, once 2-mercaptoethanol is removed, the enzyme spontaneously binds to the high molecular weight form, increasing its activity. Additionally, the presence of oxygen in the extraction buffer promotes the formation of the higher molecular weight form, enhancing enzyme activity. In contrast, GAD from 3-day-old barley root extracts had a Km value of 3.1 mmol/L for L-glutamate and exhibited a 10% increase in activity with the addition of pyridoxal phosphate [103]. These findings suggest that pyridoxal phosphate can significantly boost GAD activity in plants, and supplementing this coenzyme in enzyme reactions can enhance both GAD activity and GABA production.

4.2.4. Metal Ion Regulation of GAD

GAD activity is also influenced by various metal ions. In maize embryos, metal ions such as Mg^{2+}, Mn^{2+}, Cu^{2+}, Al^{3+}, Ag^{+}, and Zn^{2+} do not significantly affect GAD activity at a concentration of 2 mmol/L. However, Ca^{2+} has a notable impact. When the Ca^{2+} concentration is below 400 μmol/L, there is a significant activation effect, with the most pronounced increase in GAD activity (31%) observed at 400 μmol/L compared to untreated groups [101]. Similarly, in soybean, Mg^{2+} at 2 mmol/L has little effect, resulting in only a 3.5% decrease in enzyme activity. KCl causes a slight decrease of about 7%, while KI and Ag^{+} have a more substantial inhibitory effect, reducing enzyme activity by approximately 13% [100]. In rice bran, Mg^{2+}, Mn^{2+}, and Al^{3+} do not significantly affect enzyme activity, while KCl has a slight effect. However, KI and Ag^{+} significantly inhibit enzyme activity [104]. These findings suggest that certain metal ions, particularly Ag^{+} and KI, consistently inhibit GAD activity across various plant species. Reducing the presence of these ions could help prevent the inhibition of GAD activity and enhance GABA production in plants.

4.2.5. Ca^{2+}/CaM Regulation

GAD in plants is a calmodulin (CaM)-binding protein and its activity is regulated by the Ca^{2+}/CaM complex. The calmodulin-binding region is located at the C-terminal end of GAD, where Ca^{2+}/CaM acts as an activator, significantly increasing GAD activity [105]. Truncating the C-terminus of rice OsGAD2 results in an enzyme with higher activity than the wild-type at any pH, with over 40-fold higher activity at physiological pH [105]. In mung beans, calmodulin significantly enhances the relative enzyme activity of GAD at certain concentrations [106]. Studies in Petunia show that the Ca^{2+}/CaM complex must bind to the calmodulin-binding domain of GAD to form a complex that promotes GABA synthesis. Additionally, in a pea and buckwheat enzyme solution, the amount of GABA synthesis increases as Ca^{2+} concentrations rise from 0 to 0.8 mmol/L, reaching the highest GABA production at 0.8 mmol/L. Beyond this concentration, from 0.8 to 1.0 mmol/L, GABA synthesis decreases, possibly due to the inhibitory effects of excessively high Ca^{2+} levels on GAD activity [107]. These findings indicate that by optimizing the concentration of Ca^{2+}/CaM during GAD enzyme reactions, it is possible to enhance GAD activity and subsequently increase GABA production.

5. External Factors Regulating GABA

5.1. *Temperature Treatment*

Extreme temperature treatment can increase the content of GABA in plants (Figure 2B). Temperature stress in plants is categorized into high-temperature stress and low-temperature stress. Low-temperature stress includes both cold damage and freeze damage. Freeze damage occurs when ice forms within plant tissues at temperatures below 0 °C, leading to the formation of large ice crystals that severely damage the cell membrane system. This damage allows Ca^{2+}, H^{+}, and other ions to penetrate deeper into the cytoplasm, where they bind to the Ca^{2+}/CaM binding domain in glutamic acid decarboxylase (GAD). This binding increases the expression of calmodulin, which activates GAD to catalyze the conversion of L-glutamate to GABA. The disruption of the cellular membrane system also facilitates the exchange of substrates between cells, allowing more L-glutamate to bind with GAD, resulting in increased GABA accumulation [108]. For example, low-temperature treatment of germinated maize kernels at −18 °C significantly increased GABA content, peaking at 1.38 mg/g—2.3 times higher than in untreated kernels. In contrast, maize kernels exposed to 5 °C without frost damage showed the least increase in GABA content, with only a 0.91-fold increase compared to untreated kernels [109]. High-temperature stress occurs when ambient temperatures exceed the optimal range for plant growth, leading to physiological and biochemical disruptions. This stress can be categorized into short-term heat stress and long-term desiccation. High temperatures cause an accumulation of reactive oxygen species (ROS) within cells, increased membrane lipid peroxidation, and electrolyte leakage, all of which impair photosynthesis and alter antioxidant enzyme activity [110]. Numerous studies have demonstrated that GABA enhances plant stress tolerance by stabilizing cell membrane structures, regulating cytoplasmic pH, inducing ethylene production, reducing ROS damage, and promoting the synthesis of biomolecules [111]. For instance, under prolonged drying treatment at 42 °C, the GABA content in maize seedlings significantly increased, peaking at 48 h before rapidly declining [112]. Similarly, in tea plants exposed to day/night temperatures of 42 °C/40 °C, endogenous GABA levels rose significantly, increasing by 75.56%, 88.99%, and 60.13% at 4, 8, and 24 h, respectively, before also decreasing [2].

5.2. Other Treatments

It has been found that in the pre-germination period of brown rice, with the help of ultrasonic treatment, the use of soaking germination can increase the GABA content in brown rice (Figure 2B). With the prolongation of germination time, GAD activity had a tendency to increase, then decrease and then increase [113]. Similarly, the combination of soaking buckwheat in EFW (electric field water) with high-voltage electric field (HVEF) treatment has been shown to significantly increase GABA accumulation during germination (Figure 2B). At 24 h, the combined treatment reached a peak GABA content of 198.72 ± 0.75 mg/100 g, showing a significant difference ($p < 0.05$) compared to other treatment groups [114]. This suggests that the combined use of EFW soaking and HVEF treatment is more effective at enhancing GABA content in sweet buckwheat than either treatment alone. Pulsed electric field (PEF) treatment is another method that influences membrane permeability, enhancing the interaction between substrates and enzymes, and thereby promoting GABA synthesis through electroporation (Figure 2B). The effectiveness of PEF treatment depends on carefully designed factors, including pulse intensity, pulse duration, and pulse type [115].

6. Summary and Outlook

GABA plays an important role in the resistance to biotic and abiotic stresses. The main GABA transporters on the cell membrane are ALMTs and GATs, among which ALMTs is also considered to be GABA receptors. CAT9 is a cationic transporter, which can also participate in GABA transport. BAT1 is a bidirectional GABA transporter on mitochondria, and all the above transporters can participate in GABA transport. The metabolism of GABA in plants primarily occurs through the GABA shunt and the polyamine degradation pathway. Regulation of key enzymes within these pathways is likely crucial for increasing GABA content in plants, with the GABA shunt being the dominant metabolic route in most plants and the primary target for regulation. Enhancing GABA in plants can be approached in several ways. First, the key enzyme genes involved in GABA synthesis can be regulated through gene overexpression, transcription factor manipulation, or RNA interference technology to enhance the expression of GAD or inhibit the expression of GABA-T. Second, identifying and optimizing the conditions that maximize GAD enzyme activity can further promote GABA synthesis. Additionally, external factors such as extreme temperature treatment, ultrasonic treatment, and other innovative methods have been extensively studied and can be used to regulate GABA levels. Moreover, regulating glutamate synthesis by targeting key enzymes in nitrogen metabolism, such as GOGAT and GS, may increase glutamate content, thereby providing more substrate for the GABA shunt and promoting GABA synthesis. Although GAD genes are part of multigene families, further research is required to understand the forms of GAD present in plants, as well as the gene interactions and localization of the main effector genes within these families. In particular, research on the GAD gene family in crops like barley, to identify optimal conditions for GAD enzyme activity, represents a substantial direction for future studies.

Author Contributions: G.Y. and Y.H. designed the review. Y.H., Q.X. and X.H. analyzed the data and wrote the manuscript. X.W., Q.T., W.M., N.S., Y.L. and H.Z., improve the figures and manuscript writing. Z.F. and G.Y., read and approved the contents. All authors have read and agreed to the published version of the manuscript.

Funding: This work was supported by the China Agriculture Research System (CARS-05-10B), Natural Science Foundation of Sichuan province in China (2024NSFSC0325) and China Postdoctoral Science Foundation (2024M752292).

Data Availability Statement: Not applicable.

Acknowledgments: We would like to express our gratitude to Nishbah Mughal for taking time from their busy schedule to enhance the language and overall robustness of the manuscript.

Conflicts of Interest: The authors declare that the review was written in the absence of any commercial or financial relationships that could be construed as a potential conflict of interest.

References

1. Li, L.; Dou, N.; Zhang, H.; Wu, C. The versatile GABA in plants. *Plant Signal. Behav.* **2021**, *16*, 1862565. [CrossRef] [PubMed]
2. Ren, T. Effect of Exogenous γ-Aminobutyric Acid on Physiological Metabolism of Tea Plants under Heat-Stress Condition. Master's Thesis, Nanjing Agricultural University, Nanjing, China, 2021.
3. Li, N.; Li, T.; Hou, C.; Wang, D. Role of γ-aminobutyric acid in plant resistance to stress. *Plant Physiol. J.* **2020**, *56*, 134–140. [CrossRef]
4. Tarkowski, Ł.P.; Signorelli, S.; Höfte, M. γ-Aminobutyric acid and related amino acids in plant immune responses: Emerging mechanisms of action. *Plant Cell Environ.* **2020**, *43*, 1103–1116. [CrossRef] [PubMed]
5. Mirabella, R.; Rauwerda, H.; Struys, E.A.; Jakobs, C.; Triantaphylidès, C.; Haring, M.A.; Schuurink, R.C. The Arabidopsis her1 mutant implicates GABA in E-2-hexenal responsiveness. *Plant J.* **2008**, *53*, 197–213. [CrossRef] [PubMed]
6. Yang, R.; Guo, Y.; Wang, S.; Gu, Z. Ca^{2+} and aminoguanidine on γ-aminobutyric acid accumulation in germinating soybean under hypoxia-NaCl stress. *J. Food Drug Anal.* **2015**, *23*, 287–293. [CrossRef] [PubMed]
7. Li, W.; Liu, J.; Ashraf, U.; Li, G.; Li, Y.; Lu, W.; Gao, L.; Han, F.; Hu, J. Exogenous γ-aminobutyric Acid (GABA) Application Improved Early Growth, Net Photosynthesis, and Associated Physio-Biochemical Events in Maize. *Front. Plant Sci.* **2016**, *7*, 919. [CrossRef]
8. Liu, T.; Liu, Z.; Li, Z.; Peng, Y.; Zhang, X.; Ma, X.; Huang, L.; Liu, W.; Nie, G.; He, L. Regulation of Heat Shock Factor Pathways by γ-aminobutyric Acid (GABA) Associated with Thermotolerance of Creeping *Bentgrass*. *Int. J. Mol. Sci.* **2019**, *20*, 4713. [CrossRef]
9. Suhel, M.; Husain, T.; Prasad, S.M.; Singh, V.P. GABA requires nitric oxide for alleviating arsenate stress in tomato and brinjal seedlings. *J. Plant Growth Regul.* **2023**, *42*, 670–683. [CrossRef]
10. Li, Y.; Fan, Y.; Ma, Y.; Zhang, Z.; Yue, H.; Wang, L.; Li, J.; Jiao, Y. Effects of exogenous γ-aminobutyric acid (GABA) on photosynthesis and antioxidant system in pepper (*Capsicum annuum* L.) seedlings under low light stress. *J. Plant Growth Regul.* **2017**, *36*, 436–449. [CrossRef]
11. Akçay, N.; Bor, M.; Karabudak, T.; Ozdemir, F.; Türkan, I. Contribution of Gamma amino butyric acid (GABA) to salt stress responses of *Nicotiana sylvestris* CMSII mutant and wild type plants. *J. Plant Physiol.* **2012**, *169*, 452–458. [CrossRef]
12. Carillo, P. GABA Shunt in Durum Wheat. *Front. Plant Sci.* **2018**, *9*, 100. [CrossRef] [PubMed]
13. Shi, S.Q.; Shi, Z.; Jiang, Z.P.; Qi, L.W.; Sun, X.M.; Li, C.X.; Liu, J.F.; Xiao, W.F.; Zhang, S.G. Effects of exogenous GABA on gene expression of Caragana intermedia roots under NaCl stress: Regulatory roles for H_2O_2 and ethylene production. *Plant Cell Environ.* **2010**, *33*, 149–162. [CrossRef]
14. Luo, H.Y.; Yang, L.W.; Gao, H.B.; Wu, X.L.; Liu, H.H. Physiological mechanism of GABA soaking to tomato seed germination and seedling development under NaCl stress. *Acta Bot. Boreal.-Occident. Sin.* **2011**, *31*, 2235–2242.
15. Li, X.; Wei, Q.; Qin, Z.; Liang, L.; Li, Y. Effect of seed priming with γ-aminobutyric acid (GABA) on seed germination and seedling growth of pepper under salt stress. *J. Gansu Agric. Univ.* **2023**, *59*, 154–164. [CrossRef]
16. Peng, D.; Gao, Q.; Cui, K.; Qi, N.; Zhu, W.; Tang, J.; Hua, Z. Effects of Spraying GABA on Growth of Rice Seedlings under Salt Stress. *China Rice* **2022**, *28*, 43–48. [CrossRef]
17. Wang, Y.; Zhang, Y.; Yan, D. Physiological role of γ-aminobutyric acid in protecting the photosynthetic system of maize seedlings under drought stress. *Acta Pratacult. Sin.* **2020**, *29*, 191–203. [CrossRef]

18. Al-Quraan, N.A.; Samarah, N.H.; Tanash, A.A. Effect of drought stress on wheat (*Triticum durum*) growth and metabolism: Insight from GABA shunt, reactive oxygen species and dehydrin genes expression. *Funct. Plant Biol.* **2024**, *51*, 1–22. [CrossRef]
19. Mishra, V.; Gahlowt, P.; Singh, S.; Dubey, N.K.; Singh, S.P.; Tripathi, D.K.; Singh, V.P. GABA: A key player of abiotic stress regulation. *Plant Signal. Behav.* **2023**, *18*, 2163343. [CrossRef] [PubMed]
20. Shang, H.; Cao, S.; Yang, Z.; Cai, Y.; Zheng, Y. Effect of exogenous γ-aminobutyric acid treatment on proline accumulation and chilling injury in peach fruit after long-term cold storage. *J. Agric. Food Chem.* **2011**, *59*, 1264–1268. [CrossRef]
21. Kinnersley, A.M.; Turano, F.J. Gamma aminobutyric acid (GABA) and plant responses to stress. *Crit. Rev. Plant Sci.* **2000**, *19*, 479–509. [CrossRef]
22. Islam, S.N.U.; Kouser, S.; Hassan, P.; Asgher, M.; Shah, A.A.; Khan, N.A. Gamma-aminobutyric acid interactions with phytohormones and its role in modulating abiotic and biotic stress in plants. *Stress Biol.* **2024**, *4*, 36. [CrossRef] [PubMed]
23. Roberts, M.R. Does GABA Act as a Signal in Plants? Hints from Molecular Studies. *Plant Signal. Behav.* **2007**, *2*, 408–409. [CrossRef] [PubMed]
24. Ahmad, S.; Fariduddin, Q. Deciphering the enigmatic role of gamma-aminobutyric acid (GABA) in plants: Synthesis, transport, regulation, signaling, and biological roles in interaction with growth regulators and abiotic stresses. *Plant Physiol. Biochem.* **2024**, *208*, 108502. [CrossRef]
25. Shan, Q.; Liu, M.; Li, R.; Shi, Q.; Li, Y.; Gong, B. γ-Aminobutyric acid (GABA) improves pesticide detoxification in plants. *Sci. Total Environ.* **2022**, *835*, 155404. [CrossRef]
26. Xu, B.; Long, Y.; Feng, X.; Zhu, X.; Sai, N.; Chirkova, L.; Betts, A.; Herrmann, J.; Edwards, E.J.; Okamoto, M.; et al. GABA signalling modulates stomatal opening to enhance plant water use efficiency and drought resilience. *Nat. Commun.* **2021**, *12*, 1952. [CrossRef]
27. De Angeli, A.; Zhang, J.; Meyer, S.; Martinoia, E. AtALMT9 is a malate-activated vacuolar chloride channel required for stomatal opening in Arabidopsis. *Nat. Commun.* **2013**, *4*, 1804. [CrossRef]
28. Balfagón, D.; Gómez-Cadenas, A.; Rambla, J.L.; Granell, A.; de Ollas, C.; Bassham, D.C.; Mittler, R.; Zandalinas, S.I. γ-Aminobutyric acid plays a key role in plant acclimation to a combination of high light and heat stress. *Plant Physiol.* **2022**, *188*, 2026–2038. [CrossRef] [PubMed]
29. Xu, B.; Sai, N.; Gilliham, M. The emerging role of GABA as a transport regulator and physiological signal. *Plant Physiol.* **2021**, *187*, 2005–2016. [CrossRef]
30. Chen, Z.H.; Chen, G.; Dai, F.; Wang, Y.; Hills, A.; Ruan, Y.L.; Zhang, G.; Franks, P.J.; Nevo, E.; Blatt, M.R. Molecular Evolution of Grass Stomata. *Trends Plant Sci.* **2017**, *22*, 124–139. [CrossRef]
31. Caspi, Y.; Pantazopoulou, C.K.; Prompers, J.J.; Pieterse, C.M.J.; Hulshoff Pol, H.; Kajala, K. Why did glutamate, GABA, and melatonin become intercellular signalling molecules in plants? *eLife* **2023**, *12*, e83361. [CrossRef]
32. Wu, F.; Wang, G.; Bai, Y.; Zhang, Y.; Guo, Y.; Sun, Z. Effects of γ-Aminobutyric Acid on Seed Germination Respiratory Metabolism of *Glycyrrhiza uralensis*. *Seed* **2023**, *42*, 18–24. [CrossRef]
33. Qiang, B.; Jin, X.; Zhou, W.; Liu, J.; Chu, P.; Zhang, Y. Effects of γ-aminobutyric acid on seed germination and root morphogenesis of soybean. *Agric. Res. Arid Areas* **2023**, *41*, 69–77. [CrossRef]
34. Samarah, N.H.; Al-Quraan, N.A.; Al-Wraikat, B.S. Ultrasonic treatment to enhance seed germination and vigour of wheat (*Triticum durum*) in association with γ-aminobutyric acid (GABA) shunt pathway. *Funct. Plant Biol.* **2023**, *50*, 277–293. [CrossRef] [PubMed]
35. Ramesh, S.A.; Tyerman, S.D.; Gilliham, M.; Xu, B. γ-Aminobutyric acid (GABA) signalling in plants. *Cell. Mol. Life Sci.* **2017**, *74*, 1577–1603. [CrossRef] [PubMed]
36. Meyer, A.; Eskandari, S.; Grallath, S.; Rentsch, D. AtGAT1, a high affinity transporter for gamma-aminobutyric acid in *Arabidopsis thaliana*. *J. Biol. Chem.* **2006**, *281*, 7197–7204. [CrossRef]
37. Dündar, E.; Bush, D.R. BAT1, a bidirectional amino acid transporter in Arabidopsis. *Planta* **2009**, *229*, 1047–1056. [CrossRef]
38. Yuan, D.; Wu, X.; Gong, B.; Huo, R.; Zhao, L.; Li, J.; Lü, G.; Gao, H. GABA Metabolism, Transport and Their Roles and Mechanisms in the Regulation of Abiotic Stress (Hypoxia, Salt, Drought) Resistance in Plants. *Metabolites* **2023**, *13*, 347. [CrossRef]
39. Ramesh, S.A.; Kamran, M.; Sullivan, W.; Chirkova, L.; Okamoto, M.; Degryse, F.; McLaughlin, M.; Gilliham, M.; Tyerman, S.D. Aluminum-Activated Malate Transporters Can Facilitate GABA Transport. *Plant Cell* **2018**, *30*, 1147–1164. [CrossRef]
40. Ma, H.; Cao, X.; Shi, S.; Li, S.; Gao, J.; Ma, Y.; Zhao, Q.; Chen, Q. Genome-wide survey and expression analysis of the amino acid transporter superfamily in potato (*Solanum tuberosum* L.). *Plant Physiol. Biochem.* **2016**, *107*, 164–177. [CrossRef]
41. Gilliham, M.; Tyerman, S.D. Linking Metabolism to Membrane Signaling: The GABA-Malate Connection. *Trends Plant Sci.* **2016**, *21*, 295–301. [CrossRef]
42. Long, Y.; Tyerman, S.D.; Gilliham, M. Cytosolic GABA inhibits anion transport by wheat ALMT1. *New Phytol.* **2020**, *225*, 671–678. [CrossRef] [PubMed]
43. Bush, D.R. Inhibitors of the proton-sucrose symport. *Arch. Biochem. Biophys.* **1993**, *307*, 355–360. [CrossRef] [PubMed]
44. Ortiz-Lopez, A.; Chang, H.; Bush, D.R. Amino acid transporters in plants. *Biochim. Biophys. Acta* **2000**, *1465*, 275–280. [CrossRef]
45. Frezzini, M.; Guidoni, L.; Pascarella, S. Conformational transitions induced by γ-amino butyrate binding in GabR, a bacterial transcriptional regulator. *Sci. Rep.* **2019**, *9*, 19319. [CrossRef]
46. Batushansky, A.; Kirma, M.; Grillich, N.; Pham, P.A.; Rentsch, D.; Galili, G.; Fernie, A.R.; Fait, A. The transporter GAT1 plays an important role in GABA-mediated carbon-nitrogen interactions in Arabidopsis. *Front. Plant Sci.* **2015**, *6*, 785. [CrossRef]

47. Young, G.B.; Jack, D.L.; Smith, D.W.; Saier, M.H., Jr. The amino acid/auxin: Proton symport permease family. *Biochim. Biophys. Acta* **1999**, *1415*, 306–322. [CrossRef]
48. Wipf, D.; Ludewig, U.; Tegeder, M.; Rentsch, D.; Koch, W.; Frommer, W.B. Conservation of amino acid transporters in fungi, plants and animals. *Trends Biochem. Sci.* **2002**, *27*, 139–147. [CrossRef]
49. Dinkeloo, K.; Boyd, S.; Pilot, G. Update on amino acid transporter functions and on possible amino acid sensing mechanisms in plants. *Semin. Cell Dev. Biol.* **2018**, *74*, 105–113. [CrossRef] [PubMed]
50. Kishor, P.; Hong, Z.; Miao, G.H.; Hu, C.; Verma, D. Overexpression of [delta]-Pyrroline-5-Carboxylate Synthetase Increases Proline Production and Confers Osmotolerance in Transgenic Plants. *Plant Physiol.* **1995**, *108*, 1387–1394. [CrossRef]
51. Nounjan, N.; Nghia, P.T.; Theerakulpisut, P. Exogenous proline and trehalose promote recovery of rice seedlings from salt-stress and differentially modulate antioxidant enzymes and expression of related genes. *J. Plant Physiol.* **2012**, *169*, 596–604. [CrossRef]
52. Xu, Z. Identification and Functional Analyses of Genes Encoding Proline Transporters in Rice (*Oryza sativa* L.). Master's Thesis, Agricultural University of Hunan, Changsha, China, 2020. [CrossRef]
53. Snowden, C.J.; Thomas, B.; Baxter, C.J.; Smith, J.A.; Sweetlove, L.J. A tonoplast Glu/Asp/GABA exchanger that affects tomato fruit amino acid composition. *Plant J.* **2015**, *81*, 651–660. [CrossRef] [PubMed]
54. Carter, C.; Pan, S.; Zouhar, J.; Avila, E.L.; Girke, T.; Raikhel, N.V. The vegetative vacuole proteome of *Arabidopsis thaliana* reveals predicted and unexpected proteins. *Plant Cell* **2004**, *16*, 3285–3303. [CrossRef]
55. Schmidt, U.G.; Endler, A.; Schelbert, S.; Brunner, A.; Schnell, M.; Neuhaus, H.E.; Marty-Mazars, D.; Marty, F.; Baginsky, S.; Martinoia, E. Novel tonoplast transporters identified using a proteomic approach with vacuoles isolated from cauliflower buds. *Plant Physiol.* **2007**, *145*, 216–229. [CrossRef] [PubMed]
56. Whiteman, S.A.; Serazetdinova, L.; Jones, A.M.; Sanders, D.; Rathjen, J.; Peck, S.C.; Maathuis, F.J. Identification of novel proteins and phosphorylation sites in a tonoplast enriched membrane fraction of *Arabidopsis thaliana*. *Proteomics* **2008**, *8*, 3536–3547. [CrossRef]
57. Koike, S.; Matsukura, C.; Takayama, M.; Asamizu, E.; Ezura, H. Suppression of γ-aminobutyric acid (GABA) transaminases induces prominent GABA accumulation, dwarfism and infertility in the tomato (*Solanum lycopersicum* L.). *Plant Cell Physiol.* **2013**, *54*, 793–807. [CrossRef]
58. Feng, L. Cloning and Functional Analysis of CAT Amino Acid Transporter Genes in Tea Plant (*Camellia sinesis* L.). Master's Thesis, Anhui Agricultural University, Hefei, China, 2017. [CrossRef]
59. Yang, H.; Stierhof, Y.D.; Ludewig, U. The putative Cationic Amino Acid Transporter 9 is targeted to vesicles and may be involved in plant amino acid homeostasis. *Front. Plant Sci.* **2015**, *6*, 212. [CrossRef] [PubMed]
60. Chen, L.; Ortiz-Lopez, A.; Jung, A.; Bush, D.R. ANT1, an aromatic and neutral amino acid transporter in Arabidopsis. *Plant Physiol.* **2001**, *125*, 1813–1820. [CrossRef]
61. Tian, R.; Yang, Y.; Chen, M. Genome-wide survey of the amino acid transporter gene family in wheat (*Triticum aestivum* L.): Identification, expression analysis and response to abiotic stress. *Int. J. Biol. Macromol.* **2020**, *162*, 1372–1387. [CrossRef]
62. Michaeli, S.; Fait, A.; Lagor, K.; Nunes-Nesi, A.; Grillich, N.; Yellin, A.; Bar, D.; Khan, M.; Fernie, A.R.; Turano, F.J.; et al. A mitochondrial GABA permease connects the GABA shunt and the TCA cycle, and is essential for normal carbon metabolism. *Plant J.* **2011**, *67*, 485–498. [CrossRef]
63. Sun, X.; Wang, W.; Wei, C.; Lu, S.; Zheng, Z.; Zhang, H.; Xu, C.; Wang, C. Research progress on improving GABA content in tomato by gene editing technology. *J. Shanghai Norm. Univ. Nat. Sci.* **2022**, *51*, 87–93. [CrossRef]
64. Zhou, L. Study on the High Levels of GABA Accumulation in Transgenic Rice Grains. Master's Thesis, Nanjing Agricultural University, Nanjing, China, 2015.
65. Li, M.; Liu, Y.; Yang, B.; Zhang, F.; Song, J.; Su, C.; Zhang, W.; Wu, Y. Cloning and Expression Analysis of Glutamate Decarboxylase GADs from Watermelon. *Acta Agric. Boreali-Sin.* **2023**, *38*, 18–25. [CrossRef]
66. Li, L. Analysis of GAD Polymorphism in Chenopodium Quinoa and Gene Mapping of a Dewarf Mutant of *Arabidopsis thaliana*. Master's Thesis, Shandong Normal University, Jinan, China, 2019. [CrossRef]
67. Liu, K. Study on Enrichment Technology of γ-Aminobutyric Acid in Wheat Grain and Development of Germinated Whole Wheat Flour. Master's Thesis, Nanjing Agricultural University, Nanjing, China, 2019. [CrossRef]
68. Wang, Y. Process Study and Quality Analysis of Enrichment of Corn Laozao Residue γ-Aminobutyric Acid. Master's Thesis, Inner Mongolia Agricultural University, Hohhot, China, 2023. [CrossRef]
69. Jiang, Z.; Gu, Z. Biosynthesis, Catabolism and Physiological Roles of γ-Aminobutyric Acid in Higher Plants. *Plant Physiol. J.* **2003**, *39*, 249–254. [CrossRef]
70. Qiu, X. Molecular Mechanism of Glutamine Dehydrogenase Genes for Primary Study of the Rice Nitrogen Metabolite. Ph.D. Thesis, Huazhong Agricultural University, Wuhan, China, 2009.
71. Son, H.F.; Kim, I.K.; Kim, K.J. Structural insights into domain movement and cofactor specificity of glutamate dehydrogenase from *Corynebacterium glutamicum*. *Biochem. Biophys. Res. Commun.* **2015**, *459*, 387–392. [CrossRef] [PubMed]
72. Niu, C.; Liu, G.; Qu, C. Summary of glutamate synthetase genes and their regulatory roles in plant nitrogen metabolism. *Jiangsu Agric. Sci.* **2018**, *46*, 10–16. [CrossRef]
73. Liao, H.S.; Chung, Y.H.; Hsieh, M.H. Glutamate: A multifunctional amino acid in plants. *Plant Sci.* **2022**, *318*, 111238. [CrossRef]
74. Lu, J.; Bai, S.; Zheng, X. Research progress on the metabolism of γ-aminobutyric acid and the mechanism of stress accumulation in germinated grains. *Food Ferment. Ind.* **2024**, *50*, 380–386. [CrossRef]

75. Xing, S.G.; Jun, Y.B.; Hau, Z.W.; Liang, L.Y. Higher accumulation of gamma-aminobutyric acid induced by salt stress through stimulating the activity of diamine oxidases in *Glycine max* (L.) Merr. roots. *Plant Physiol. Biochem.* **2007**, *45*, 560–566. [CrossRef]
76. Shelp, B.J.; Bozzo, G.G.; Trobacher, C.P.; Zarei, A.; Deyman, K.L.; Brikis, C.J. Hypothesis/review: Contribution of putrescine to 4-aminobutyrate (GABA) production in response to abiotic stress. *Plant Sci.* **2012**, *193–194*, 130–135. [CrossRef]
77. Shelp, B.J.; Zarei, A. Subcellular compartmentation of 4-aminobutyrate (GABA) metabolism in *Arabidopsis*: An update. *Plant Signal. Behav.* **2017**, *12*, e1322244. [CrossRef]
78. Zik, M.; Arazi, T.; Snedden, W.A.; Fromm, H. Two isoforms of glutamate decarboxylase in *Arabidopsis* are regulated by calcium/calmodulin and differ in organ distribution. *Plant Mol. Biol.* **1998**, *37*, 967–975. [CrossRef]
79. Turano, F.J.; Fang, T.K. Characterization of two glutamate decarboxylase cDNA clones from *Arabidopsis*. *Plant Physiol.* **1998**, *117*, 1411–1421. [CrossRef] [PubMed]
80. Miyashita, Y.; Good, A.G. Contribution of the GABA shunt to hypoxia-induced alanine accumulation in roots of *Arabidopsis thaliana*. *Plant Cell Physiol.* **2008**, *49*, 92–102. [CrossRef] [PubMed]
81. Li, C. Spatiotemporal Expression of GAD Family Genes and Theirmultiple Mutant Phenotype Analysis in *Arabidopsis thaliana*. Bachelor's Thesis, South-Central Minzu University, Wuhan, China, 2021.
82. Pires, N.; Dolan, L. Origin and diversification of basic-helix-loop-helix proteins in plants. *Mol. Biol. Evol.* **2010**, *27*, 862–874. [CrossRef]
83. Kazan, K.; Manners, J.M. MYC2: The master in action. *Mol. Plant* **2013**, *6*, 686–703. [CrossRef] [PubMed]
84. Xu, Y.H.; Liao, Y.C.; Lv, F.F.; Zhang, Z.; Sun, P.W.; Gao, Z.H.; Hu, K.P.; Sui, C.; Jin, Y.; Wei, J.H. Transcription Factor AsMYC2 Controls the Jasmonate-Responsive Expression of ASS1 Regulating Sesquiterpene Biosynthesis in *Aquilaria sinensis* (Lour.) Gilg. *Plant Cell Physiol.* **2017**, *58*, 1924–1933. [CrossRef] [PubMed]
85. Oikawa, T.; Maeda, H.; Oguchi, T.; Yamaguchi, T.; Tanabe, N.; Ebana, K.; Yano, M.; Ebitani, T.; Izawa, T. The Birth of a Black Rice Gene and Its Local Spread by Introgression. *Plant Cell* **2015**, *27*, 2401–2414. [CrossRef]
86. Chen, S.; Zhao, H.; Luo, T.; Liu, Y.; Nie, X.; Li, H. Characteristics and Expression Pattern of MYC Genes in *Triticum aestivum*, *Oryza sativa*, and *Brachypodium distachyon*. *Plants* **2019**, *8*, 274. [CrossRef]
87. El-Kereamy, A.; Bi, Y.M.; Ranathunge, K.; Beatty, P.H.; Good, A.G.; Rothstein, S.J. The rice R2R3-MYB transcription factor OsMYB55 is involved in the tolerance to high temperature and modulates amino acid metabolism. *PLoS ONE* **2012**, *7*, e52030. [CrossRef]
88. Chen, Y. The Function of MaGAD1 in Ethylene Biosynthesis and Signal Transduction Pathways. Master's Thesis, University of Hainan, Haikou, China, 2011.
89. Chen, W.; Cheng, T.; Ji, J.; Wu, Y.; Xie, T.; Jiang, Z.; Shi, S. Identification of three gene families in the GABA shunt and their expression analysis in poplar. *J. Nanjing For. Univ. Nat. Sci. Ed.* **2020**, *44*, 67–77. [CrossRef]
90. Zhou, L.; Shen, B.; Bai, S.; Liu, X.; Jiang, L.; Zhai, H.; Wan, J. RNA Interference of OsGABA-T1 Gene Expression Induced GABA Accumulation in Rice Grain. *Acta Agron. Sin.* **2015**, *41*, 1305–1312. [CrossRef]
91. Sun, L.; Wang, Y.; Bai, Y.; Luo, H.; Yao, B.; Xu, T. Structure and catalytic mechanism of glutamate decarboxylase: A review. *Microbiol. China* **2020**, *47*, 2236–2244. [CrossRef]
92. Dai, F.; Miao, Y.; Chen, J.; Mi, Z. Change Pattern of the Glutamate Decarboxylase Activity in Banana During Maturation. *J. Chin. Inst. Food Sci. Technol.* **2010**, *10*, 81–86. [CrossRef]
93. Satyanarayan, V.; Nair, P.M. Purification and characterization of glutamate decarboxylase from *Solanum tuberosum*. *Eur. J. Biochem.* **1985**, *150*, 53–60. [CrossRef] [PubMed]
94. Liu, Q.; Yao, H.; Yuan, H. Studies on Some Properties of Glutamate Decarboxylase from Lactic Acid Bacterium. *Food Sci.* **2005**, *26*, 100–104.
95. Shu, C.; Wang, C.; Guo, X.; Shan, S.; Xue, Y.; Ou, K.; Li, Y. Enzymatic Characteristics of Pumpkin Glutamate Decarboxylase and Accumulation Conditions of γ-Aminobutyric Acid. *Sci. Technol. Food Ind.* **2018**, *39*, 191–194.
96. Jia, S.; Liu, J.; Li, G. Recent Advances in Gamma-Aminobutyric Acid Biosynthesis. *Food Res. Dev.* **2023**, *44*, 174–181. [CrossRef]
97. Capitani, G.; De Biase, D.; Aurizi, C.; Gut, H.; Bossa, F.; Grütter, M.G. Crystal structure and functional analysis of *Escherichia coli* glutamate decarboxylase. *EMBO J.* **2003**, *22*, 4027–4037. [CrossRef]
98. Lyu, J.-F.; Lyu, C.-J.; Cao, J.-R.; Mei, J.-Q.; Hu, S.; Zhao, W.-R.; Xu, T.-Y.; Wang, Y.-T.; Wang, D.-L.; Huang, J. High level production of γ-aminobutyric acid in engineered *Escherichia coli* by refactoring the glutamate decarboxylase. *Process Biochem.* **2022**, *118*, 243–251. [CrossRef]
99. Liu, Q.; Cheng, H.; Ma, X.; Xu, N.; Liu, J.; Ma, Y. Expression, characterization and mutagenesis of a novel glutamate decarboxylase from *Bacillus megaterium*. *Biotechnol. Lett.* **2016**, *38*, 1107–1113. [CrossRef]
100. Yao, Q.; Zhang, Y. Isolation and properties of soybean glutamate decarboxylase. *China Oils Fats* **2011**, *36*, 26–30.
101. Li, N.; Wang, L.; Wu, Z.; Wang, W. Some Properties of Glutamate Decarboxylase from Corn Germ. *Food Ferment. Ind.* **2011**, *37*, 34–38. [CrossRef]
102. Zhang, H.; Yao, H.-Y.; Chen, F.; Wang, X. Purification and characterization of glutamate decarboxylase from rice germ. *Food Chem.* **2007**, *101*, 1670–1676. [CrossRef]
103. Inatomi, K.; Slaughter, J.C. Glutamate decarboxylase from barley embryos and roots. General properties and the occurrence of three enzymic forms. *Biochem. J.* **1975**, *147*, 479–484. [CrossRef]

104. Lv, Y.; Zhang, H.; Xu, D.; Fan, L.; Yao, H. Studies on Some Properties of Glutamate Decarboxylase from Rice Bran. *Food Ferment. Ind.* **2008**, *34*, 24–28.
105. Akama, K.; Takaiwa, F. C-terminal extension of rice glutamate decarboxylase (OsGAD2) functions as an autoinhibitory domain and overexpression of a truncated mutant results in the accumulation of extremely high levels of GABA in plant cells. *J. Exp. Bot.* **2007**, *58*, 2699–2707. [CrossRef] [PubMed]
106. Wang, X.; Wei, T.; Shi, Y. Extraction and purification of mung bean calmodulin and its regulation on glutamate decarboxylase. *Cereals Oils* **2020**, *33*, 39–43.
107. Arazi, T.; Baum, G.; Snedden, W.A.; Shelp, B.J.; Fromm, H. Molecular and biochemical analysis of calmodulin interactions with the calmodulin-binding domain of plant glutamate decarboxylase. *Plant Physiol.* **1995**, *108*, 551–561. [CrossRef]
108. Mazzucotelli, E.; Tartari, A.; Cattivelli, L.; Forlani, G. Metabolism of gamma-aminobutyric acid during cold acclimation and freezing and its relationship to frost tolerance in barley and wheat. *J. Exp. Bot.* **2006**, *57*, 3755–3766. [CrossRef]
109. Yin, Y.; Wu, J.; Liu, C.; Li, D.; Yang, R.; Gu, Z. Effects of Hypoxia and Cold Stress on γ-Aminobutyric Acid Accumulation in Germinating Maize. *Food Sci.* **2015**, *36*, 89–93. [CrossRef]
110. Yi, X.; Li, M.; Huang, J.; Chi, H. Effects of Salicylic Acid on Membrane Lipid Peroxidation in Purple Majesty (*Pennisetum glaucum*) Seedlings under Drought, High Temperature or the Dual Stress. *J. Southwest Univ. Nat. Sci. Ed.* **2013**, *35*, 41–45. [CrossRef]
111. Fait, A.; Fromm, H.; Walter, D.; Galili, G.; Fernie, A.R. Highway or byway: The metabolic role of the GABA shunt in plants. *Trends Plant Sci.* **2008**, *13*, 14–19. [CrossRef] [PubMed]
112. Zhang, H.; Cui, L.; Dong, S.; Gao, R.; Sun, A. Physiological Role of Hot Stress—Induced GABA Accumulation in Maize Seedlings. *Shandong Agric. Sci.* **2011**, *07*, 35–37. [CrossRef]
113. Cao, J. Study on the Change of γ-Aminobutyric Acid in Germinated Brown Rice and Its Products Processing. *Chin. Acad. Agric. Sci.* **2018**, *39*, 47–52.
114. Yang, S. Effects of Electrolyzed Functional Water Combined with High Voltage Electric Field Treatment on the Accumulation of γ-Aminobutyric Acid in Germinated Buckwheat. Master's Thesis, Shanxi Agricultural University, Taiyuan, China, 2022.
115. Wang, S.; Liu, X.; Hu, Y.; Jin, Y.; Bai, T.; Zhu, M.; Zhang, Y. Metabolism and Enrichment Mechanism of γ-aminobutyric Acid in Plants. *J. Anhui Agric. Sci.* **2020**, *48*, 9–12. [CrossRef]

plants

MDPI

Article

Comparative Study on the Response of Hyssop (*Hyssopus officinalis* L.), Salvia (*Salvia officinalis* L.), and Oregano (*Origanum vulgare* L.) to Drought Stress Under Foliar Application of Selenium

Liubov Skrypnik [1,*], Pavel Maslennikov [2], Maria Antipina [2], Dmitriy Katserov [2] and Pavel Feduraev [2]

1 Laboratory of Natural Antioxidants, Research and Education Center "Industrial Biotechnologies", Immanuel Kant Baltic Federal University, 236040 Kaliningrad, Russia

2 Scientific and Educational Cluster MEDBIO, Immanuel Kant Baltic Federal University, 236040 Kaliningrad, Russia; pfeduraev@kantiana.ru (P.F.)

* Correspondence: lskrypnik@kantiana.ru

Abstract: Drought is one of the most important abiotic factors limiting plant productivity. Although the aromatic plants of the Lamiaceae family often grow in arid regions, drought tolerance varies greatly among the different species of this family. The effect of induced drought stress can be reduced by the application of selenium. The current study aims to compare the growth and biochemical responses of three species of the Lamiaceae family (hyssop, salvia, and oregano) to drought stress and the possibility of reducing the effect of stress in these plants by foliar treatment with selenium. Drought stress reduced the fresh and dry biomass of hyssop (by 35% and 15%), salvia (by 45% and 41%), and oregano (by 51% and 32%). Se treatment did not affect the growth of plants under drought stress, but it improved relative water content in hyssop and salvia under moderate drought conditions. A reduction in the content of chlorophyll *a* and chlorophyll *b* (in hyssop and salvia). In addition, an increase in the content of hydrogen peroxide (in oregano and salvia), malondialdehyde, and proline in plants cultivated under drought conditions was observed. Se treatment led to reduced levels of hydrogen peroxide and malondialdehyde, along with an increase in chlorophyll *a* content (in hyssop and oregano) and proline content. The response of the antioxidant system depended on the plant species. Hyssop exhibited a significant increase in glutathione peroxidase, superoxide dismutase, and peroxidase activities. Oregano showed enhanced catalase activity. Salvia experienced a sharp increase in ascorbic acid content. Se treatment stimulated the accumulation of phenolic compounds and increased glutathione peroxidase activity in all studied species.

Keywords: abiotic stress; water-deficit stress; aromatic plants; reactive oxygen species; ascorbate–glutathione cycle; osmolyte; photosynthesis; irrigation

1. Introduction

Drought is one of the most critical limiting factors, significantly affecting plant growth, development, and productivity, thus impeding sustainable agriculture worldwide [1]. As a natural climatic phenomenon, drought occurs in almost all climatic zones, with variations in frequency, intensity, and duration [2]. Given the anticipated climate changes, it is evident that the frequency and duration of droughts will increase [3], positioning this as one of the major challenges of the 21st century and emphasizing the importance of this study.

The Lamiaceae family comprises over 6000 species of herbs and shrubs, which are globally distributed and have significant practical applications in pharmaceutical, food, and cosmetic industries [4]. This family includes plant species such as mint, thyme, rosemary, basil, lavender, oregano, salvia, and hyssop. Hyssop (*Hyssopus officinalis* L.) is a notable medicinal plant and a typical xerophyte, well adapted to drought and low-input conditions [5]. Salvia (*Salvia officinalis* L.) is a perennial woody shrub native to the

Mediterranean region but is currently cultivated all over the world [6]. While various *Salvia* species exhibit relative drought tolerance, *S. officinalis* is particularly sensitive to water stress [7]. Oregano (*Origanum vulgare* L.), a perennial herbaceous species also originating from the Mediterranean, not only serves as a consumable plant but also contributes to sustainable agroecosystems and the reclamation of cultivated land [8]. Under water stress, oregano experiences a reduction in fresh weight, dry weight, relative water content (RWC), and total chlorophyll [9]. In this study, we selected three species from the Lamiaceae family, distinguished by their resistance to drought stress based on the literature data.

Plant responses to stress factors, including drought, are highly complex and occur at multiple levels, ranging from molecular changes to structural alterations in various parts of the plant. One of the most sensitive indicators of plants' response to stress is the redox balance, which refers to the equilibrium between the production of reactive oxygen species (ROS) and the components of the antioxidant defense system [10]. Typically, exposure to abiotic stress factors such as salinity, drought, heavy metals, and metalloids triggers increased ROS production, along with changes in the activity of antioxidant enzymes and levels of non-enzymatic antioxidants [11–13]. Drought stress activates various signaling pathways, leading to changes in gene expression, which results in the production of functional compounds such as proline, glycine betaine, soluble sugars, late embryogenesis-abundant proteins, and aquaporins. These compounds are involved in plant metabolism and contribute to stress tolerance. Plants utilize two main signaling pathways—abscisic acid (ABA)-dependent and ABA-independent—to detect and respond to drought stress [14]. Regulatory gene products such as calcium-dependent protein kinases, mitogen-activated protein kinases, HD-ZIP/bZIP, AP2/ERF, NAC, MYB, and WRKY play key roles in regulating signal transduction pathways or acting as transcription factors, allowing plants to adapt to arid conditions by altering their morphology and physiology [15].

Although selenium (Se) has not yet been officially recognized as an essential microelement for plants, its significance has been demonstrated in numerous studies. A growing body of experimental evidence shows that selenium can enhance plant resistance to various abiotic stressors [16–19]. Notably, selenium treatment has been shown to improve the drought resistance of several plant species [20–24]. The protective effects of selenium are primarily linked to the activation of plants' antioxidant system, particularly enzymes such as glutathione peroxidase (GPX), ascorbate peroxidase (APX), and catalase (CAT). Additionally, selenium helps regulate water balance, supports the photosynthetic apparatus, and stimulates the biosynthesis of osmoprotectants and secondary metabolites [21,25,26].

Due to the significant economic value of various Lamiaceae species, many studies have focused on the effects of drought stress on their growth and phytochemical composition, particularly on essential oils and phenolic compounds [27]. However, the data on selenium's impact on the drought resistance of Lamiaceae species are limited and often contradictory. The goal of the present study was to verify the hypothesis according to which the difference in drought tolerance demonstrated for different Lamiaceae species is associated with species-specific reactions of the plant antioxidant system and the protective effect of selenium in plants under stress conditions determined by the regulation of cell redox metabolism. To test this hypothesis (1), a comparative analysis of the responses of three Lamiaceae species—hyssop, salvia, and oregano—to varying degrees of drought stress by examining changes in growth, photosynthetic pigment content, oxidative stress intensity, and antioxidant system activity, and (2) an assessment of the potential of foliar selenium treatment on the drought resistance of these species through its effect on the antioxidant system were conducted. The findings from this study will provide deeper insights into the mechanisms driving Lamiaceae plants' response to drought stress and the role of selenium in regulating their stress resistance

2. Results

2.1. Plant Growth Characteristics and Se Accumulation

Drought had a pronounced impact on the growth of hyssop, salvia, and oregano. The study revealed a significant reduction in shoot height (by 30%), fresh weight (by 35 to 50%), and dry weight (by 15 to 40%), depending on the species (Table 1). However, selenium treatment did not result in notable changes in growth indicators for any of the three species, regardless of the drought severity.

Table 1. Effect of Selenium on Plant Growth under Drought Stress.

Level of Drought	Plant Height [cm]		Fresh Weight [g]		Dry Weight [g]		RWC [1] [%]	
	Control	Se	Control	Se	Control	Se	Control	Se
				Hyssop				
25%	23.46 ± 1.04 a	23.60 ± 1.47 a	5.32 ± 0.25 a	5.77 ± 0.29 a	0.58 ± 0.03 a	0.65 ± 0.03 a	80.06 ± 0.68 a	80.20 ± 0.57 a
50%	23.17 ± 1.09 a	24.34 ± 1.45 a	4.81 ± 0.27 a	5.01 ± 0.25 a	0.57 ± 0.02 a	0.63 ± 0.03 a	74.14 ± 0.61 b	78.68 ± 0.59 a *
75%	16.41 ± 1.24 b	18.07 ± 1.20 b	3.44 ± 0.21 b	2.92 ± 0.18 b	0.50 ± 0.02 a	0.41 ± 0.01 b	66.09 ± 0.33 c	64.87 ± 0.94 b
				Salvia				
25%	40.10 ± 2.25 a [2]	42.86 ± 2.87 a	8.47 ± 0.29 a	8.58 ± 0.27 a	0.93 ± 0.07 a	0.96 ± 0.03 a	77.43 ± 0.67 a	77.54 ± 0.39 a
50%	35.53 ± 2.61 ab	40.63 ± 2.80 ab	6.88 ± 0.36 b	7.51 ± 0.28 b	0.85 ± 0.06 a	0.94 ± 0.03 a	71.98 ± 0.35 b	76.54 ± 0.34 a *
75%	29.91 ± 1.58 b	31.91 ± 0.99 b	4.63 ± 0.30 c	4.72 ± 0.19 c	0.55 ± 0.03 b	0.63 ± 0.02 b	68.71 ± 0.81 c	69.38 ± 0.55 b
				Oregano				
25%	52.78 ± 2.98 a	54.41 ± 2.46 a	9.41 ± 0.49 a	8.87 ± 0.48 a	1.65 ± 0.11 a	1.75 ± 0.14 a	90.13 ± 0.47 a	88.13 ± 0.64 a *
50%	42.24 ± 2.43 b	43.19 ± 3.09 b	8.25 ± 0.55 a	8.52 ± 0.60 a	1.55 ± 0.10 a	1.80 ± 0.12 a	77.05 ± 0.39 b	75.84 ± 1.06 b
75%	35.77 ± 1.72 b	36.75 ± 1.61 b	4.63 ± 0.31 b	5.10 ± 0.24 b	1.11 ± 0.07 b	1.03 ± 0.08 b	64.95 ± 0.74 c	63.07 ± 0.45 c

[1] RWC, relative water content. [2] Different letters indicate significant differences between plants by drought level for control, Se-treatment, and each species separately, and asterisks * indicate significant differences between control and Se-treatment at $p \leq 0.05$ based on post hoc Tukey's test (n = 4).

The RWC in hyssop, salvia, and oregano decreased significantly as drought stress intensified (Table 1). Selenium treatment, however, improved RWC in hyssop and salvia under moderate drought conditions. In oregano, Se did not affect RWC under moderate and severe drought stress, but significantly reduced it under control conditions.

The foliar application of exogenous selenium led to an increase in selenium concentration in the plant shoots (Table 2), with hyssop showing the highest selenium accumulation. The selenium concentration in hyssop shoots reached around 8–9 μg g^{-1}, which was 2.3–3.5 times higher than in salvia and oregano. A three-way analysis of variance showed that the level of drought stress and the interaction between drought and selenium treatment did not influence selenium concentration in the plants (Supplementary Materials Table S1A).

Table 2. Selenium concentration in plants (μg g^{-1} DW).

Level of Drought	Hyssop		Salvia		Oregano	
	Control	Se	Control	Se	Control	Se
25%	0.024 ± 0.001 a [1]	9.53 ± 0.61 a *	0.0504 ± 0.003 a	2.93 ± 0.17 a *	0.184 ± 0.011 a	3.36 ± 0.16 a *
50%	0.024 ± 0.001 a	8.89 ± 0.56 a *	0.0522 ± 0.002 a	2.94 ± 0.16 a *	0.197 ± 0.007 a	3.22 ± 0.21 a *
75%	0.023 ± 0.001 a	7.86 ± 0.24 a *	0.0489 ± 0.002 a	3.30 ± 0.12 a *	0.173 ± 0.011 a	2.71 ± 0.15 a *

[1] Different letters indicate significant differences between plants by drought level for control, Se-treatment, and each species separately, and asterisks * indicate significant differences between control and Se-treatment at $p \leq 0.05$ based on post hoc Tukey's test (n = 4).

2.2. Photosynthetic Pigments

Cultivating plants under drought conditions resulted in a reduction in chlorophyll *a* content (Table 3). At the highest drought stress level, the chlorophyll *a* content decreased by 1.6 times in hyssop, 1.4 times in salvia, and 1.3 times in oregano. However, treatment with exogenous selenium under drought conditions led to an increase in chlorophyll *a* content in hyssop and salvia (under moderate drought), as well as in oregano (under moderate and severe drought). Interestingly, in salvia under severe drought, selenium treatment resulted in a lower chlorophyll *a* content compared to untreated plants.

Table 3. Effect of Selenium on Photosynthetic Pigments under Drought Stress.

Level of Drought	Chl *a* [mg g^{-1} DW [1]]		Chl *b* [mg g^{-1} DW]		Chl *a/b*		Total Chl [mg g^{-1} DW]		Car [mg g^{-1} DW]	
	Control	Se	Control	Se	Control	Se	Control	Se	Control	Se
					Hyssop					
25%	13.13 ± 0.63 a [2]	12.92 ± 0.44 a	6.17 ± 0.18 a	5.83 ± 0.11 a	2.14 ± 0.16 b	2.22 ± 0.11 a	19.30 ± 0.50 a	18.75 ± 0.37 a	2.76 ± 0.16 a	2.87 ± 0.09 a
50%	9.95 ± 0.29 b	12.25 ± 0.33 a *	3.66 ± 0.12 b	4.32 ± 0.12 b *	2.73 ± 0.13 a	2.84 ± 0.13 a	13.61 ± 0.29 b	16.58 ± 0.32 b *	2.67 ± 0.19 a	2.54 ± 0.20 a
75%	8.13 ± 0.26 c	9.38 ± 0.52 b	3.06 ± 0.13 c	3.66 ± 0.12 c *	2.67 ± 0.13 ab	2.58 ± 0.22 a	11.20 ± 0.29 c	13.04 ± 0.43 c *	3.13 ± 0.26 a	2.88 ± 0.12 a
					Salvia					
25%	8.35 ± 0.42 a	8.30 ± 0.32 a	3.74 ± 0.13 a	3.62 ± 0.25 a	2.23 ± 0.13 a	2.33 ± 0.22 a	12.09 ± 0.45 a	11.93 ± 0.31 a	2.34 ± 0.02 a	2.16 ± 0.13 ab
50%	6.18 ± 0.18 b	7.55 ± 0.12 a *	2.58 ± 0.06 b	3.20 ± 0.05 a *	2.40 ± 0.10 a	2.36 ± 0.03 a	8.76 ± 0.16 b	10.74 ± 0.16 a *	2.14 ± 0.17 a	2.02 ± 0.13 b
75%	6.02 ± 0.36 b	4.89 ± 0.31 b *	2.34 ± 0.11 b	1.91 ± 0.13 b *	2.60 ± 0.26 a	2.56 ± 0.14 a	8.36 ± 0.28 b	6.79 ± 0.41 b *	2.31 ± 0.11 a	2.60 ± 0.15 a
					Oregano					
25%	5.91 ± 0.31 a	5.87 ± 0.12 a	2.26 ± 0.14 a	2.31 ± 0.15 a	2.62 ± 0.02 a	2.57 ± 0.14 a	8.17 ± 0.45 a	8.18 ± 0.22 a	2.99 ± 0.15 ab	3.07 ± 0.15 a
50%	5.12 ± 0.26 ab	5.37 ± 0.18 ab	2.57 ± 0.13 a	2.15 ± 0.09 a	2.00 ±0.03 b	2.51 ± 0.10 a *	7.69 ± 0.39 a	7.52 ± 0.24 a	2.55 ± 0.10 b	3.23 ± 0.15 a
75%	4.46 ± 0.17 b	5.27 ± 0.16 b *	2.48 ± 0.06 a	2.40 ± 0.07 a	1.80 ±0.07 c	2.19 ± 0.09 b *	6.94 ± 0.19 a	7.66 ± 0.17 a	3.28 ± 0.17 a	3.22 ± 0.21 a

[1] Chl *a*, chlorophyll *a*; Chl *b*, chlorophyll *b*; Total Chl, total chlorophylls; Chl *a*/*b*, chlorophyll *a*/*b* ratio; Car, carotenoids; DW, dry weight. [2] Different letters indicate significant differences between plants by drought level for control, Se-treatment, and each species separately, and asterisks * indicate significant differences between control and Se-treatment at $p \leq 0.05$ based on post hoc Tukey's test (n = 4).

Drought conditions also caused a reduction in chlorophyll *b* content in hyssop and salvia (Table 3). In contrast, the chlorophyll *b* content in oregano was unaffected by water availability or selenium treatment. Selenium treatment of hyssop cultivated under drought stress increased chlorophyll *b* content by 1.2 times compared to untreated plants. In salvia, selenium treatment led to an increase in chlorophyll *b* content under moderate drought, but a decrease under severe drought conditions.

According to the results of the 3-way analysis of variance, the chlorophyll *a*/chlorophyll *b* ratio was influenced by both plant species and the interaction between species and drought conditions (Supplementary Materials Table S1B). The most notable differences in this ratio were observed in oregano, where it decreased by 1.3 to 1.5 times under drought compared to control plants. Selenium treatment, however, significantly increased the chlorophyll *a*/chlorophyll *b* ratio in oregano under drought conditions.

The total chlorophyll content decreased in hyssop and salvia under drought conditions compared to the control, while no significant changes were observed in oregano (Table 3). Selenium treatment also had no effect on the total chlorophyll content in oregano. Notably, the effect of selenium on the total chlorophyll content in hyssop and salvia under drought followed the same pattern as the changes observed in chlorophyll *a* and chlorophyll *b* levels.

The effect of drought on carotenoid content was significant only in oregano (Table 3), where the highest carotenoid levels were recorded under maximum stress conditions. Selenium treatment did not significantly affect carotenoid content in any of the studied plant species.

2.3. Oxidative Stress Parameters

Cultivation of salvia and oregano under drought conditions resulted in a significant increase in hydrogen peroxide levels, rising by 1.2 to 1.4 times in salvia and 1.5 to 1.8 times in oregano (Figure 1). In oregano, selenium treatment reduced hydrogen peroxide levels under drought conditions. However, in salvia, selenium treatment resulted in a significant decrease in hydrogen peroxide levels only in control plants, whereas under severe drought stress, hydrogen peroxide content increased significantly. In hyssop, no significant changes in hydrogen peroxide levels were observed in response to either increased drought stress or selenium treatment.

Cultivating hyssop, salvia, and oregano under drought conditions resulted in the accumulation of malondialdehyde, a marker of oxidative stress, in the plants (Figure 2). In hyssop and salvia, significantly higher levels were detected only under severe drought conditions (75%), while in oregano, the increase in malondialdehyde was observed earlier, at moderate drought levels (50%). Selenium treatment under drought conditions led to a reduction in malondialdehyde content in hyssop and oregano by 1.1 to 1.2 times and 1.5

to 1.7 times, respectively. However, in salvia, selenium treatment caused an increase in malondialdehyde levels compared to untreated plants under the same conditions.

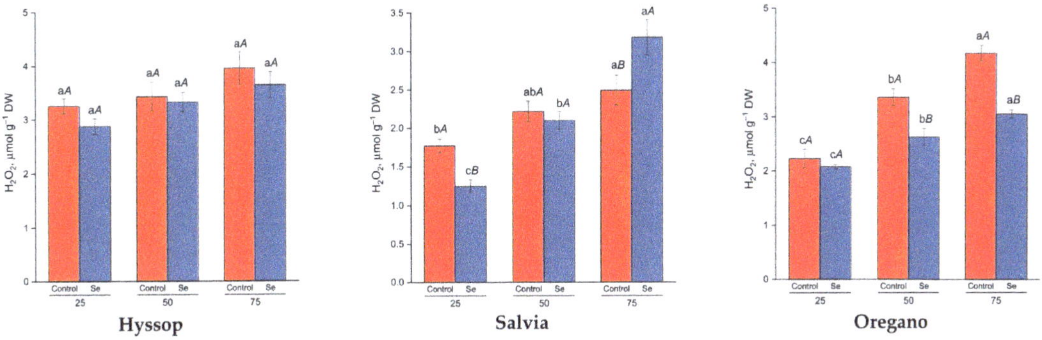

Figure 1. Effect of selenium on hydrogen peroxide content (H_2O_2) under drought stress. Different lower-case letters indicate significant differences between plants by drought level for control, Se-treatment, and each species separately, and upper-case letters indicate significant differences between control and Se-treatment at $p \leq 0.05$ based on post hoc Tukey's test (n = 4).

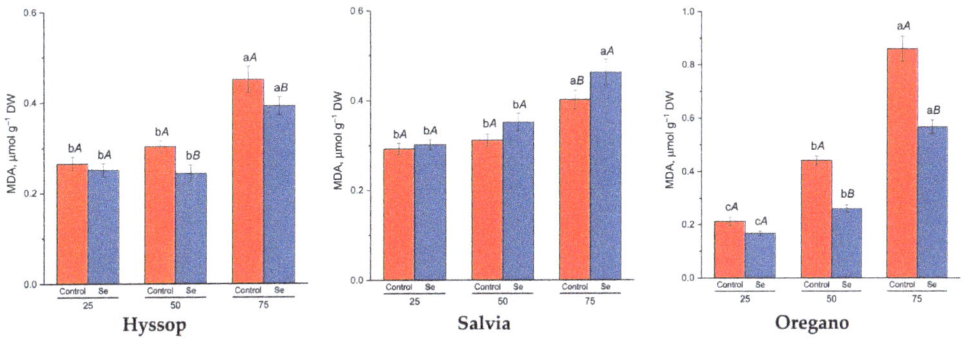

Figure 2. Effect of selenium on malondialdehyde (MDA) content under drought stress. Different lower-case letters indicate significant differences between plants by drought level for control, Se-treatment, and each species separately, and upper-case letters indicate significant differences between control and Se-treatment at $p \leq 0.05$ based on post hoc Tukey's test (n = 4).

2.4. Proline and Non-Enzymatic Antioxidants

A significant increase in proline content, by approximately 2.5 to 3.0 times, was found in all three studied species in response to the augmenting drought level (Figure 3). Treatment of plants with selenium provided an even greater increase in proline content compared to untreated plants. However, it should be noted that in hyssop and salvia, a significant increase was revealed only under moderate drought conditions. Treatment of oregano with selenium led to an increase in proline content in all examined variants including control plants.

The changes in the total content of phenolic compounds in response to increasing drought levels varied depending on the plant species (Figure 4). In hyssop, the highest phenolic compound content was observed under moderate drought conditions, while the lowest was detected at the highest drought level. In salvia, there was a linear increase in phenolic compound content as drought stress intensified. In oregano, phenolic compound content at moderate and severe drought levels was significantly higher than in the control, but there was no difference between these two stress levels. Selenium treatment stimulated the accumulation of phenolic compounds in both control plants and those subjected to drought. The exception was salvia under severe drought conditions, where selenium treatment led to a 1.4-fold decrease in phenolic compound content compared to untreated plants.

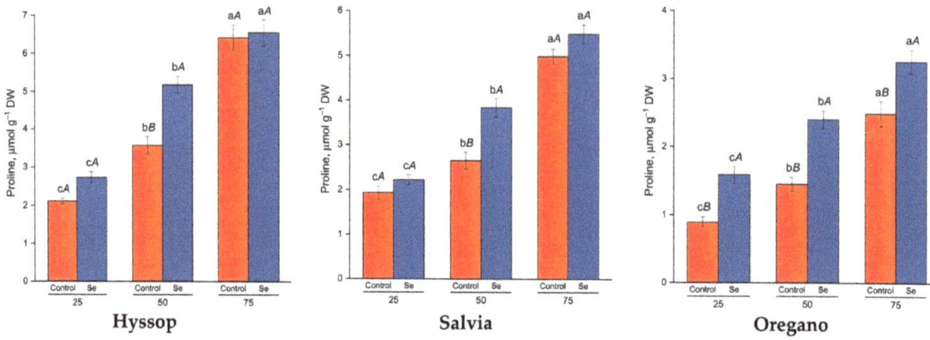

Figure 3. Effect of selenium on proline content under drought stress. Different lower-case letters indicate significant differences between plants by drought level for control, Se-treatment, and each species separately, and upper-case letters indicate significant differences between control and Se-treatment at $p \leq 0.05$ based on post hoc Tukey's test (n = 4).

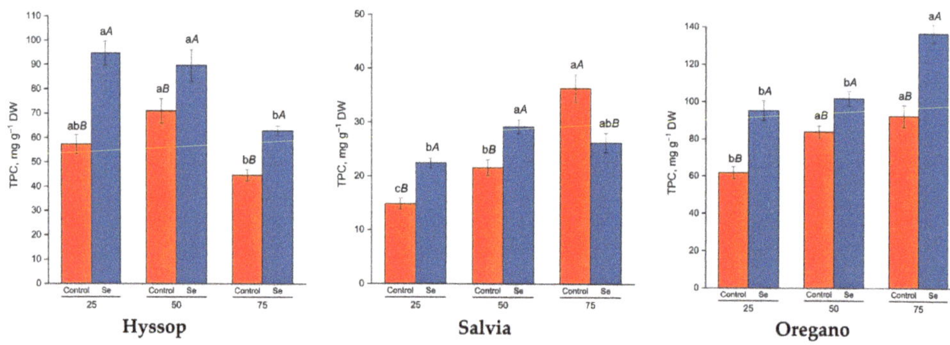

Figure 4. Effect of selenium on total phenolics content (TPC) under drought stress. Different lower-case letters indicate significant differences between plants by drought level for control, Se-treatment, and each species separately, and upper-case letters indicate significant differences between control and Se-treatment at $p \leq 0.05$ based on post hoc Tukey's test (n = 4).

To evaluate the effects of drought and selenium on glutathione metabolism in the studied plants, we measured the content of both reduced and oxidized glutathione, as well as their ratio and total content (Table 4). Drought conditions in all three plant species resulted in a significant decrease in reduced glutathione and an increase in its oxidized form. However, the total glutathione content remained unchanged, while the ratio of reduced to oxidized glutathione significantly decreased. Selenium treatment did not affect glutathione redox metabolism under drought conditions, except in hyssop plants exposed to severe drought, where a significant decrease in oxidized glutathione content was observed.

The investigation of a drought's impact on ascorbic acid metabolism revealed species-specific responses (Table 5). In hyssop and salvia, the highest content of reduced ascorbic acid was observed at moderate drought stress levels. In contrast, oregano showed a linear increase in ascorbic acid content as drought stress intensified. Additionally, the content of the oxidized form of ascorbic acid (dehydroascorbic acid) increased by 1.2 to 2.1 times in plants grown under drought conditions.

Table 4. Effect of Selenium on Glutathione Content under Drought Stress.

Level of Drought	GSH [μmol g^{-1} DW [1]]		GSSG [μmol g^{-1} DW]		GSH/GSSG		GSH + GSSG [μmol g^{-1} DW]	
	Control	Se	Control	Se	Control	Se	Control	Se
				Hyssop				
25%	254.01 ± 14.94 a [2]	273.21 ± 15.13 a	42.29 ± 0.86 b	41.25 ± 2.11 b	6.03 ± 0.47 a	6.64 ± 0.31 a	296.30 ± 14.22 a	314.46 ± 16.41 a
50%	233.77 ± 17.00 ab	245.86 ± 8.55 ab	50.10 ± 1.60 b	48.30 ± 1.79 b	4.66 ± 0.26 b	5.13 ± 0.36 b	283.88 ± 18.20 a	294.16 ± 6.88 a
75%	198.11 ± 5.64 b	208.87 ± 8.45 b	67.63 ± 4.05 a	60.07 ± 1.90 a *	2.97 ± 0.22 c	3.48 ± 0.14 c	265.73 ± 5.42 a	268.94 ± 9.41 a
				Salvia				
25%	192.89 ± 13.57 a	207.96 ± 10.13 a	36.39 ± 2.05 b	38.18 ± 0.90 b	5.32 ± 0.33 a	5.45 ± 0.31 a	229.27 ± 14.89 a	246.13 ± 10.04 a
50%	194.67 ± 11.47 a	185.75 ± 8.57 a	44.02 ± 2.10 b	39.89 ± 1.84 b	4.44 ± 0.28 a	4.71 ± 0.40 a	238.70 ± 12.28 a	225.64 ± 7.39 ab
75%	160.65 ± 6.15 b	148.92 ± 3.09 b	59.77 ± 4.77 a	64.24 ± 1.64 a	2.74 ± 0.26 b	2.32 ± 0.03 b	220.42 ± 7.96 a	213.17 ± 4.57 b
				Oregano				
25%	86.74 ± 3.92 a	76.05 ± 3.27 a	14.54 ± 0.96 b	13.66 ± 1.28 b	6.06 ± 0.53 a	5.66 ± 0.35 a	101.28 ± 4.01 a	89.72 ± 4.50 a
50%	77.42 ± 4.66 ab	78.54 ± 3.36 a	22.20 ± 1.16 a	20.89 ± 1.48 a	3.50 ± 0.21 b	3.83 ± 0.38 b	99.62 ± 5.26 a	99.42 ± 3.01 a
75%	67.41 ± 3.04 b	70.49 ± 3.57 a	24.87 ± 1.68 a	23.74 ± 1.61 a	2.77 ± 0.29 b	3.00 ± 0.22 b	92.29 ± 2.06 a	94.23 ± 4.48 a

[1] GSH, reduced glutathione; GSSG, oxidized glutathione; DW, dry weight. [2] Different letters indicate significant differences between plants by drought level for control, Se-treatment, and each species separately, and asterisks * indicate significant differences between control and Se-treatment at $p \leq 0.05$ based on post hoc Tukey's test (n = 4).

Table 5. Effect of Selenium on Ascorbic Acid Content under Drought Stress.

Level of Drought	AsA [mg g^{-1} DW [1]]		DHA [mg g^{-1} DW]		AsA/DHA		AsA + DHA [mg g^{-1} DW]	
	Control	Se	Control	Se	Control	Se	Control	Se
				Hyssop				
25%	15.32 ± 0.81 b [2]	21.28 ± 1.63 a *	6.72 ± 0.31 b	6.34 ± 0.41 b	2.30 ± 0.16 a	3.43 ± 0.45 a *	22.04 ± 0.85 c	27.62 ± 1.36 ab *
50%	25.17 ± 0.94 a	23.34 ± 1.24 a	11.22 ± 1.57 a	9.50 ± 0.58 a	2.41 ± 0.42 a	2.47 ± 0.15 ab	36.39 ± 0.96 a	32.84 ± 1.62 a
75%	16.47 ± 1.16 b	14.34 ± 1.51 b	11.34 ± 0.49 a	9.85 ± 0.65 a	1.46 ± 0.12 a	1.48 ± 0.18 b	27.82 ± 1.30 b	24.19 ± 1.59 b
				Salvia				
25%	6.74 ± 0.35 b	9.59 ± 0.57 b *	3.54 ± 0.15 c	3.32 ± 0.21 c	1.92 ± 0.14 a	2.94 ± 0.34 a *	10.27 ± 0.36 b	12.93 ± 0.42 c *
50%	12.23 ± 0.65 a	12.45 ± 0.44 a	5.99 ± 0.36 b	6.47 ±0.32 b	2.05 ± 0.11 a	1.94 ± 0.12 b	18.22 ± 0.92 a	18.92 ±0.51 a
75%	8.31 ± 0.38 b	6.45 ± 0.42 c *	7.50 ± 0.32 a	9.84 ±0.28 a *	1.11 ± 0.07 b	0.64 ± 0.06 c	15.80 ± 0.50 a	16.30 ± 0.30 b
				Oregano				
25%	18.77 ± 1.42 c	24.51 ± 1.20 b *	5.68 ± 0.24 b	6.16 ± 0.11 b	3.30 ± 0.16 b	3.98 ± 0.20 a	24.44 ± 1.63 c	30.67 ± 1.21 b *
50%	24.31 ± 1.51 b	29.64 ± 1.74 ab *	6.32 ± 0.23 ab	7.26 ± 0.45 ab *	3.84 ± 0.13 ab	4.16 ± 0.44 a	30.63 ± 1.72 b	36.90 ± 1.32 a *
75%	30.25 ± 1.19 a	33.95 ± 1.37 a	6.75 ± 0.26 a	7.99 ± 0.43 a *	4.52 ± 0.33 a	4.29 ± 0.29 a	37.00 ± 0.96 a	41.94 ± 1.40 a *

[1] AsA, ascorbic acid; DHA, dehydroascorbic acid; DW, dry weight. [2] Different letters indicate significant differences between plants by drought level for control, Se-treatment, and each species separately, and asterisks * indicate significant differences between control and Se-treatment at $p \leq 0.05$ based on post hoc Tukey's test (n = 4).

Selenium treatment led to an increase in ascorbic acid content, as well as the total content of ascorbic and dehydroascorbic acids, in all three species cultivated in control conditions (Table 5). However, the response of the plants to selenium treatment under drought conditions varied. In hyssop, no significant effect of selenium was observed. In salvia, selenium treatment under severe drought resulted in a decrease in ascorbic acid content and an increase in dehydroascorbic acid content compared to untreated plants. In oregano, selenium treatment significantly increased ascorbic acid content at moderate drought levels, while both dehydroascorbic acid and the total content of both forms increased significantly at moderate and severe drought levels.

2.5. Antioxidative Enzymes

Drought cultivation led to increased superoxide dismutase (SOD) activity in hyssop and salvia by 2.5 and 1.5 times, respectively, whereas SOD activity in oregano remained unchanged (Table 6). Selenium treatment reduced SOD activity in hyssop under moderate drought and in salvia under severe drought. In contrast, selenium treatment caused a significant increase in SOD activity in oregano across all watering conditions.

Changes in catalase (CAT) activity under drought conditions varied by plant species (Table 6). Drought led to a decrease in CAT activity in hyssop and salvia by 1.6 to 2.1 times and 1.5 to 1.6 times, respectively, while in oregano, enzyme activity increased by 1.6 to 2.6 times. Selenium treatment significantly increased CAT activity in hyssop under moderate drought and in oregano under both moderate and severe drought conditions. However, selenium treatment did not have a significant effect on CAT activity in salvia.

Table 6. Effect of Selenium on the Activities of Antioxidant Enzymes under Drought Stress.

Level of Drought	SOD [1] [U mg^{-1} Protein]		CAT [µmol H_2O_2 mg^{-1} Protein min^{-1}]		APX [µmol AsA mg^{-1} Protein min^{-1}]		GPX [µmol GSH mg^{-1} Protein min^{-1}]		POD [µmol Guaiacol mg^{-1} Protein min^{-1}]	
	Control	Se	Control	Se	Control	Se	Control	Se	Control	Se
					Hyssop					
25%	1.84 ± 0.05 b [2]	1.69 ± 0.10 c	614.08 ± 50.91 a	636.06 ± 30.08 a	8.08 ± 0.45 a	8.41 ± 0.65 a	0.083 ± 0.006 c	0.144 ± 0.009 c *	0.83 ± 0.04 c	1.09 ± 0.06 b*
50%	3.97 ± 0.35 a	2.84 ± 0.15 b *	391.58 ± 28.61 b	507.24 ± 20.98 b *	7.19 ± 0.45 a	8.34 ± 0.59 a	0.224 ± 0.007 b	0.297 ± 0.020 b *	1.38 ± 0.05 b	1.23 ± 0.06 b
75%	4.42 ± 0.42 a	4.65 ± 0.31 a	291.46 ± 22.16 b	319.33 ± 11.33 c	6.78 ± 0.30 a	7.06 ± 0.41 a	0.426 ± 0.027 a	0.547 ± 0.015 a *	2.05 ± 0.11 a	1.66 ± 0.09 a *
					Salvia					
25%	2.10 ± 0.09 b	2.18 ± 0.15 a	378.38 ± 25.37 a	359.64 ± 16.08 a	8.68 ± 0.43 ab	6.34 ± 0.37 b *	0.482 ± 0.038 a	0.695 ± 0.018 a *	2.70 ± 0.08 a	2.74 ± 0.16 b
50%	2.31 ± 0.11 b	2.31 ± 0.11 a	239.06 ± 14.63 b	266.09 ± 6.11 b	7.48 ± 0.46 b	10.09 ± 0.45 a *	0.277 ± 0.009 b	0.398 ± 0.028 b *	2.69 ± 0.11 a	2.86 ± 0.19 b
75%	3.04 ± 0.20 a	2.57 ± 0.14 a *	248.31 ± 10.80 b	219.63 ±16.31 b	9.65 ± 0.70 a	11.34 ± 0.48 a *	0.103 ±0.008 c	0.192 ± 0.013 c *	2.74 ± 0.16 a	4.00 ± 0.22 a *
					Oregano					
25%	0.89 ± 0.05 a	1.13 ± 0.08 ab *	121.44 ± 6.25 c	114.02 ± 6.85 c	13.87 ± 0.71 b	15.58 ± 0.80 c	0.071 ± 0.005 a	0.131 ± 0.009 ab *	1.09 ± 0.07 b	0.73 ± 0.04 c *
50%	0.77 ± 0.03 a	1.00 ± 0.07 b *	191.36 ± 12.37 b	303.63 ± 23.13 b *	15.20 ± 1.14 b	22.24 ± 1.25 b *	0.084 ± 0.005 a	0.153 ± 0.007 a *	1.33 ± 0.05 ab	1.04 ± 0.07 b *
75%	0.94 ± 0.04 a	1.42 ± 0.08 a *	317.49 ± 20.78 a	386.62 ± 19.78 a *	21.49 ± 1.57 a	28.78 ± 1.71 a *	0.051 ± 0.004 b	0.122 ± 0.006 b *	1.53 ± 0.08 a	1.41 ± 0.08 a

[1] SOD, superoxide dismutase; U, units; CAT, catalase; APX, ascorbate peroxidase; AsA, ascorbic acid; GPX, glutathione peroxidase; GSH, glutathione; POD, peroxidase. [2] Different letters indicate significant differences between plants by drought level for control, Se-treatment, and each species separately, and asterisks * indicate significant differences between control and Se-treatment at $p \leq 0.05$ based on post hoc Tukey's test ($n = 4$).

The activity of ascorbate peroxidase (APX) in hyssop remained unchanged regardless of the drought level (Table 6). In contrast, salvia and oregano exhibited the highest APX activity under severe drought conditions, with APX activity in oregano being 1.5 times higher than in control plants. Selenium treatment led to a decrease in APX activity in salvia under control conditions but an increase in activity in both salvia and oregano under drought stress.

Changes in glutathione peroxidase (GPX) activity in response to drought were species-specific (Table 6). In hyssop, GPX activity increased sharply by 2.7 to 5.1 times, while in salvia, it decreased dramatically by 1.7 to 4.7 times under the same conditions. In oregano, a reduction in GPX activity was observed only under severe drought, with a decrease of 1.4 times. Selenium treatment stimulated GPX activity in all studied species, both in control plants and those grown under drought conditions.

Peroxidase (POD) activity increased in hyssop and oregano under drought conditions, while in salvia, POD activity remained unaffected by watering conditions (Table 6).

Selenium treatment led to an increase in POD activity in hyssop in control conditions and in salvia grown under severe drought conditions. However, in oregano in control and moderate drought stress conditions, as well as in hyssop under severe drought stress, selenium treatment caused a decrease in POD activity.

2.6. Cluster Analysis

To construct the heat map and perform cluster analysis, the results were normalized to the control value (25% drought, no selenium treatment) separately for each plant species. In addition to the control group, cluster analysis identified four additional plant groups (Figure 5).

The first group consisted of selenium-treated hyssop, salvia, and oregano plants grown under non-stress conditions. For this group, most of the measured indicators were similar to those of the control plants.

The second group comprised salvia plants grown under drought conditions, both with and without selenium treatment. This group was characterized by a significant increase in dehydroascorbic acid, ascorbic acid, and their total content.

The third group consisted of oregano plants, which were characterized by elevated malondialdehyde content, an increased AsA/DHA ratio, and higher CAT activity.

The fourth group, the most distinct from all the others, comprised of hyssop plants cultivated under drought conditions, including both selenium-treated and untreated plants. This group was notable for a sharp increase in the activities of GPX, SOD, and POD enzymes compared to the control plants.

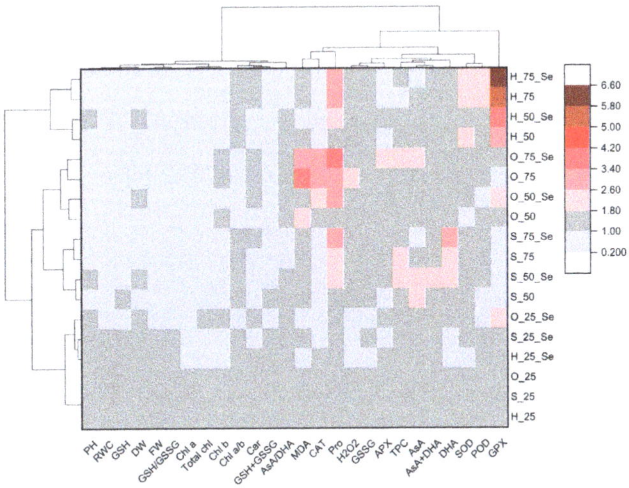

Figure 5. Heat map with clusters for the studied parameters (at the top) and *Lamiaceae* species cultivated under different levels of drought without and with Se-treatment (on the left). PH, plant height; FW, fresh weight; DW, dry weight; RWC, relative water content; MDA, malondialdehyde; H_2O_2, hydrogen peroxide; Chl *a*, chlorophyll *a*; Chl *b*, chlorophyll *b*; Chl *a*/*b*, chlorophyll *a*/*b* ratio; Total chl, total chlorophylls; Car, carotenoid; Pro, proline; GSH, reduced glutathione; GSSG, oxidized glutathione; AsA, ascorbic acid; DHA, dehydroascorbic acid; TPC, total phenolic compounds; SOD, superoxide dismutase; CAT, catalase; APX, ascorbate peroxidase; GPX, glutathione peroxidase; POD, peroxidase. H_75, H_75_Se, H_50, H_50_Se, H_25, H_25_Se, S_75, S_75_Se, S_50, S_50_Se, S_25, S_25_Se and O_75, O_75_Se, O_50, O_50_Se, O_25, O_25_Se, hyssop, salvia, and oregano plants cultivated at 75%, 50%, and 25% of drought stress without and with Se-treatment, respectively.

3. Discussion

3.1. Differences in the Response of Hyssop, Salvia, and Oregano to Drought Stress

Drought, as one of the most significant abiotic stresses, has a detrimental impact on plants. Natural populations of aromatic plants from the Lamiaceae family are found in regions such as the Mediterranean, the Massif Central in France, the Alps, the Pyrenees, the Carpathians, the Balkan Peninsula, and the Crimean Peninsula [28]. These regions are characterized by water scarcity, high solar insolation, and elevated temperatures, particularly in summer—conditions that contribute to drought stress [29]. Such environmental factors can negatively affect the growth, yield, and quality of aromatic Lamiaceae plants. However, many of these plants possess adaptive mechanisms that enable them to respond to drought stress through morphological, anatomical, physiological, biochemical, and molecular changes, which help them mitigate the effects of stress and enhance their resistance [4]. At the same time, the response to drought stress and the level of resistance can be highly species-specific, varying significantly even within the same genus [30–32].

This study focused on three species—hyssop, salvia, and oregano—that differ in their drought resistance based on the literature data. Drought stress had the least impact on hyssop's growth, with a 35% reduction in fresh weight and a 15% decrease in dry weight under severe drought conditions. As a typical xerophyte, hyssop exhibits high resistance to water scarcity [5]. In contrast, salvia and oregano, which are more sensitive to drought, showed a larger decrease in fresh weight (45–50%) and dry weight (30–40%) under similar conditions.

The reduction in plant productivity under drought stress can be attributed to several factors, many of which are related to changes in the functioning of the photosynthetic apparatus. For example, drought may reduce the absorption of photosynthetically active radiation due to delayed leaf expansion, temporary wilting or twisting of leaves during periods of high stress, and premature leaf senescence. Additionally, drought stress can impair the efficiency of utilizing absorbed photosynthetically active radiation for the production of new dry matter [33]. In our study, drought conditions in hyssop and salvia resulted in a reduction in chlorophylls *a* and *b*, as well as their total content in the leaves. This decline in chlorophyll content under water-deficient conditions may be linked to increased chlorophyllase activity, as its gene expression is known to be induced by stress. Furthermore, the inhibition of chlorophyll biosynthesis enzymes during the light stage and the degradation of chlorophylls by ROS, which are generated under stress, could also contribute to this decrease [34–36]. It is important to note that one of the plant adaptation mechanisms to drought involves regulating the size of the light-harvesting complex, which impacts the chlorophyll *a*/*b* ratio [37]. In our study, only oregano showed a reduction in this ratio by 1.3 to 1.5 times compared to control plants under drought conditions.

During drought stress, the production of ROS typically intensifies in various cellular compartments of plants [38]. In this study, a significant increase in hydrogen peroxide content was observed in salvia and oregano under drought conditions. However, in hyssop, no heightened hydrogen peroxide production was detected compared to control plants under water deficit. Despite this, all three species showed increased levels of malondialdehyde, a marker of lipid peroxidation, under drought stress. In addition to their pro-oxidant role, ROS are also involved in signal transduction, initiating a wide range of molecular, biochemical, physiological, and morphological responses to stress [14]. A cDNA microarray study identified in *Arabidopsis* 113 genes that were upregulated and 62 that were downregulated by H_2O_2, highlighting its key role in modulating plant drought responses. This includes the regulation of Ca^{2+} signaling, MAPK cascades, and gene expression [39].

Plants under water stress have developed several mechanisms to mitigate this challenge. The antioxidant system and osmotic regulation are the primary defense mechanisms that enhance crop resistance under such conditions [40]. Proline, a key osmolyte and signaling molecule, predominantly accumulates in the cytosol and plays a crucial role in protecting membranes and proteins, including enzymes, against various stresses [41]. In this study, a linear increase in proline content was observed as drought stress levels intensified across all three species. These findings are consistent with previous research showing that proline accumulation in response to drought stress is a common adaptive response in a variety of species, including members of the Lamiaceae family [42].

In addition to osmoprotectants, the antioxidant system plays a crucial role in protecting plants from drought stress. This system includes non-enzymatic antioxidants such as ascorbic acid, glutathione, phenolic compounds, and tocopherols, as well as antioxidant enzymes like SOD, CAT, APX, and GPX [14,43]. Our experiments demonstrated that the effects of drought stress on this system were species-specific. In salvia and oregano, there was an increase in the total content of phenolic compounds, ascorbic acid, and dehydroascorbic acid in response to increasing drought stress levels. In contrast, hyssop showed maximum values for these indicators only at moderate drought stress. Under severe drought, hyssop exhibited a sharp increase in GPX activity (by 5-fold), along with an increase in SOD and POD activities (by 2.5-fold). Meanwhile, in salvia, the activity of antioxidant enzymes either decreased significantly (GPX by 4.7-fold and CAT by 1.5-fold) or remained similar to control levels (POD, APX, SOD), depending on the drought intensity. In oregano, the most notable change was a 2.6-fold increase in CAT activity compared to the control under drought stress. Previous studies have shown that the activation of specific antioxidant components under drought stress depends on the stress level, duration, and plant species. Differences have been observed between drought-sensitive and drought-tolerant species, with some drought-sensitive plants exhibiting only moderate induction or even down-regulation of certain antioxidant system components [44].

Responses to drought stress are extremely different according to the genetic background. In fact, inter- and intra-species variations in drought resistance are known [45]. RWC is one of the most reliable and widely used indicators for defining both the sensitivity and the tolerance to water stress in plants [46]. In our experiments, drought stress reduced

RWC in all three species, namely in hyssop by 7.5 to 17.5%, in salvia by 7.8 to 11.6%, and in oregano by 14.4 to 28.8%. Based on these results, it can be concluded that oregano is a more drought-sensitive species compared to hyssop and salvia. However, it is worth noting that RWC may not be considered a reliable indicator of drought resistance for all species. Previous experiments on rye and wheat have shown that RWC does not correlate with the stress tolerance index [47,48]. This could explain the lack of differences in the changes in RWC between the more resistant hyssop and the more sensitive salvia under drought stress conditions. Another important indicator used as a measure of sensitivity and tolerance to water stress is proline content [49]. In our study, hyssop, being the more resistant species, exhibited a higher content of proline under control conditions (Supplements, Table S2F). Furthermore, under stress conditions, proline content in hyssop increased by 1.7 to 3.0 times, while in salvia and oregano, proline content increased by 1.4 to 2.6 and 1.6 to 2.8 times, respectively. It is also important to note the high glutathione content in hyssop under control conditions, which sharply decreased under drought conditions and was accompanied by a significant increase in glutathione peroxidase activity (Supplementary Materials Table S2H). GSH plays an important role in the antioxidant and glyoxalase systems that are vital for ROS and methylglyoxal detoxification, respectively [50]. Previously, it was shown that methylglyoxal (MG) homeostasis played an essential role in promoting plant growth, development, metabolic adaptation, signal transduction, and thereby responses to drought stress [51]. Thus, the higher drought stress resistance of hyssop may be related to the more intense biosynthesis of glutathione and its more active involvement in the cell's protective systems under stress conditions. Recent experimental data suggest that drawing conclusions about the sensitivity or resistance of plants to drought stress based on a single indicator (e.g., RWC) is not accurate. Generally, it is necessary to consider agronomic traits, as well as physiological, biochemical, and molecular markers for a comprehensive assessment. In our study, the cluster analysis conducted based on 25 indicators revealed that the response of hyssop to drought stress differed from that of salvia and oregano. However, for a more accurate interpretation, additional studies aimed at identifying differences among the studied species using molecular technologies are needed. Furthermore, expanding the research subjects to include various genotypes of a single species will allow for a deeper understanding of the mechanisms of drought stress resistance in the Lamiaceae family.

3.2. *Features of the Selenium Effect on Growth and Biochemical Changes in Hyssop, Salvia, and Oregano Under Drought Stress Conditions*

It is well established that most plants, including medicinal and aromatic species of the Lamiaceae family, are non-accumulators of selenium [52]. In this study, foliar treatment of hyssop, salvia, and oregano with selenium resulted in a significant increase in selenium content in all three species. Interestingly, the drought stress level did not influence the plants' ability to accumulate selenium, contrasting with previous findings in sunflowers, where increased drought levels led to more active selenium absorption [53].

Although selenium treatment did not have a noticeable effect on growth parameters—such as height, fresh, and dry weight—in hyssop, salvia, and oregano grown under drought stress, biochemical changes were observed. Specifically, under moderate drought stress, the RWC in hyssop and salvia increased in response to selenium treatment. Similar results have been reported in other species, such as *Triticum aestivum*, *Trifolium repens*, *Sorghum*, and *Fragaria* × *ananassa* Duch. [24,54].

Selenium treatment under drought conditions had a significant impact on the levels of hydrogen peroxide, malondialdehyde, and chlorophylls in the plants. In hyssop and oregano, foliar selenium treatment during drought led to reduced levels of hydrogen peroxide and malondialdehyde, alongside an increase in chlorophyll *a* content compared to untreated plants. However, in salvia, particularly under severe drought stress, selenium treatment resulted in elevated levels of hydrogen peroxide and malondialdehyde but a decrease in chlorophyll content. Several studies have shown that selenium can enhance plant defense systems by detoxifying intracellular free radicals and boosting both enzymatic and non-enzymatic antioxidant activities, helping plants mitigate ROS and oxidative stress [55]. The observed differences in salvia's response, though inconsistent with previous findings, could be attributed to species-specific reactions to selenium, particularly in non-accumulator plants, where selenium may act as a pro-oxidant [56]. In salvia, it is possible that selenium treatment exacerbated oxidative stress under severe drought conditions. However, further research on selenium metabolism in salvia is necessary to better interpret these results.

All species studied showed an increase in proline and phenolic compound accumulation in response to selenium treatment. Previous studies have demonstrated that selenium treatment can stimulate proline biosynthesis under stress conditions by increasing glutamyl kinase activity and reducing proline oxidase activity [56]. Additionally, selenium treatment in water-deficient plants has been found to enhance the activity of phenylalanine-ammonia-lyase, a key enzyme involved in the biosynthesis of phenolic compounds [57].

This study found that selenium treatment did not significantly alter glutathione metabolism in any of the three plant species. However, selenium did influence the levels of ascorbic and dehydroascorbic acids, particularly in salvia and oregano. Additionally, an increase in APX activity was observed in these species under drought stress conditions. Previous research has shown that APX activity can either increase or remain unchanged in response to selenium treatment, depending on the plant species, genotype, selenium concentration, and the form applied. It is worth noting that the regulation of APX activity may occur at both the transcriptional level, through increased gene expression, and at the post-translational level, via the replacement of Cys with SeCys during APX synthesis [58]. Despite the absence of a significant effect of selenium on glutathione redox metabolism in all three species, GPX activity was higher in selenium-treated plants compared to untreated ones. This may be due to selenium activating various signaling pathways that affect the expression of not only the genes encoding GPX but also glutathione reductase (GR), which aids in GSH regeneration. Furthermore, selenium treatment has been shown to increase the activity of both GPX and GR in plants under water deficiency conditions [59]. It is worth noting that for some plant species, an increase in the activity of enzymes involved in the regeneration of oxidized forms of glutathione and ascorbate has been shown (in particular, MDHAR, DHAR, and GR) [60], while changes in the activity of APX, GPX and in the level of substrates (AsA and GSH) may not be observed. Some studies demonstrate that certain conditions in different plant species activate different detoxification mechanisms within the AsA-GSH cycle, possibly due to the chemical form and concentration of applied Se and the plant species itself [58,61,62]. In addition, selenium metabolism and its transformation into organic forms in plants occur with the participation of glutathione [63]. Therefore, the peculiarities of the effect of selenium on the AsA-GSH cycle in different plant species may be associated with different intensities and rates of transformations of inorganic forms of selenium into organic ones. In this regard, for a more correct interpretation of the data obtained, it is necessary to conduct additional studies, both using different forms and concentrations of selenium, and studying the genes and enzymes involved in selenium metabolism.

4. Materials and Methods

4.1. Experimental Design and Plant Material

The study was conducted as a pot experiment under greenhouse conditions in the Institute of Living Systems at Immanuel Kant Baltic Federal University (Kaliningrad, Russia) in 2023–2024. For the experiment, seeds of three Lamiaceae plants (*Hyssopus officinalis* L. cv Lekar; *Salvia officinalis* L. cv. Kubanec; *Oregano vulgare* L. cv. Feya) were used. The seeds were sown into seed trays in the greenhouse. After germination, seedlings were grown for two months. Homogeneous seedlings of each species were transplanted into culture pots filled with the soil having the following physicochemical properties: soil texture—clay clam, pH—6.8, organic matter—2.21%, total N content—0.27%, available P—279 mg kg^{-1}, available K—206 mg kg^{-1}. The pots were 29 cm in height and 20 cm in radius. Each pot contained 10 kg of soil. There were four plants in each pot. During the growing experiment, the following conditions were maintained: humidity 70%; day—22 °C, 16 h; night—18 °C, 8 h; flux density of photosynthetically active radiation quanta—320 μmol/(m^2·s).

After transplantation to pots, plants were grown for two weeks under normal irrigation (75% of the soil field capacity (SFC)) for root adaptation and stabilization in the soil. Following this acclimatization period, the experiment involving drought stress and selenium-treatment began.

Drought stress was applied using a reduced irrigation strategy. Treatments were subjected to three levels of irrigation: (i) control—plants were irrigated up to 75% of SFC (this corresponds to 25% of the drought stress level); (ii) moderate drought stress—plants were irrigated up to 50% of SFC (this corresponds to 50% of the drought stress level); (iii) severe drought stress—plants were irrigated up to 25% of SFC (this corresponds to 75% of the drought stress level). The weight method was used to calculate the SFC. The pots were weighed every day, and the plants were irrigated with distilled water up to the corresponding target SFC.

The foliar treatment of plants with selenium was carried out using a hand atomizer once a week. As a source of selenium, an aqueous solution of sodium selenate (Na_2SeO_4) was used. The concentration of selenium was 50 μM. This concentration of selenium was selected based on a preliminary study on the effect of selenium on growth and secondary metabolite accumulation in hyssop, salvia, and oregano.

The factors of the experiment consisted of: (i) three plant species (hyssop, salvia, oregano); (ii) three levels of drought stress (25% (control), 50% (moderate drought stress), and 75% (severe drought stress)); (iii) foliar application of selenium (control and 50 μM). The experiment was conducted in a completely randomized design with four replications.

The plants were exposed to drought stress and selenium treatment for four weeks and then the harvesting was performed.

4.2. Determination of Growth Characteristics and Relative Water Content

On the day of plant collection, shoot height and fresh weight were measured. To determine dry weight, shoots were dried at 75 °C for 48 h and then weighed.

To determine the relative water content (RWC%), leaves were cut at the base of the blade at the time of harvesting the experimental plants. The weight of the fresh leaves was determined. Then, they were soaked in distilled water for 24 h at low light and low temperature (4 °C) levels. Following this, the leaves were removed from the water, blotted with a filter cloth, and the turgid weight was determined. The dry weight was measured after drying at 75 °C for 48 h. RWC% was calculated according to the formula presented in [64].

4.3. Determination of Selenium Concentration in Plant

Selenium concentration in plants was determined by hydride generation atomic absorption spectrometry (SpectrAA 220 FS, Varian, Palo Alto, CA, USA). The plant material, dried at 40 °C and crushed, was mineralized using a mixture of HNO_3 and $HClO_4$ acids. Se (VI) was reduced to Se (IV) with 6M HCl. Hydride generation was performed using $NaBH_4$. The flow rate of fine gas (argon) was 200 mL per minute, and the temperature of the quartz cuvette was 900 °C. The intensity of light absorption was measured at a wavelength of 196.0 nm [65].

4.4. Determination of the Content of Photosynthetic Pigments

The content of photosynthetic pigments (chlorophyll *a*, chlorophyll *b*, and carotenoids) was determined in plant samples frozen in liquid nitrogen and stored at −80 °C by spectrophotometry after extraction with 80% acetone. Optical absorption of the extracts was measured at wavelengths of 663 nm, 645 nm, and 470 nm (UV-3600, Shimadzu, Tokyo,

Japan). The concentration of photosynthetic pigments was calculated using the formulas presented in [66].

4.5. Determination of Oxidative Stress Parameters

Hydrogen peroxide (H_2O_2) content was determined in plant samples frozen in liquid nitrogen and stored at −80 °C by spectrophotometry using potassium iodide (KI) according to the protocol described previously [67]. Hydrogen peroxide concentration was calculated from the standard curve and converted to gram dry weight.

The determination of malondialdehyde (MDA) content was carried out by measuring the products of its reaction with thiobarbituric acid (TBA) using the spectrophotometry method. A weighed portion of plant samples frozen in liquid nitrogen and stored at −80 °C (0.3 g) was homogenized with 0.1% trichloroacetic acid (TCA). The homogenate was centrifuged at 1000 g for 10 min, and then 0.5% TBA in 20% TCA was added to the supernatant. The mixture was heated in a boiling water bath for 30 min, then it was very quickly cooled in ice and centrifuged at 1000 g for 15 min. The optical density in the samples was measured at wavelengths of 532 and 600 nm. The MDA concentration was calculated after subtracting the non-specific absorbance at 600 nm using the extinction coefficient $\varepsilon = 1.56 \times 10^5\ M^{-1}\ cm^{-1}$ [68].

4.6. Determination of Proline and Non-Enzymatic Antioxidant Content

Free proline was extracted twice from a weighed portion (0.3 g) of plant material frozen in liquid nitrogen and stored at −80 °C with 3% sulfosalicylic acid. The homogenate was centrifuged at $1000\times g$ for 15 min. The proline content in the supernatant was determined with the ninhydrin reagent as described previously [56].

The total content of phenolic compounds was determined in lyophilized and crushed plant materials by spectrophotometry using the Folin–Ciocalteu reagent. The extraction of phenolic compounds was performed with a 70% aqueous solution of ethanol three times. Optical absorption was measured at a wavelength of 765 nm. Gallic acid was used as a standard [69]. The total content of phenolic compounds was expressed as mg of gallic acid equivalents per gram of dry weight.

The content of reduced and oxidized glutathione was determined in plant samples frozen in liquid nitrogen and stored at −80 °C by spectrophotometry according to the protocol presented in [70]. The content of oxidized glutathione was determined using the oxidation reaction of 5,5′-dithio-bis (2-nitrobenzoic acid). Then, the total content of glutathione was determined by the enzymatic reduction in oxidized glutathione with NADPH in the presence of glutathione reductase. The content of reduced glutathione was calculated as the difference between the total content of glutathione and oxidized glutathione. The content of glutathione was expressed in µmol per gram of dry weight.

The determination of ascorbic acid and its oxidized form (dehydroascorbic acid) was carried out in plant samples frozen in liquid nitrogen and stored at −80 °C by spectrophotometry according to the study [71]. The homogenization of plant material was carried out in 5% metaphosphoric acid. Determination of ascorbic acid was based on its ability to reduce Fe^{3+} to Fe^{2+}, which in a complex with 2,2′-bipyridyl gives a colored compound with an absorption maximum at 530 nm wavelength. Further, dehydroascorbic acid was determined by its ability to interact with 2,4-dinitrophenylhydrazine in the presence of 85% sulfuric acid with the production of a colored compound with an absorption maximum at 540 nm wavelength.

4.7. Determination of Antioxidative Enzyme Activity

To determine the activity of antioxidant enzymes, the plant samples frozen in liquid nitrogen and stored at −80 °C were homogenized in an ice-cooled mortar using 10 mM potassium phosphate buffer (pH 7.0) containing 1% polyvinylpyrrolidone and 0.1 mM ethylenediaminetetraacetic acid (EDTA). To prevent enzyme inactivation during the determination of ascorbate peroxidase activity, 1 mM ascorbic acid and 2 mM β-mercaptoethanol were added to the homogenization buffer [60]. The homogenate was centrifuged at 15,000× *g* for 15 min at 4 °C. The supernatant was used for enzyme activity measuring.

SOD activity was determined by its ability to inhibit the photochemical reduction in nitroblue tetrazolium according to a previously described protocol [72]. CAT activity was determined by the decrease in optical absorption at a wavelength of 240 nm due to the decomposition of hydrogen peroxide and was calculated using an extinction coefficient of 39.4 mM^{-1} cm^{-1} [73]. APX activity was measured by the decrease in absorbance at a wavelength of 290 nm related to the oxidation of ascorbate with hydrogen peroxide and calculated using an extinction coefficient of 2.8 mM^{-1} cm^{-1} [74]. GPX activity was assessed by the decrease in reduced glutathione due to its oxidation by hydrogen peroxide [75]. POD activity was determined by spectrophotometry using guaiacol and hydrogen peroxide as substrates and calculated using an extinction coefficient of 26.6 mM^{-1} cm^{-1} (at 470 nm) [76].

To convert the activity of all enzymes per mg of protein, the protein concentration in samples was measured by spectrophotometry using the Bradford assay [77]. Bovine serum albumin was used as a standard.

4.8. Statistical Analysis

All the biochemical assays were carried out in triplicate. Statistical analysis of data were performed only with biological replications corresponding to the number of pots in each treatment (n = 4). The obtained experimental data were analyzed using a three-way ANOVA, where factor 1 was the plant species; factor 2 was the drought level; factor 3 was selenium treatment (Supplementary Materials Table S1). Since a significant relationship between a species and drought level was revealed on the studied biochemical parameters, the experimental data were additionally processed using a two-way ANOVA, where factor 1 was the drought level; factor 2 was selenium treatment, separately for each species. The significance of differences between means was determined by the post hoc Tukey's test at $p \leq 0.05$. An analysis of variance was performed using the software SigmaPlot 12.5 (Systat Software GmbH, Erkrath, Germany). Plotting of graphs, heat maps, and cluster analysis were performed using OriginPro 2019b (OriginLab Corporation, Northampton, MA, USA). The Euclidean distance was used as a measure of similarity in cluster analysis.

5. Conclusions

The study highlighted differences in the drought stress response among three plant species from the Lamiaceae family: hyssop, salvia, and oregano. Under drought conditions, hyssop exhibited a significant increase in GPX, SOD, and POD activities, oregano showed enhanced CAT activity, and salvia experienced a sharp increase in ascorbic acid content under moderate stress. Selenium treatment of all three species in drought-free conditions did not lead to notable changes in growth or biochemical characteristics. However, under drought stress, foliar selenium treatment stimulated the accumulation of proline and phenolic compounds and increased GPX activity in hyssop, salvia, and oregano. Based on the outcome of the present investigation, the foliar application of selenium to the hyssop, salvia, and oregano cultivated in arid and semiarid areas and regions suffering from temporary drought can be recommended. However, further studies aimed at evaluating the metabolism of selenium compounds in each of the studied species and exploring how selenium regulates biochemical processes at the transcriptomic and post-translational levels are required. The latter will provide new insights into the molecular mechanisms responsible for differences in the response of closely related plant species to drought and selenium treatment.

Supplementary Materials: The following supporting information can be downloaded at: https://www.mdpi.com/article/10.3390/plants13212986/s1, Table S1: Results of 3-factorial ANOVA; Table S2: Results of 2-factorial ANOVA.

Author Contributions: Conceptualization, L.S.; methodology, P.F. and P.M.; validation, P.F. and P.M.; investigation, M.A. and D.K.; resources, L.S.; writing—original draft preparation, P.F.; writing—review and editing, P.F.; visualization, M.A. and D.K.; supervision, L.S.; funding acquisition, L.S. All authors have read and agreed to the published version of the manuscript.

Funding: This research was funded by the Russian Science Foundation, grant number 22-14-00106, https://rscf.ru/en/project/22-14-00106/ (accessed on 18 September 2024).

Data Availability Statement: The original contributions presented in this study are included in the article and Supplementary Materials. Further inquiries can be directed to the corresponding author.

Conflicts of Interest: The authors declare no conflicts of interest. The funders had no role in the design of the study; in the collection, analyses, or interpretation of data; in the writing of the manuscript; or in the decision to publish the results.

References

1. Wahab, A.; Abdi, G.; Saleem, M.H.; Ali, B.; Ullah, S.; Shah, W.; Mumtaz, S.; Yasin, G.; Muresan, C.C.; Marc, R.A. Plants' Physio-Biochemical and Phyto-Hormonal Responses to Alleviate the Adverse Effects of Drought Stress: A Comprehensive Review. *Plants* **2022**, *11*, 1620. [CrossRef]
2. Osmolovskaya, N.; Shumilina, J.; Kim, A.; Didio, A.; Grishina, T.; Bilova, T.; Keltsieva, O.A.; Zhukov, V.; Tikhonovich, I.; Tarakhovskaya, E.; et al. Methodology of Drought Stress Research: Experimental Setup and Physiological Characterization. *Int. J. Mol. Sci.* **2018**, *19*, 4089. [CrossRef]
3. Seleiman, M.F.; Al-Suhaibani, N.; Ali, N.; Akmal, M.; Alotaibi, M.; Refay, Y.; Dindaroglu, T.; Abdul-Wajid, H.H.; Battaglia, M.L. Drought Stress Impacts on Plants and Different Approaches to Alleviate Its Adverse Effects. *Plants* **2021**, *10*, 259. [CrossRef]
4. Avasiloaiei, D.I.; Calara, M.; Brezeanu, P.M.; Murariu, O.C.; Brezeanu, C. On the Future Perspectives of Some Medicinal Plants within Lamiaceae Botanic Family Regarding Their Comprehensive Properties and Resistance Against Biotic and Abiotic Stresses. *Genes* **2023**, *14*, 955. [CrossRef]
5. Kizil, S.; Haşimi, N.; Tolan, V.; Kilinc, E.; Karataş, H. Chemical composition, antimicrobial and antioxidant activities of hyssop (*Hyssopus officinalis* L.) essential oil. *Not. Bot. Horti Agrobot. Cluj-Napoca* **2010**, *38*, 99–103. [CrossRef]
6. Aslani, Z.; Hassani, A.; Mandoulakani, B.A.; Barin, M.; Maleki, R. Effect of drought stress and inoculation treatments on nutrient uptake, essential oil and expression of genes related to monoterpenes in sage (*Salvia officinalis*). *Sci. Hortic.* **2023**, *309*, 111610. [CrossRef]
7. Soltanbeigi, A.; Yıldız, M.; Dıraman, H.; Terzi, H.; Sakartepe, E.; Yıldız, E. Growth responses and essential oil profile of *Salvia officinalis* L. Influenced by water deficit and various nutrient sources in the greenhouse. *Saudi J. Biol. Sci.* **2021**, *28*, 7327–7335. [CrossRef]
8. Gonceariuc, M.; Muntean, M.V.; Butnaraş, V.; Duda, M.M.; Benea, A.; Jelezneac, T.; Vornicu, Z.; Cotelea, L.; Botnarenco, P. Quality Variation of the Moldovan *Origanum vulgare* L. ssp. *vulgare* L. and *Origanum vulgare* L. ssp. *hirtum* (Link) Ietsw. Varieties in Drought Conditions. *Agriculture* **2021**, *11*, 1211. [CrossRef]
9. Emrahi, R.; Morshedloo, M.R.; Ahmadi, H.; Javanmard, A.; Maggi, F. Intraspecific divergence in phytochemical characteristics and drought tolerance of two carvacrol-rich *Origanum vulgare* subspecies: Subsp. *hirtum* and subsp. *gracile*. *Ind. Crops Prod.* **2021**, *168*, 113557. [CrossRef]
10. Paciolla, C.; Paradiso, A.; De Pinto, M. Cellular redox homeostasis as central modulator in plant stress response. In *Redox State as a Central Regulator of Plant-Cell Stress Responses*; Springer: Berlin/Heidelberg, Germany, 2016; pp. 1–23. [CrossRef]
11. Huang, H.; Ullah, F.; Zhou, D.-X.; Yi, M.; Zhao, Y. Mechanisms of ROS Regulation of Plant Development and Stress Responses. *Front. Plant Sci.* **2019**, *10*, 800. [CrossRef]
12. Mittler, R.; Zandalinas, S.I.; Fichman, Y.; Van Breusegem, F. Reactive oxygen species signalling in plant stress responses. *Nat. Rev. Mol. Cell Biol.* **2022**, *23*, 663–679. [CrossRef]
13. Das, K.; Roychoudhury, A. Reactive oxygen species (ROS) and response of antioxidants as ROS-scavengers during environmental stress in plants. *Front. Environ. Sci.* **2014**, *2*, 53. [CrossRef]
14. Oguz, M.C.; Aycan, M.; Oguz, E.; Poyraz, I.; Yildiz, M. Drought Stress Tolerance in Plants: Interplay of Molecular, Biochemical and Physiological Responses in Important Development Stages. *Physiologia* **2022**, *2*, 180–197. [CrossRef]
15. Yang, X.; Lu, M.; Wang, Y.; Wang, Y.; Liu, Z.; Chen, S. Response Mechanism of Plants to Drought Stress. *Horticulturae* **2021**, *7*, 50. [CrossRef]
16. Moulick, D.; Mukherjee, A.; Das, A.; Roy, A.; Majumdar, A.; Dhar, A.; Pattanaik, B.K.; Chowardhara, B.; Ghosh, D.; Upadhyay, M.K.; et al. Selenium—An Environmentally Friendly Micronutrient in Agroecosystem in the Modern Era: An Overview of 50-Year Findings. *Ecotoxicol. Environ. Saf.* **2021**, *270*, 115832. [CrossRef]
17. Feng, R.; Wei, C.; Tu, S. The roles of selenium in protecting plants against abiotic stresses. *Environ. Exp. Bot.* **2013**, *87*, 58–68. [CrossRef]
18. Nazir, F.; Kumari, S.; Mahajan, M.; Khan, M.I.R. An explicit story of plant abiotic stress resilience: Overtone of selenium, plant hormones and other signaling molecules. *Plant Soil.* **2023**, *486*, 135–163. [CrossRef]

19. Liu, H.; Xiao, C.; Qiu, T.; Deng, J.; Cheng, H.; Cong, X.; Cheng, S.; Rao, S.; Zhang, Y. Selenium Regulates Antioxidant, Photosynthesis, and Cell Permeability in Plants under Various Abiotic Stresses: A Review. *Plants* **2023**, *12*, 44. [CrossRef]
20. Jóźwiak, W.; Politycka, B. Effect of Selenium on Alleviating Oxidative Stress Caused by a Water Deficit in Cucumber Roots. *Plants* **2019**, *8*, 217. [CrossRef]
21. Ahmad, R.; Waraich, E.A.; Nawaz, F.; Ashraf, M.Y.; Khalid, M. Selenium (Se) improves drought tolerance in crop plants–a myth or fact? *J. Sci. Food Agric.* **2016**, *96*, 372–380. [CrossRef]
22. Adnan, M. Application of selenium a useful way to mitigate drought stress: A review. *Open Access J. Biog. Sci. Res.* **2020**, *3*, 1–4. [CrossRef]
23. Raza, M.A.S.; Aslam, M.U.; Valipour, M.; Iqbal, R.; Haider, I.; Mustafa, A.E.-Z.M.A.; Elshikh, M.S.; Ali, I.; Roy, R.; Elshamly, A.M.S. Seed priming with selenium improves growth and yield of quinoa plants suffering drought. *Sci. Rep.* **2024**, *14*, 886. [CrossRef]
24. Sapna; Sinky; Rani, B.; Pooja Pazhany, A.S.; Sharma, K.D.; Kumar, A. Selenium Mediated Amelioration of Drought in Crop Plants: A Review. In *Salinity and Drought Tolerance in Plants*; Kumar, A., Dhansu, P., Mann, A., Eds.; Springer: Singapore, 2023. [CrossRef]
25. Wang, Y.; Rao, C.; Huang, L.; Wu, J.; Sun, P.; Zhan, J.; Wu, J.; Liu, S.; Zhou, C.; Hu, L.; et al. Effects of Organic Selenium and Nanoselenium on Drought Stress of Pak Choi (*Brassica chinensis* var. *pekinensis*. cv. 'Suzhouqing') and Its Transcriptomic Analysis. *Agronomy* **2024**, *14*, 78. [CrossRef]
26. Han, D.; Tu, S.; Dai, Z.; Huang, W.; Jia, W.; Xu, Z.; Shao, H. Comparison of selenite and selenate in alleviation of drought stress in *Nicotiana tabacum* L. *Chemosphere* **2021**, *287*, 132136. [CrossRef]
27. Mansinhos, I.; Gonçalves, S.; Romano, A. How Climate Change-Related Abiotic Factors Affect the Production of Industrial Valuable Compounds in Lamiaceae Plant Species: A Review. *Front. Plant Sci.* **2024**, *15*, 1370810. [CrossRef]
28. Cheminal, A.; Kokkoris, I.P.; Strid, A.; Dimopoulos, P. Medicinal and Aromatic Lamiaceae Plants in Greece: Linking Diversity and Distribution Patterns with Ecosystem Services. *Forests* **2020**, *11*, 661. [CrossRef]
29. Munné-Bosch, S.; Mueller, M.; Schwarz, K.; Alegre, L. Diterpenes and antioxidative protection in drought-stressed *Salvia officinalis* plants. *J. Plant Physiol.* **2001**, *158*, 1431–1437. [CrossRef]
30. Kulak, M. Recurrent drought stress effects on essential oil profile of Lamiaceae plants: An approach regarding stress memory. *Ind. Crops Prod.* **2020**, *154*, 112695. [CrossRef]
31. García-Caparrós, P.; Romero, M.J.; Llanderal, A.; Cermeño, P.; Lao, M.T.; Segura, M.L. Effects of Drought Stress on Biomass, Essential Oil Content, Nutritional Parameters, and Costs of Production in Six Lamiaceae Species. *Water* **2019**, *11*, 573. [CrossRef]
32. Stefanakis, M.K.; Giannakoula, A.E.; Ouzounidou, G.; Papaioannou, C.; Lianopoulou, V.; Philotheou-Panou, E. The Effect of Salinity and Drought on the Essential Oil Yield and Quality of Various Plant Species of the Lamiaceae Family (*Mentha spicata* L., *Origanum dictamnus* L., *Origanum onites* L.). *Horticulturae* **2024**, *10*, 265. [CrossRef]
33. Sharafzadeh, S.; Zare, M. Effect of drought stress on qualitative and quantitative characteristics of some medicinal plants from Lamiaceae family: A review. *Adv. Environ. Biol.* **2011**, *5*, 2058–2062.
34. Shin, Y.K.; Bhandari, S.R.; Jo, J.S.; Song, J.W.; Lee, J.G. Effect of Drought Stress on Chlorophyll Fluorescence Parameters, Phytochemical Contents, and Antioxidant Activities in Lettuce Seedlings. *Horticulturae* **2021**, *7*, 238. [CrossRef]
35. Nikolaeva, M.K.; Maevskaya, S.N.; Shugaev, A.G.; Bukhov, N.G. Effect of drought on chlorophyll content and antioxidant enzyme activities in leaves of three wheat cultivars varying in productivity. *Russ. J. Plant Physiol.* **2010**, *57*, 87–95. [CrossRef]
36. Zonouri, M.; Javadi, T.; Ghaderi, N.; Saba, M.K. Effect of foliar spraying of ascorbic acid on chlorophyll *a*, chlorophyll *b*, total chlorophyll, carotenoids, hydrogen peroxide, leaf temperature and leaf relative water content under drought stress in grapes. *Bull. Environ. Pharmacol. Life Sci.* **2014**, *3*, 178–184.
37. Goto, K.; Yabuta, S.; Ssenyonga, P.; Tamaru, S.; Sakagami, J.I. Response of leaf water potential, stomatal conductance and chlorophyll content under different levels of soil water, air vapor pressure deficit and solar radiation in chili pepper (*Capsicum chinense*). *Sci. Hortic.* **2021**, *281*, 109943. [CrossRef]
38. Samanta, S.; Seth, C.S.; Roychoudhury, A. The molecular paradigm of reactive oxygen species (ROS) and reactive nitrogen species (RNS) with different phytohormone signaling pathways during drought stress in plants. *Plant Physiol. Biochem.* **2023**, *206*, 108259. [CrossRef]
39. Desikan, R.; Mackerness, S.A.H.; Hancock, S.J.T.; Neill, S.J. Regulation of the *Arabidopsis* transcriptome by oxidative stress. *Plant Physiol.* **2001**, *127*, 9–172. [CrossRef]
40. Bashir, S.S.; Hussain, A.; Hussain, S.J.; Wani, O.A.; Zahid Nabi, S.; Dar, N.A.; Baloch, F.S.; Mansoor, S. Plant Drought Stress Tolerance: Understanding Its Physiological, Biochemical and Molecular Mechanisms. *Biotechnol. Biotechnol. Equip.* **2021**, *35*, 1912–1925. [CrossRef]
41. Singh, M.; Kumar, J.; Singh, S.; Singh, V.P.; Prasad, S.M. Roles of osmoprotectants in improving salinity and drought tolerance in plants: A review. *Rev. Environ. Sci. Bio/Technol.* **2015**, *14*, 407–426. [CrossRef]
42. Moradi, P.; Ford-Lloyd, B.; Pritchard, J. Metabolomic Approach Reveals the Biochemical Mechanisms Underlying Drought Stress Tolerance in Thyme. *Anal. Biochem.* **2017**, *527*, 49–62. [CrossRef]
43. Mukarram, M.; Choudhary, S.; Kurjak, D.; Petek, A.; Khan, M.M.A. Drought: Sensing, signalling, effects and tolerance in higher plants. *Physiol. Plant.* **2021**, *172*, 1291–1300. [CrossRef]
44. Laxa, M.; Liebthal, M.; Telman, W.; Chibani, K.; Dietz, K.-J. The Role of the Plant Antioxidant System in Drought Tolerance. *Antioxidants* **2019**, *8*, 94. [CrossRef]

45. Rampino, P.; Pataleo, S.; Gerardi, C.; Mita, G.; Perrotta, C. Drought stress response in wheat: Physiological and molecular analysis of resistant and sensitive genotypes. *Plant Cell Environ.* **2006**, *29*, 2143–2152. [CrossRef]
46. Maghsoodi, M.; Razmjoo, J. Identify Physiological Markers for Drought Tolerance in Alfalfa. *Agron. J.* **2015**, *107*, 149–157. [CrossRef]
47. Geravandi, M.; Farshadfar, E.; Kahrizi, D. Evaluation of some physiological traits as indicators of drought tolerance in bread wheat genotypes. *Russ. J. Plant Physiol.* **2011**, *58*, 69–75. [CrossRef]
48. Farshadfar, E.; Mohammadi, R.; Aghaee, M.; Sutka, J. Identification of QTLs involved in physiological and agronomic indicators of drought tolerance in rye using a multiple selection index. *Acta Agron. Hung.* **2003**, *51*, 419–428. [CrossRef]
49. Parida, A.K.; Dagaonkar, V.S.; Phalak, M.S.; Aurangabadkar, L.P. Differential responses of the enzymes involved in proline biosynthesis and degradation in drought tolerant and sensitive cotton genotypes during drought stress and recovery. *Acta Physiol. Plant.* **2008**, *30*, 619–627. [CrossRef]
50. Nahar, K.; Hasanuzzaman, M.; Alam, M.M.; Fujita, M. Roles of exogenous glutathione in antioxidant defense system and methylglyoxal detoxification during salt stress in mung bean. *Biol. Plant.* **2015**, *59*, 745–756. [CrossRef]
51. Askari-Khorasgani, O.; Pessarakli, M. Manipulation of plant methylglyoxal metabolic and signaling pathways for improving tolerance to drought stress. *J. Plant Nutr.* **2019**, *42*, 1268–1275. [CrossRef]
52. Pilon-Smits, E.A.H. On the Ecology of Selenium Accumulation in Plants. *Plants* **2019**, *8*, 197. [CrossRef]
53. Ameen, M.; Zia, M.A.; Najeeb Alawadi, H.F.; Naqve, M.; Mahmood, A.; Shahzad, A.N.; Khan, B.A.; Alhammad, B.A.; Aljabri, M.; Seleiman, M.F. Exogenous application of selenium on sunflower (*Helianthus annuus* L.) to enhance drought stress tolerance by morpho-physiological and biochemical adaptations. *Front. Plant Sci.* **2024**, *15*, 1427420. [CrossRef]
54. Zahedi, S.M.; Moharrami, F.; Sarikhani, S.; Padervand, M. Selenium and silica nanostructure-based recovery of strawberry plants subjected to drought stress. *Sci. Rep.* **2020**, *10*, 17672. [CrossRef]
55. Lanza, M.G.D.B.; Reis, A.R.D. Roles of selenium in mineral plant nutrition: ROS scavenging responses against abiotic stresses. *Plant Physiol. Biochem.* **2021**, *164*, 27–43. [CrossRef]
56. Khan, M.I.R.; Nazir, F.; Asgher, M.; Per, T.S.; Khan, N.A. Selenium and sulfur influence ethylene formation and alleviate cadmium-induced oxidative stress by improving proline and glutathione production in wheat. *J. Plant Physiol.* **2015**, *173*, 9–18. [CrossRef]
57. Neysanian, M.; Iranbakhsh, A.; Ahmadvand, R.; Ardebili, Z.O.; Ebadi, M. Selenium nanoparticles conferred drought tolerance in tomato plants by altering the transcription pattern of *microRNA-172* (*miR-172*), *bZIP*, and *CRTISO* genes, upregulating the antioxidant system, and stimulating secondary metabolism. *Protoplasma* **2024**, *261*, 735–747. [CrossRef]
58. Popović, A.V.; Čamagajevac, I.; Vuković, R.; Matić, M.; Velki, M.; Gupta, D.K.; Galić, V.; Lončarić, Z. Biochemical and molecular responses of the ascorbate-glutathione cycle in wheat seedlings exposed to different forms of selenium. *Plant Physiol. Biochem.* **2024**, *208*, 108460. [CrossRef]
59. Hasanuzzaman, M.; Fujita, M. Selenium Pretreatment Upregulates the Antioxidant Defense and Methylglyoxal Detoxification System and Confers Enhanced Tolerance to Drought Stress in Rapeseed Seedlings. *Biol. Trace Elem. Res.* **2011**, *143*, 1758–1776. [CrossRef]
60. Wang, C.Q.; Xu, H.J.; Liu, T. Effect of selenium on ascorbate–glutathione metabolism during peg-induced water deficit in *Trifolium repens* L. *J. Plant Growth Regul.* **2011**, *30*, 436–444. [CrossRef]
61. Silva, V.M.; Tavanti, R.F.R.; Gratão, P.L.; Alcock, T.D.; dos Reis, A.R. Selenate and selenite affect photosynthetic pigments and ROS scavenging through distinct mechanisms in cowpea (*Vigna unguiculata* (L.) walp) plants. *Ecotoxicol. Environ. Saf.* **2020**, *201*, 110777. [CrossRef]
62. Das, D.; Seal, P.; Biswas, S.; Biswas, A.K. Modulation of ascorbate-glutathione cycle by selenate and sulphate treatments in the seedlings of two rice (*Oryza sativa* L.) cultivars. *Plant Sci. Today* **2020**, *7*, 441–455. [CrossRef]
63. Chauhan, R.; Awasthi, S.; Srivastava, S.; Dwivedi, S.; Pilon-Smits, E.A.H.; Dhankher, O.P.; Tripathi, R.D. Understanding selenium metabolism in plants and its role as a beneficial element. *Crit. Rev. Environ. Sci. Technol.* **2019**, *49*, 1937–1958. [CrossRef]
64. Khaleghi, A.; Naderi, R.; Brunetti, C.; Maserti, B.E.; Salami, S.A.; Babalar, M. Morphological, physiochemical and antioxidant responses of *Maclura pomifera* to drought stress. *Sci. Rep.* **2019**, *9*, 19250. [CrossRef]
65. Kurkova, T.; Skrypnik, L.; Zalieckienėe, E. Features of plant material pre-treatment for selenium determination by atomic absorption and fluorimetric methods. *Chemija* **2008**, *19*, 40–43.
66. Asghari, J.; Mahdavikia, H.; Rezaei-Chiyaneh, E.; Banaei-Asl, F.; Amani Machiani, M.; Harrison, M.T. Selenium Nanoparticles Improve Physiological and Phytochemical Properties of Basil (*Ocimum basilicum* L.) under Drought Stress Conditions. *Land* **2023**, *12*, 164. [CrossRef]
67. Sharma, S.; Anand, G.; Singh, N.; Kapoor, R. Arbuscular mycorrhiza augments arsenic tolerance in wheat (*Triticum aestivum* L.) by strengthening antioxidant defense system and thiol metabolism. *Front. Plant Sci.* **2017**, *8*, 906. [CrossRef]
68. Akcay, U.C.; Ercan, O.; Kavas, M.; Yildiz, L.; Yilmaz, C.; Oktem, H.A.; Yucel, M. Drought-induced oxidative damage and antioxidant responses in peanut (*Arachis hypogaea* L.) seedlings. *Plant Growth Regul.* **2010**, *61*, 21–28. [CrossRef]
69. Aalipour, H.; Nikbakht, A.; Sabzalian, M.R. Essential oil composition and total phenolic content in *Cupressus arizonica* G. in response to microbial inoculation under water stress conditions. *Sci. Rep.* **2023**, *13*, 1209. [CrossRef]

70. Lwalaba, J.L.W.; Louis, L.T.; Zvobgo, G.; Richmond, M.E.A.; Fu, L.; Naz, S.; Mwamba, M.; Mundende, R.P.M.; Zhang, G. Physiological and molecular mechanisms of cobalt and copper interaction in causing phyto-toxicity to two barley genotypes differing in Co tolerance. *Ecotoxicol. Environ. Saf.* **2020**, *187*, 109866. [CrossRef]
71. Kaur, E.; Bhardwaj, R.D.; Kaur, S.; Grewal, S.K. Drought stress-induced changes in redox metabolism of barley (*Hordeum vulgare* L.). *Biol. Futur.* **2021**, *72*, 347–358. [CrossRef] [PubMed]
72. Giannopolitis, C.N.; Ries, S.K. Superoxide dismutases: I. Occurrence in higher plants. *Plant Physiol.* **1977**, *59*, 309–314. [CrossRef] [PubMed]
73. Aebi, H. Catalase In Vitro. *Methods Enzymol.* **1984**, *105*, 121–126.
74. Nakano, Y.; Asada, K. Hydrogen peroxide is scavenged by ascorbate specific peroxidase in spinach chloroplasts. *Plant Cell Physiol.* **1981**, *22*, 867–880.
75. Hartikainen, H.; Xue, T.; Piironen, V. Selenium as an anti-oxidant and pro-oxidant in ryegrass. *Plant Soil.* **2000**, *225*, 193–200. [CrossRef]
76. Goharrizi, K.J.; Riahi-Madvar, A.; Rezaee, F.; Pakzad, R.; Bonyad, F.J.; Ahsaei, M.G. Effect of Salinity Stress on Enzymes' Activity, Ions Concentration, Oxidative Stress Parameters, Biochemical Traits, Content of Sulforaphane, and *CYP79F1* Gene Expression Level in *Lepidium draba* Plant. *J. Plant Growth Regul.* **2020**, *39*, 1075–1094. [CrossRef]
77. Bradford, M.M. A rapid and sensitive method for the quantitation of microgram quantities of protein utilizing the principle of protein-dye binding. *Anal. Biochem.* **1976**, *72*, 248–254. [CrossRef]

plants

Article

Physiology and Metabolism Alterations in Flavonoid Accumulation During Buckwheat (*Fagopyrum esculentum* Moench.) Sprouting

Meixia Hu [1], Jia Yang [2], Jing Zhang [1], Weiming Fang [1] and Yongqi Yin [1,*]

[1] College of Food Science and Engineering, Yangzhou University, Yangzhou 225009, China; mz120222068@stu.yzu.edu.cn (M.H.); mz120222090@stu.yzu.edu.cn (J.Z.); wmfang@stu.yzu.edu.cn (W.F.)
[2] Yangzhou Center for Food and Drug Control, Yangzhou 225000, China; jiajia82112001@163.com
* Correspondence: yqyin@yzu.edu.cn

Abstract: In this research, we investigated the physiological modifications, flavonoid metabolism, and antioxidant systems of two buckwheat (*Fagopyrum esculentum* Moench.) cultivars, Pintian and Suqiao, during germination. The results demonstrated an initial increase followed by a subsequent decline in the flavonoid content of the buckwheat sprouts throughout germination. On the third day of germination, the highest flavonoid concentrations were observed, with the Pintian and Suqiao varieties reaching 996.75 and 833.98 μg/g fresh weight, respectively. Both the activity and relative gene expression level of the flavonoid metabolizing enzyme showed a significant rise in 3-day-old buckwheat sprouts, which was strongly correlated with the flavonoid content. The correlation analysis revealed that the buckwheat sprouts accumulated flavonoids by enhancing the activities and gene expression levels of flavonoid synthases. The antioxidant capacity and the activities and gene expression profiles of the antioxidant enzymes in both buckwheat cultivars notably increased after three days of germination. The correlation analysis indicated a significant positive link between antioxidant capacity and the activity and gene expression levels of the antioxidant enzymes, flavonoid content, and total phenol content. This research demonstrated that germination treatment can significantly boost the accumulation of flavonoids and total phenols, thereby enhancing the antioxidant properties of buckwheat sprouts, despite variations among different buckwheat varieties.

Keywords: buckwheat; germination; flavonoids; antioxidant capacity; gene expression

1. Introduction

Buckwheat (*Fagopyrum esculentum* Moench.)is cultivated and eaten worldwide [1]. Recently, with increasing awareness of healthy eating, buckwheat has gained attention for its nutrients and bioactive substances. Buckwheat is a pseudocereal that, unlike grains with nutrients (carbohydrates, proteins, and vitamins), is high in protein, containing between 7 and 21% [2]. Buckwheat is also a significant source of essential minerals. It contains high levels of copper, manganese, and zinc, which are crucial for maintaining various bodily functions [3]. Meanwhile, buckwheat, as a cereal crop abundant in bioactive compounds, features a remarkable chemical composition where total flavonoids hold a significant position [4]. Flavonoids, a class of polyphenolic secondary metabolites widely found in the plant kingdom, not only bestow buckwheat with beneficial properties such as antioxidant [5], anti-inflammatory [6], and blood pressure-lowering [7] properties. Meanwhile, clinical studies have shown that flavonoids extracted from buckwheat mainly inhibit the growth of human acute myeloid leukemia HL 60 cells by increasing flow cytometry annexin binding capacity and activating caspase 3 protein, thereby achieving anti-cancer properties [8]. In addition, flavonoids have become a focal point for the scientific community to conduct in-depth research into their functional mechanisms and potential applications. Therefore, the synthesis of flavonoids in plants has received more and more attention. However, the content of flavonoids is influenced by various environmental factors.

As an inherent bioprocessing methodology, germination significantly enhances the biological activity and functional attributes of plants [9]. Post-germination, the macromolecules of starch and protein within the seeds are decomposed into simpler nutrients, thereby facilitating easier absorption and utilization by the human body [10]. Concurrently, during the germination process, plants synthesize an array of bioactive constituents, encompassing phenolic acids [11] and polyphenols [12]. Liu et al. [13] concurrently germinated 17 types of seeds and observed a substantial accumulation of phenolic acids and flavonoids in the resulting sprouts. Ji et al. [14] established a significant relationship between the accumulation of flavonoids and the primary enzymes responsible for their biosynthesis. The germination of seeds initiates a series of intricate physiological and biochemical transformations, with the activation of the antioxidant enzyme system being of particular importance. Peng et al. [15] found that the contents of total phenols, total flavonoids, and antioxidant capacity in buckwheat sprouts were higher than those in buckwheat seeds. Zhang et al. [16] studied the germination process of seeds and found that flavonoid content, total phenol, and the activity of four antioxidant enzymes were enhanced. Research has confirmed that there are significant differences in the composition and content of flavonoids among various buckwheat cultivars [17–20] and different tissues [21,22] of buckwheat sprouts. Generally, the flavonoid content in bitter buckwheat sprouts is higher compared to that in common buckwheat, particularly in the cotyledons, where flavonoid levels surpass those in other tissue regions. In recent years, an increasing number of studies have focused on controlling germination conditions to enhance flavonoid accumulation in buckwheat sprouts. Appropriate light intensity [23,24], light sources [25,26], ultraviolet irradiation [14,27,28], laser treatment [10], exogenous compounds [29–32], and microwave treatment [33] have all been shown to significantly increase flavonoid content in buckwheat sprouts. Consequently, the antioxidant capacity of these flavonoid-rich buckwheat sprouts has also been markedly enhanced, establishing a foundational basis for developing nutritional foods or dietary supplements derived from buckwheat sprouts [34]. Furthermore, the enzymes involved in flavonoid metabolism are crucial for the synthesis and accumulation of flavonoids during the process of plant germination. Plant flavonoids are primarily synthesized via the phenylpropanoid metabolic pathway. The enzyme activities of phenylalanine ammonia lyase (PAL), cinnamic acid 4-hydroxylase (C4H), 4-coumarate coenzyme A ligase (4CL), and chalcone isomerase (CHI) within the flavonoid biosynthetic pathway are intricately linked to the gene expression of these respective enzymes, as well as to the synthesis and accumulation of flavonoids [35]. Ren et al. [36] demonstrated a significant positive linear relationship between PAL activity and the accumulation of flavonoids in both common and bitter varieties of buckwheat ($p < 0.01$). Chen et al. [37] observed an increase in the concentration of phenolic compounds in sprouting buckwheat, a phenomenon that was modulated by the enzymatic activities of PAL and CHI. Similarly, Li et al. [38] reported that during the germination of barley, there was an elevation in the activities of PAL, C4H, and 4CL, which corresponded with the accumulation of phenolic compounds. In addition, Ma et al. [39] investigated the impact of microwave irradiation on the expression levels of phenylalanine ammonia-lyase (PAL), chalcone isomerase (CHI), and flavanol synthase (FLS), as well as the concentration of flavonoid substances in buckwheat sprouts. Furthermore, Rao et al. [40] utilized quantitative proteomics to reveal that mildly acidic electrolyzed water induces the accumulation of flavonoid compounds by increasing the abundance of enzyme proteins in the phenylpropanoid biosynthetic pathway within buckwheat sprouts. These studies, however, did not examine the enzymatic activity and comparative gene expression profiles of flavonoid biosynthetic enzymes across the complete metabolic pathway of phenylpropanoids during the germination process of genetically engineered buckwheat sprouts.

This research aimed to explore the regulatory mechanisms of antioxidant activity and flavonoid accumulation in two cultivars (Pintian and Suqiao) during germination.

Initially, the physiological and biochemical dynamics throughout the growth of buckwheat sprouts were examined. Subsequently, the alterations in the antioxidant system during germination were disclosed. Finally, the molecular mechanisms behind flavonoid accumulation in buckwheat were elucidated by assessing the functional activity and transcriptional levels of flavonoid synthases throughout germination. The significance of this study lies in providing a theoretical foundation for comprehending the mechanism of flavonoid accumulation during buckwheat germination and offering technical support for the production of functional foods enriched with flavonoids.

2. Results

2.1. Changes in Buckwheat Sprouts in Biochemical and Physiological Indexes

The biochemical and physiological transformations of buckwheat sprouts during the germination phase are illustrated in Figure 1. The sprout growth is visibly apparent (Figure 1I). The fresh weight of the sprouts also increases during the growth phase (Figure 1II). Throughout germination, the overall phenolic and flavonoid concentrations present in buckwheat sprouts initially rise and then fall, with peak values occurring after 3 days of germination (Figure 1III,IV). Compared to ungerminated buckwheat (0 d), the flavonoid contents of the Pintian sprouts and Suqiao sprouts increased by 53.2% and 49.3%, respectively. Moreover, a notable disparity in the total phenolic and flavonoid content was identified between the Pintian and Suqiao sprouts following 3 days of germination ($p < 0.05$). At this point, the total phenolic and flavonoid concentrations in the Pintian sprouts were found to be 1.3 and 1.2 times greater, respectively, than those in the Suqiao sprouts. These findings suggest that while the overall trend in total phenolic and flavonoid contents during germination is similar across different buckwheat cultivars, the extent of change varies.

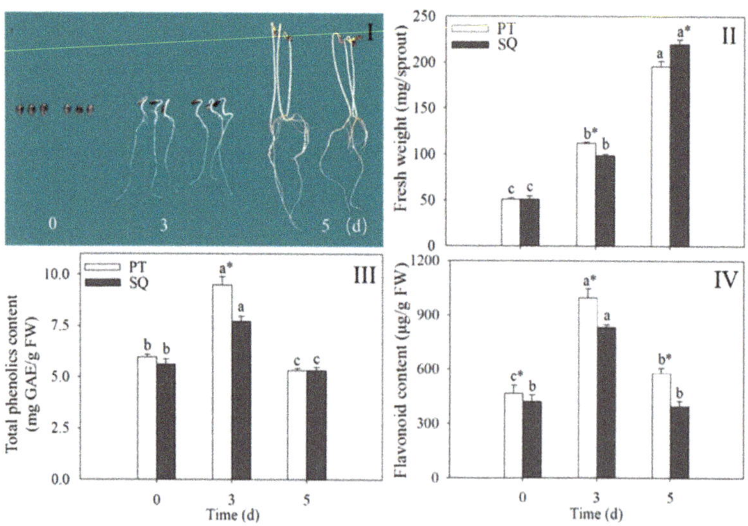

Figure 1. Changes in growth morphology (**I**), fresh weight (**II**), total phenolic content (**III**), and flavonoid content (**IV**) in buckwheat sprouts. Distinct lowercase letters signify statistically significant differences in germination times for the same buckwheat cultivar ($p < 0.05$). Asterisks (*) denote significant differences among buckwheat cultivars at identical germination times ($p < 0.05$). PT: Pintian. SQ: Suqiao.

2.2. Changes in Antioxidant Capacity in Buckwheat Sprouts

Throughout the germination phase, the antioxidant capacity initially increased, peaking at three days, and subsequently declined in the buckwheat sprouts (Figure 2). On the third day following the initiation of germination, the antioxidant activity exhibited a statistically significant increase in comparison with both 0 d and 5 d ($p < 0.05$). Following a three-day period of germination, the antioxidant capacities, as measured by ABTS and FRAP, were notably greater in the Pintian sprouts compared with the Suqiao sprouts—exceeding them by factors of 1.1 and 1.2, respectively. However, no notable variation in DPPH activity was detected between the Pintian and Suqiao sprouts after three days ($p > 0.05$). This suggests that the antioxidant capacities among different varieties of buckwheat vary throughout their growth stages.

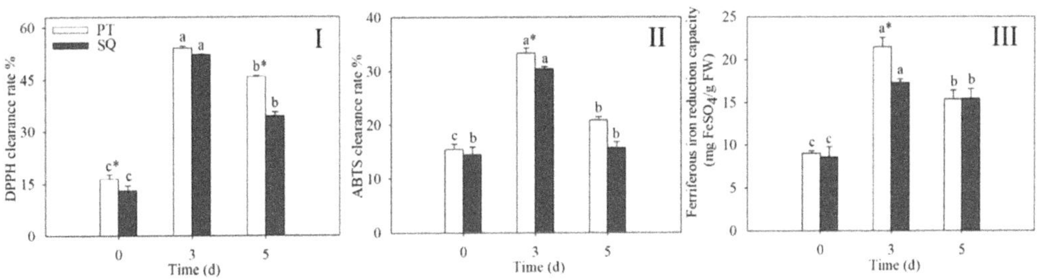

Figure 2. Changes in the capacities of DPPH (**I**), ABTS (**II**), and FRAP (**III**) in buckwheat sprouts are shown. Distinct lowercase letters signify statistically significant differences in germination times for the same buckwheat cultivar ($p < 0.05$). Asterisks (*) denote significant differences among buckwheat cultivars at identical germination times ($p < 0.05$). PT: Pintian. SQ: Suqiao.

2.3. *Changes in the Activity and Gene Expression Levels of the Flavonoid Synthase Enzyme in Buckwheat Sprouts*

Figure 3 depicts the changes in both the activity and gene expression levels of the flavonoid synthetase. The observed trend exhibits a preliminary rise succeeded by a decline, with the peak value occurring at 3 days post-germination. Compared to ungerminated buckwheat (0 d), a notable increase ($p < 0.05$) was observed in both the activity and comparative gene expression of flavonoid-metabolizing enzymes at 3 days of germination. However, different buckwheat cultivars displayed variable enzymatic activities and gene expression profiles at the identical germination phase. Specifically, the activity of PAL in the Suqiao sprouts was observed to be 1.2 times greater than that in the Pintian sprouts after 3 days of germination (Figure 3I). Conversely, the Pintian sprouts exhibited elevated activities of C4H, 4CL, and CHI, with increases of 1.3, 1.6, and 1.2 times, respectively, when compared to the Suqiao sprouts (Figure 3II–IV). The enzymatic activity and associated gene expression levels related to flavonoid metabolism in both cultivars exhibited statistically significant differences on the third day of germination ($p < 0.05$). Additionally, the findings in Figure 3I were corroborated by Figure 3V, demonstrating a marked increase in PAL expression levels in the Suqiao sprouts compared to the Pintian sprouts at 3 days of germination. However, at 3 days of germination, the corresponding gene expression levels of the flavonoid-metabolizing enzymes (C4H, 4CL, CHI, Chalcone synthase (CHS) and Flavanone 3-hydroxylase (F3H)) in the Pintian sprouts were significantly increased compared to the Suqiao sprouts ($p < 0.05$) (Figure 3VI–X).

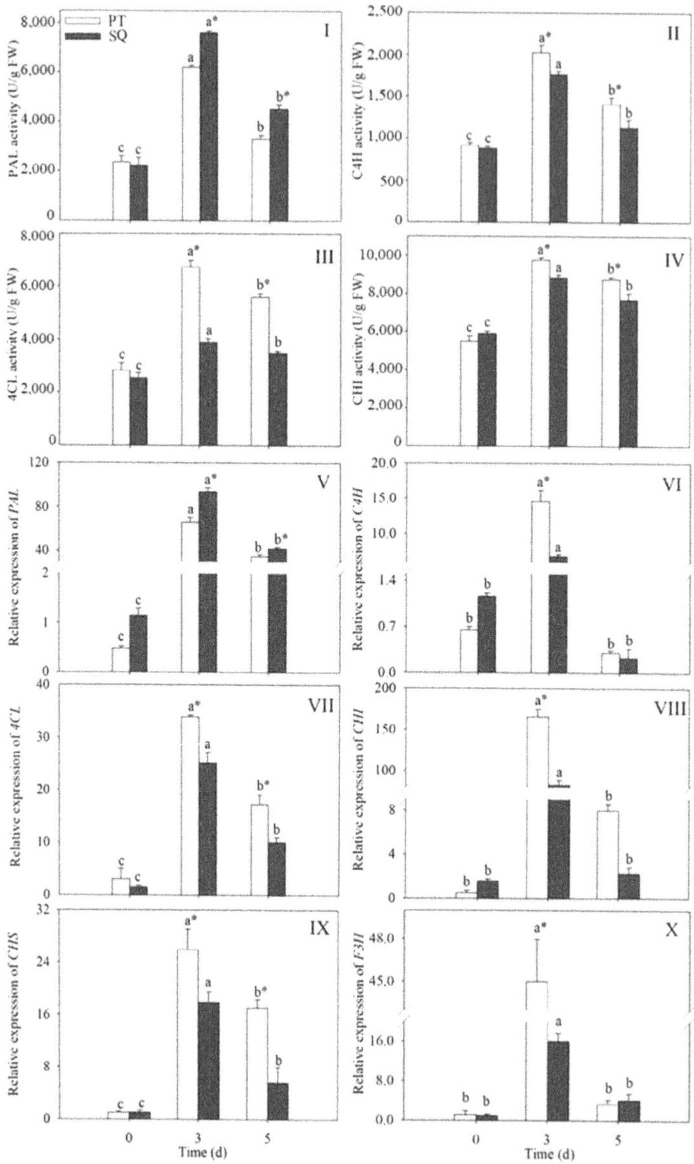

Figure 3. Changes in the activities of PAL (**I**), C4H (**II**), 4CL (**III**), and CHI (**IV**) and the expression levels of *PAL* (**V**), *C4H* (**VI**), *4CL* (**VII**), *CHI* (**VIII**), *CHS* (**IX**), and *F3H* (**X**) in buckwheat sprouts. Distinct lowercase letters signify statistically significant differences in germination times for the same buckwheat cultivar ($p < 0.05$). Asterisks (*) denote significant differences among buckwheat cultivars at identical germination times ($p < 0.05$). PT: Pintian. SQ: Suqiao.

2.4. Changes in the Activities and Gene Expression Levels of Antioxidant Enzymes in Buckwheat Sprouts

Throughout the germination phase, the activities and gene expression patterns of the antioxidant enzymes in the buckwheat sprouts exhibited an initial increase followed by a subsequent decline, with the highest levels observed at 3 days of germination. Compared to

0 d and 5 d, the activities and gene expression levels of the antioxidant enzymes in the buckwheat sprouts exhibited a significant increase following three days of germination ($p < 0.05$). Furthermore, following a three-day germination period, there were notable differences in both the activity and gene expression levels of the antioxidant enzymes between the Pintian sprouts and Suqiao sprouts ($p < 0.05$). Specifically, at three days post-germination, the activities of POD, CAT, SOD, and APX in the Pintian sprouts were observed to be 1.1, 1.4, 1.2, and 1.3 times greater than those in the Suqiao sprouts, respectively (Figure 4I–IV). Concurrently, the expression levels of *POD*, *CAT*, *SOD*, and *APX* in the Pintian sprouts were 1.1, 3.1, 1.5, and 5.1 times those of the Suqiao sprouts, respectively, after 3 days of germination (Figure 4V–VIII).

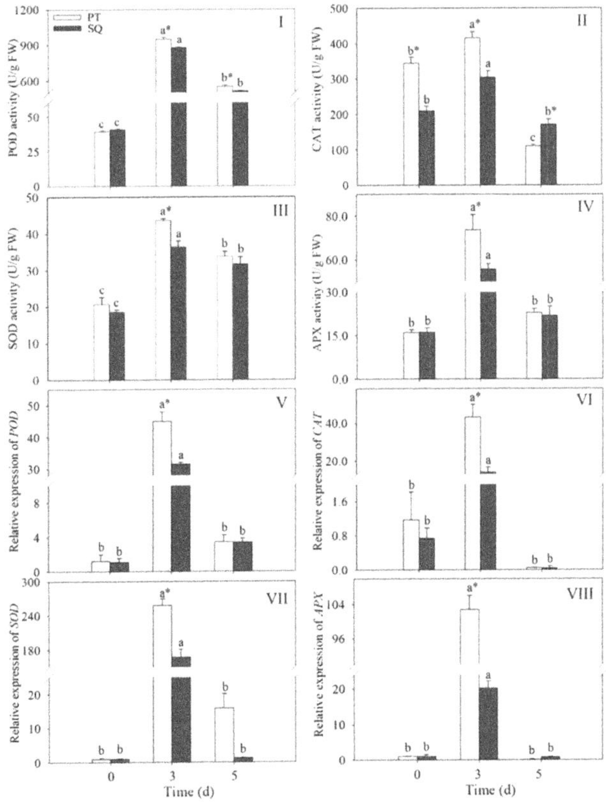

Figure 4. Changes in the activity of POD (**I**), CAT (**II**), SOD (**III**), and APX (**IV**) and the expression levels of *POD* (**V**), *CAT* (**VI**), *SOD* (**VII**), and *APX* (**VIII**) in buckwheat sprouts. Distinct lowercase letters signify statistically significant differences in germination times for the same buckwheat cultivar ($p < 0.05$). Asterisks (*) denote significant differences among buckwheat cultivars at identical germination times ($p < 0.05$). PT: Pintian. SQ: Suqiao.

2.5. The Pearson Correlation Analysis of Flavonoids with Various Indexes

The Pearson correlation analysis examining the relationship between flavonoid content and the activities and gene expression levels of antioxidant enzymes, as well as the flavonoid-metabolizing enzyme, during the germination of buckwheat is presented (Figure 5). The aim of this study was to elucidate the correlation between flavonoid levels in sprouts and various biological markers. In both the Pintian and Suqiao sprouts, the flavonoid content demonstrated a notable positive correlation with the activities of PAL,

C4H, 4CL, and CHI, and the expression levels of *PAL*, *C4H*, *4CL*, *CHI*, and *CHS*. Additionally, in the Pintian sprouts, a significant positive correlation was observed between the flavonoid content and *F3H* ($p < 0.01$), but no significant correlation was found between the flavonoid content and *F3H* in the Suqiao sprouts. The flavonoid content in the Pintian sprouts and Suqiao sprouts exhibited a substantial positive correlation with the activities of POD and APX, as well as the expression levels of *POD*, *CAT*, *SOD*, and *APX*. In the interim, a significant positive correlation was observed between the flavonoid content in the Pintian and Suqiao sprouts and their respective capacities for DPPH, ABTS, and FRAP. In the Pintian sprouts, the flavonoid concentration exhibited a notable positive association with SOD activity ($p < 0.01$) but not with CAT activity. Conversely, in the Suqiao sprouts, the flavonoid concentration exhibited a notable positive association with CAT activity ($p < 0.01$) but showed no significant correlation with SOD activity.

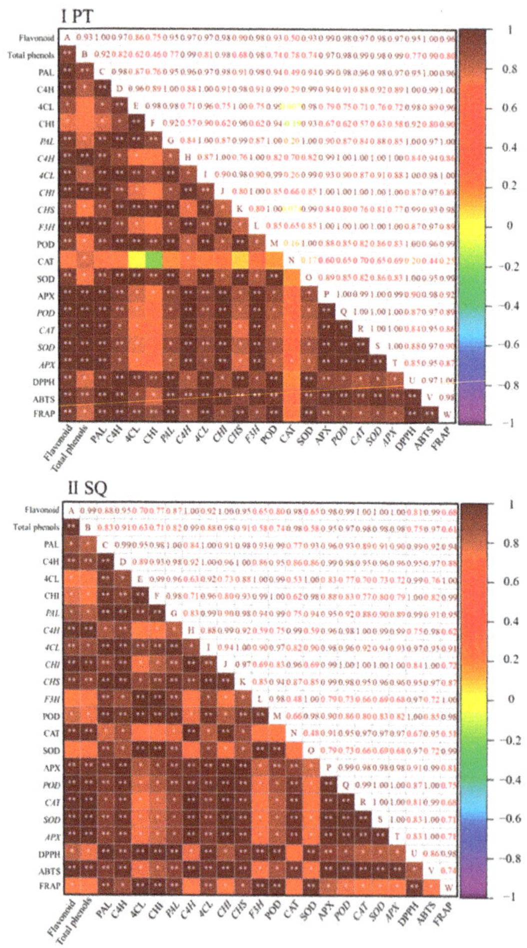

Figure 5. The examination of the correlation analysis of flavonoids with various indexes in Pintian sprouts and Suqiao sprouts is represented by (I) and (II), respectively. Negative correlations are indicated by shades of purple, blue, and green, whereas positive correlations are represented by shades of red. The correlation coefficients can vary from −1 to 1. Asterisks (*) and (**) denote that the correlation coefficient is statistically significant at *p*-value thresholds of 0.05 and 0.01, respectively. PT: Pintian; SQ: Suqiao.

2.6. Antioxidant Changes and Regulatory Mechanism of Flavonoid Metabolism in Buckwheat During Germination

Figure 6 depicts the predicted molecular process of the flavonoid production in the two buckwheat cultivars after three days of germination. At this moment, the flavonoid concentration in both cultivars had peaked. The higher flavonoid content in these cultivars is due to flavonoid production being regulated by flavonoid-metabolizing enzyme activity and gene expression levels. However, after three days of germination, the flavonoid content differed between the two cultivars, owing mostly to changes in C4H, 4CL, and CHI activity, as well as expression levels of *C4H*, *4CL*, *CHI*, *CHS*, and *F3H*. Furthermore, the antioxidant system in both germinates was active after three days, as evidenced by the antioxidant enzyme activity and gene expression. Interestingly, the antioxidant content of the Pintian sprouts topped that of the Suqiao sprouts.

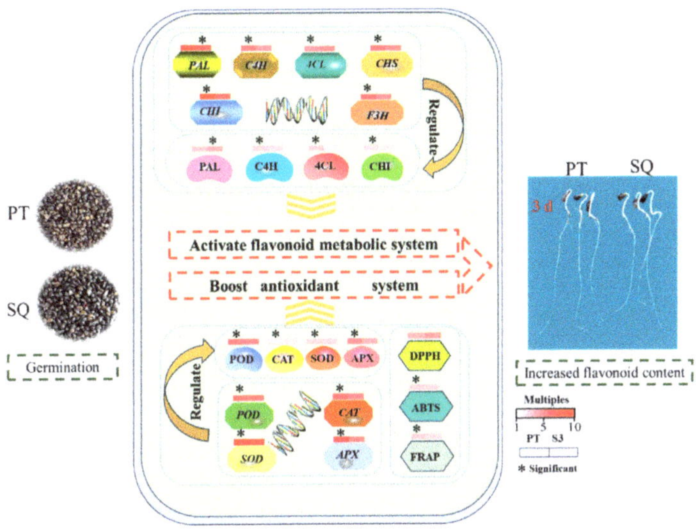

Figure 6. The mechanism diagram of germination that regulates the antioxidant system and the flavonoid metabolism system of buckwheat sprouts.

3. Discussion

Flavonoids represent the most prevalent natural secondary metabolites of polyphenols, generated through the phenylpropanoid pathway in plants during their expansion and progression [41]. Beyond facilitating the growth and development of plants, their antioxidant properties significantly contribute to human health. Flavonoids are ubiquitously present in most plant tissues [42]. Notably, there is a substantial increase in flavonoid content during the germination of seeds [43]. During the germination process, different sprout species exhibit varying levels of flavonoids. For instance, among four types of sprouts, sunflower sprouts possess the highest flavonoid concentration, having a concentration 1.2, 1.3, and 3.2 times greater than that found in broccoli sprouts, radish sprouts, and mung bean sprouts, respectively [44]. Research has confirmed that the predominant flavonoids present in buckwheat sprouts include orientin, isoorientin, vitexin, isovitexin, rutin, quercetrin, and quercetin, with varying responses to external treatments among different flavonoid monomers [17,45]. For instance, blue light treatment resulted in a notable increase in the levels of C-glycosyl flavonoids (orientin, vitexin and its isomers, and rutin and its isomers) [25]. The study by Ren et al. [36] indicated that throughout the germination phase, the concentrations of total flavonoids and rutin in both varieties of buckwheat exhibited a progressive increase, reaching their maximum levels on the 7th and 9th days of germination, respectively. In contrast, the quercetin content in bitter buckwheat begins to decline from the 5th day and becomes undetectable by the 7th day. This highlights the considerable influence of germination duration on the biosynthesis of total flavonoids and flavonoid monomers across various buckwheat cultivars. Although our current study did not assess the changes in monomeric flavonoids during the germination of the two buckwheat varieties, our research discovered that as the germination time increased, the flavonoid content of the buckwheat initially rose and then declined, peaking at three days of germination. However, the flavonoid content showed a significant difference ($p < 0.05$), with the Pintian sprouts containing 996.75 μg/g fresh weight of flavonoids, which was 1.2 times higher than that of the Suqiao sprouts. Lv et al. [46] observed that most broccoli sprouts obtained peak flavonoid content at 3 days old. The results of this investigation are consistent with the existing literature, demonstrating an initial increase in flavonoid content during the germination process, followed by a subsequent decrease. Furthermore, these fluctuations were observed to differ among the various sprout cultivars. To further investigate the variations in flavonoid biosynthesis during germination in these two buckwheat varieties, it is essential to analyze the trends in monomeric flavonoid changes in further research. Additionally, while earlier research has demonstrated an absence of a significant relationship between the leaf length of common buckwheat sprouts and their rutin content, a notable positive correlation has been identified between leaf length and rutin content in bitter buckwheat sprouts [21]. However, in our study, the physiological states of the two buckwheat seedlings (sprout length and fresh weight) observed a notably substantial positive correlation with total flavonoid content. Previous research has demonstrated that the ethanol extract of buckwheat sprouts exhibits significant reducing power, as well as notable free radical scavenging and superoxide anion scavenging activities, effectively inhibiting the production of peroxides within human liver cancer HepG2 cells and eliminating intracellular superoxide anions [19]. Research indicates that treatments utilizing blue light [25], a combination of microwave and l-phenylalanine [15], sucrose [30], and ultraviolet light [14] have led to buckwheat sprouts demonstrating not only the highest levels of total flavonoids and total phenolic compounds but also the most significant antioxidant activity. Lee et al. [47] demonstrated a significant correlation between the antioxidant capacity of buckwheat sprouts and the concentration of flavonoids present.

The research also examined fluctuations in the antioxidant potential of buckwheat throughout its developmental stages, specifically assessing DPPH, ABTS, and FRAP capacities. The results indicated that as the germination time increased, the capacities of DPPH, ABTS, and FRAP in the buckwheat initially rose before diminishing, with the highest levels recorded at three days of germination. Shabbir et al. [48] also demonstrated that the antioxidant capacity of germinated black bean sprouts significantly improved, as evidenced by DPPH, ABTS, and FRAP assays. The upward trend in antioxidant capacity in the buckwheat sprouts corresponded with increased activity of antioxidant enzymes and demonstrated a notable positive correlation with the content of flavonoids. Similarly, Lee et al. [47] demonstrated a significant correlation between the antioxidant capacity of buckwheat sprouts and the concentration of flavonoids present.

The fundamental metabolic mechanism involved in flavonoid production in buckwheat is facilitated by four enzymes: PAL, C4H, 4CL, and CHI. The enzyme-catalyzed conversion of phenylalanine by PAL, C4H, and 4CL leads to the formation of 4-coumaryl CoA. Under conditions of salt stress, the content of flavonoids is increased through the regulation of both the phenylpropanoid and flavonoid biosynthetic pathways, as well as the enzymatic activities related to these pathways [49]. Melatonin is integral to mediating flavonoid synthesis genes, significantly promoting flavonoid accumulation [50]. The principal enzymes implicated in the biosynthetic metabolism of flavonoids in buckwheat sprouts subjected to blue light and ultraviolet treatment exhibited a notable correlation with the overall flavonoid content [11]. The administration of sucrose has been shown to promote the accumulation of flavonoids in buckwheat sprouts by elevating the enzymatic activity of tyrosinase and phenylalanine ammonia-lyase [30]. The study observed that the activities of PAL, C4H, 4CL, and CHI initially increased and then decreased over the germination period. This pattern was consistent with the observation that the concentration of flavonoids diminished with increased germination duration. Notably, while the flavonoid content in the Pintian sprouts exceeded that of the Suqiao sprouts, the latter exhibited higher PAL activity compared to the Pintian sprouts. The phenylpropane metabolism pathway was catalyzed by a variety of enzymes. While the PAL activity in the Pintian sprouts was lower, the activities of C4H, 4CL, and CHI were higher compared to the Suqiao sprouts. These findings suggest that buckwheat germination promotes the accumulation of flavonoids by increasing the activities of PAL, C4H, 4CL, and CHI. The content of flavonoids exhibited a notable positive correlation with the activities of the four flavonoid synthases (PAL, C4H, 4CL, and CHI). Prior research has additionally shown a positive association between flavonoids and the effects of PAL, C4H, 4CL, and CHI [51]. Furthermore, it was proved that treatment with microwave radiation significantly enhances the transcription of multiple genes associated with flavonoid biosynthetic enzymes in bitter buckwheat sprouts, consequently facilitating the accumulation of flavonoids [39]. An analysis of gene expression at the molecular level may yield a more comprehensive understanding of the mechanisms that govern flavonoid accumulation during the germination process of buckwheat. During the germination process, the quantification of *PAL, C4H, 4CL, CHI, CHS*, and *F3H* exhibited an initial increase, peaking on the third day, after which a decline was observed. Compared to non-germinated buckwheat (0 d), the quantification of flavonoid-metabolizing enzyme genes in the buckwheat sprouts had a notable upregulation on the third day of germination. However, the relative expression levels varied across the different buckwheat cultivars. For instance, on the third day of germination, the Suqiao sprouts exhibited higher relative expression levels of *PAL* than the Pintian sprouts, which correlated with PAL activity. The same expression levels of *PAL, C4H, 4CL, CHI, CHS*, and *F3H* for flavonoids in buckwheat have been also recognized in prior research studies [52].

In the in-depth study of plant physiology and its adaptive mechanisms, a prominent area of focus is on the "dynamic regulation of the antioxidant enzyme system during the germination process". This process not only illustrates the complex reactions of living organisms to environmental fluctuations in their natural surroundings but also emphasizes the essential function of antioxidant enzymes in facilitating the transition of seeds from dormancy to active growth. With the germination of seeds, a series of complex physiological and biochemical changes are triggered, among which the activation of the antioxidant enzyme system is particularly crucial. In this investigation, the activities and quantification of antioxidant enzymes were measured throughout the developmental stages of buckwheat. The findings indicated that the gene expression levels of the antioxidant enzymes were consistent with their activities in the buckwheat sprouts. Specifically, the activities and mRNA transcription quantities of CAT, POD, and APX were increased significantly compared to those of seeds ($p < 0.05$) during the third day of germination. This trend mirrors the changes in flavonoid content during buckwheat sprout development. Additionally, the flavonoid content in buckwheat sprouts exhibited a significant positive correlation with the activities and gene expression levels of antioxidant enzymes. Similarly, prior research has shown a positive correlation between flavonoid content and antioxidant activity in vegetable sprouts [53].

4. Materials and Methods

4.1. Germination Process

Buckwheat seeds (*Fagopyrum esculentum*), the Pintian and Suqiao cultivars, were planted in 2022 and sourced from the Institute of Agricultural Sciences in Taixing City, Jiangsu Province. After the seeds were sterilized and germination was performed. The buckwheat sprouts were systematically sampled at predetermined intervals (0, 3, and 5 days) for subsequent analysis.

4.2. Measurement of Principal Physiological and Biochemical Parameters

The fresh weight of 30 buckwheat sprouts with uniform growth was measured. Every 10 buckwheat sprouts were rinsed and wiped dry to remove surface moisture before weighing. The total fresh weight of the 10 buckwheat sprouts was divided by 10 to obtain the average fresh weight of each sprout. The above experiments were repeated three times. Flavonoids were extracted and quantified according to the methodology of Yin et al. [54]. In this experiment, a total of fresh buckwheat sprouts (1.0 g) was initially subjected to grinding using 6 mL of 80% (*v*/*v*) ethanol solution, resulting in a homogenate that exhibited no discernible plant tissue. Subsequently, this homogenate underwent sonication for a duration of 25 min at a controlled temperature of 25 °C to further enhance the extraction process. The homogenate was then centrifuged. From this supernatant, an aliquot of 0.5 milliliters was meticulously withdrawn and subsequently diluted 25-fold with the same 80% (*v*/*v*) ethanol solution. The absorbance of the prepared solution was then measured at a wavelength of 260 nm.

The total phenolic compounds were extracted and quantified following Zhang et al. [55] protocol. The buckwheat sprouts (1.0 g) are ground with 5 mL of 50% (*v*/*v*) methanol, and the supernatant is collected by centrifugation. Next, 1 mL of the supernatant is mixed with 1 mL of 0.2 mM Folin–Ciocalteu reagent and 2 mL of 2% (*w*/*v*) sodium carbonate (Na_2CO_3), and then treated in the dark for 2 h. The total phenolic content is quantified by measuring the absorbance at a wavelength of 765 nm.

4.3. Assessment of Antioxidant Capacity

Antioxidant capacities were measured for 1,1-diphenyl-2-picrylhydrazyl (DPPH) and 2,2′-azino-bis(3-ethylbenzothiazoline-6-sulfonic acid) (ABTS) using the Alfuraydi et al. [56] method. To begin with, a total of 1.0 g of freshly harvested millet sprouts was subjected to extraction using 5 mL of a 50% (*v*/*v*) methanol solution. Subsequently, 2 mL of DPPH (2,2-diphenyl-1-picrylhydrazyl) solution, prepared at a concentration of 0.2 mM in ethanol, was added to 1 mL of the previously obtained extraction solution. The absorbance levels of the resulting mixture were then measured at a wavelength of 517 nm. For comparison purposes, distilled water was utilized as the control in this experimental setup.

The ABTS cation radical was produced through a chemical reaction involving a 2.45 mM solution of potassium persulfate ($K_2S_2O_8$) in combination with a 7 mM aqueous solution of ABTS, maintaining a ratio of 1:1. This mixture was allowed to stand for 12 h in a dark environment before its application. The ABTS activity assay was modified, the working solution was diluted 8-fold with absolute ethanol, and then 0.1 mL of the extract was mixed with 4.9 mL of the working solution.

Ferriferous ion reduction capacity (FRAP) was evaluated according to Zhang et al. [57]. A total of 100 µL of the extract was placed into a test tube, followed by an adjustment of the volume to 1 mL using methanol as the solvent. Subsequently, 2.5 mL of phosphate buffer was added, along with 2.5 mL of a 1% potassium ferricyanide solution. The contents of the test tube were then subjected to vortex mixing to ensure thorough mixing of the components. The resulting mixture was allowed to incubate for a duration of 20 min at a controlled temperature of 50 °C within a water bath. After the incubation, 2.5 mL of 10% (*w*/*v*) trichloroacetic acid was introduced to the mixture and centrifuged at a force of 8000× *g* for a period of 20 min. Following centrifugation, 2.5 mL of the resultant supernatant was extracted and combined with 2.5 mL of distilled water, along with 0.5 mL of a 0.1% (*w*/*v*) ferric chloride solution within another test tube. Finally, the absorbance of this solution was measured at a wavelength of 700 nm. It is important to note that a higher absorbance value is indicative of an increased reducing power of the extract being analyzed.

4.4. Enzymatic Activity Measurement of Flavonoid Biosynthesis

The activities of phenylalanine ammonia lyase (PAL), cinnamic acid 4-hydroxylase (C4H), and 4-coumarate coenzyme A ligase (4CL) were analyzed as per Yin et al. [58]. After grinding the buckwheat sprouts with Tris-HCl buffer solution, a centrifuge was used to collect the supernatant. For the purpose of this study, one unit of enzyme activity for phenylalanine ammonia-lyase (PAL), cinnamate-4-hydroxylase (C4H), and 4-coumarate-CoA ligase (4CL) was defined as a measurable change of 0.01 per minute in optical density readings at 290 nm, 340 nm, and 333 nm, respectively.

The determination of chalcone isomerase (CHI) activity was evaluated utilizing the methodology outlined by Ji et al. [14]. The supernatant obtained from the prior extraction process was utilized as a source of crude enzyme. Then, the enzyme solution was mixed with the reaction solution and allowed to react for 30 min. The evaluation of chitinase (CHI) activity was conducted by monitoring the changes in absorbance at a wavelength of 381 nm.

4.5. Evaluation of Antioxidant Enzyme Activities

Buckwheat sprouts underwent a grinding process utilizing sodium phosphate buffer at a pH of 7.0, with a concentration of 50 mM. Following this initial preparation, the resultant homogenate was subjected to centrifugation at 12,000× *g* for a duration of 15 min at 4 °C. This procedure effectively facilitated the separation of the supernatant from the solid residue.

The activities of catalase (CAT) and peroxidase (POD) were evaluated following the procedure of Zhang et al. [59]. A single unit of activity for both CAT (catalase) and POD (peroxidase) is meticulously defined as a variation or change of 0.01 per minute when measured at optical densities of 240 nm and 470 nm, respectively.

The activities of superoxide dismutase (SOD) and ascorbate peroxidase (APX) were determined using the method of Bin et al. [60]. A specific unit for measuring the activity of superoxide dismutase (SOD) and ascorbate peroxidase (APX) was established, which is characterized by a variation of 0.01 per minute at optical densities of 560 nm and 290 nm, respectively.

4.6. RNA Extraction and Quantitative Real-Time PCR Analysis

Buckwheat sprouts were treated with the E.A.N.A.TM Plant RNA Kit (R6827-01, Omega, Norcross, GA, USA) to extract total RNA. After being extracted, the RNA was reverse transcribed using the PrimeScript^{TM} RT Master Mix Kit (RR036A, Takara Bio, Kusatsu, Japan) into complementary cDNA. Each cDNA sample was evaluated in triplicate for quantitative analysis using the Light Cycler 480 II real-time PCR equipment (Roche, Basel, Switzerland) and SYBR Premix Ex-Taq^{TM} II (RR820S, Takara Bio, Kusatsu, Japan). Supplementary Table S1 provides specific information on the oligonucleotide primers used in the quantitative reverse transcription (qRT-PCR). The $2^{-\Delta\Delta Ct}$ comparative approach was conducted to ascertain the relative expression levels of genes.

4.7. Statistical Analysis

All experiments were performed in triplicate, and the results are presented as the mean ± standard deviation. The DPS data processing system software (2024) was used for statistical analysis. Statistical analysis employed one-way ANOVA and Tukey's multiple comparisons test, with significance thresholds set at $p < 0.05$ and $p < 0.01$.

5. Conclusions

In summary, two buckwheat cultivars were used to assess germination and flavonoid accumulation. Additionally, the physiology, flavonoid metabolism, and antioxidant system during buckwheat germination were investigated. The results showed that germination treatment greatly increased flavonoid and total phenol accumulation, as well as the antioxidant capacity of the buckwheat sprouts. Nevertheless, significant differences were detected between the cultivars.

Supplementary Materials: The following supporting information can be downloaded at: https://www.mdpi.com/article/10.3390/plants13233342/s1, Table S1: The sequences of the primers utilized in the study.

Author Contributions: M.H.—Writing—original draft and investigation, J.Y.—Supervision, J.Z.—Investigation, W.F.—Supervision, Y.Y.—Writing—review and editing. All authors have read and agreed to the published version of the manuscript.

Funding: This research was supported by Yangzhou Science and Technology Program (YZ2023066) and Jiangsu Provincial Administration for Market Regulation Science and Technology Program (KJ2024096).

Data Availability Statement: The data presented in this study are available on request from the corresponding author.

Conflicts of Interest: The authors declare no conflicts of interest.

References

1. Borgonovi, S.M.; Iametti, S.; Speranza, A.R.; Di Nunzio, M. Cell culture models for assessing the effects of bioactive compounds in common buckwheat (*Fagopyrum esculentum*): A systematic review. *Food Funct.* **2024**, *15*, 2799–2813. [CrossRef] [PubMed]
2. Sturza, A.; Paucean, A.; Chis, M.S.; Muresan, V.; Vodnar, D.C.; Man, S.M.; Urcan, A.C.; Rusu, I.E.; Fostoc, G.; Muste, S. Influence of buckwheat and buckwheat sprouts flours on the nutritional and textural parameters of wheat buns. *Appl. Sci.* **2020**, *10*, 7969. [CrossRef]
3. Luthar, Z.; Golob, A.; Germ, M.; Vombergar, B.; Kreft, I. Tartary buckwheat in human nutrition. *Plants* **2021**, *10*, 700. [CrossRef] [PubMed]
4. Wu, Y.L.; Li, Z.M.; Zhu, H.; Zi, R.; Xue, F.; Yu, Y. Identification of tartary buckwheat (*Fagopyrum tataricum* (L.) Gaertn) and common buckwheat (*Fagopyrum esculentum* Moench) using gas chromatography-mass spectroscopy-based untargeted metabolomics. *Foods* **2023**, *12*, 2578. [CrossRef] [PubMed]
5. Aloo, S.O.; Ofosu, F.K.; Muchiri, M.N.; Vijayalakshmi, S.; Pyo, C.G.; Oh, D.H. In vitro bioactivities of commonly consumed cereal, vegetable, and legume seeds as related to their bioactive components: An untargeted metabolomics approach using UHPLC-QTOF-MS^2. *Antioxidants* **2023**, *12*, 1501. [CrossRef] [PubMed]
6. Molska, M.; Regula, J.; Kapusta, I.; Swieca, M. Analysis of phenolic compounds in buckwheat (*Fagopyrum esculentum* Moench) sprouts modified with probiotic yeast. *Molecules* **2022**, *27*, 7773. [CrossRef]
7. Hou, Z.X.; Hu, Y.Y.; Yang, X.B.; Chen, W.S. Antihypertensive effects of tartary buckwheat flavonoids by improvement of vascular insulin sensitivity in spontaneously hypertensive rats. *Food Funct.* **2017**, *8*, 4217–4228. [CrossRef]
8. Ren, W.; Qiao, Z.; Wang, H.; Zhu, L.; Zhang, L.; Lu, Y.; Cui, Y.; Zhang, Z.; Wang, Z. Tartary buckwheat flavonoid activates caspase 3 and induces HL-60 cell apoptosis. *Method Find. Exp. Clin.* **2001**, *23*, 427–432. [CrossRef]
9. Li, R.; Li, Z.J.; Wu, N.N.; Tan, B. Effect of pre-treatment on the functional properties of germinated whole grains: A review. *Cereal Chem.* **2022**, *99*, 253–269. [CrossRef]
10. Wang, S.M.; Wang, S.M.; Wang, J.Z.; Peng, W.P. Label-free quantitative proteomics reveals the mechanism of microwave-induced Tartary buckwheat germination and flavonoids enrichment. *Food Res. Int.* **2022**, *160*, 111758. [CrossRef]
11. Abdel-Aty, A.M.; Salama, W.H.; Fahmy, A.S.; Mohamed, S.A. Impact of germination on antioxidant capacity of garden cress: New calculation for determination of total antioxidant activity. *Sci. Hortic.* **2019**, *246*, 155–160. [CrossRef]
12. Li, B.; Shen, X.T.; Shen, H.F.; Zhou, Y.; Yao, X.M. Effect of optimized germination technology on polyphenol content and hypoglycemic activity of mung bean. *Front. Nutr.* **2023**, *10*, 1138739. [CrossRef] [PubMed]
13. Liu, H.Y.; Liu, Y.; Li, M.Y.; Ge, Y.Y.; Geng, F.; He, X.Q.; Xia, Y.; Guo, B.L.; Gan, R.Y. Antioxidant capacity, phytochemical profiles, and phenolic metabolomics of selected edible seeds and their sprouts. *Front. Nutr.* **2022**, *9*, 167597. [CrossRef] [PubMed]
14. Ji, H.B.; Tang, W.; Zhou, X.L.; Wu, Y. Combined effects of blue and ultraviolet lights on the accumulation of flavonoids in tartary buckwheat sprouts. *Pol. J. Food Nutr. Sci.* **2016**, *66*, 93–98. [CrossRef]
15. Peng, W.P.; Wang, N.; Wang, S.M.; Wang, J.Z.; Bian, Z.X. Effects of microwave and exogenous l-phenylalanine treatment on phenolic constituents, antioxidant capacity and enzyme inhibitory activity of Tartary buckwheat sprouts. *Food Sci. Biotechnol.* **2023**, *32*, 11–19. [CrossRef] [PubMed]
16. Zhang, J.; Yang, J.; Yin, Y.Q. Germination promotes flavonoid accumulation of finger millet (*Eleusine coracana* L.): Response surface optimization and investigation of accumulation mechanism. *Plants* **2024**, *13*, 2191. [CrossRef]
17. Nam, T.G.; Lee, S.M.; Park, J.H.; Kim, D.O.; Baek, N.I.; Eom, S.H. Flavonoid analysis of buckwheat sprouts. *Food Chem.* **2015**, *170*, 97–101. [CrossRef]
18. Wiczkowski, W.; Szawara-Nowak, D.; Debski, H.; Mitrus, J.; Horbowicz, M. Comparison of flavonoids profile in sprouts of common buckwheat cultivars and wild tartary buckwheat. *Int. J. Food Sci. Technol.* **2014**, *49*, 1977–1984. [CrossRef]
19. Liu, C.L.; Chen, Y.S.; Yang, J.H.; Chiang, B.H. Antioxidant activity of tartary (*Fagopyrum tataricum* (L.) Gaertn.) and common (*Fagopyrum esculentum moench*) buckwheat sprouts. *J. Agric. Food Chem.* **2008**, *56*, 173–178. [CrossRef]

20. Kim, S.J.; Zaidul, I.S.M.; Suzuki, T.; Mukasa, Y.; Hashimoto, N.; Takigawa, S.; Noda, T.; Matsuura-Endo, C.; Yamauchi, H. Comparison of phenolic compositions between common and tartary buckwheat (*Fagopyrum*) sprouts. *Food Chem.* **2008**, *110*, 814–820. [CrossRef]
21. Rauf, M.; Yoon, H.; Lee, S.; Hyun, D.Y.; Lee, M.C.; Oh, S.-J.; Choi, Y.M. Evaluation of sprout growth traits and flavonoid content in common and tartary buckwheat germplasms. *Plant Breed. Biotechnol.* **2019**, *7*, 375–385. [CrossRef]
22. Watanabe, M. An anthocyanin compound in buckwheat sprouts and its contribution to antioxidant capacity. *Biosci. Biotechnol. Biochem.* **2007**, *71*, 579–582. [CrossRef]
23. Sawai, Y. Effect of light intensity on flavonoid contents of common and tartary buckwheat sprouts. *J. Jpn. Soc. Food Sci.* **2016**, *63*, 382–387. [CrossRef]
24. Shin, J.; Yang, J.H.; Yang, J.Y. Germination of tartary buckwheat at various light strengths to enhance flavonoid content and scale-up of the process using smart-farm systems. *J. Cereal Sci.* **2023**, *112*, 103727. [CrossRef]
25. Nam, T.G.; Kim, D.O.; Eom, S.H. Effects of light sources on major flavonoids and antioxidant activity in common buckwheat sprouts. *Food Sci. Biotechnol.* **2018**, *27*, 169–176. [CrossRef]
26. Nam, T.G.; Lim, Y.J.; Eom, S.H. Flavonoid accumulation in common buckwheat (*Fagopyrum esculentum*) sprout tissues in response to light. *Hortic. Environ. Biotechnol.* **2018**, *59*, 19–27. [CrossRef]
27. Yang, W.; Qian, G.T.; Chen, Y.L.; Liu, T.X.; Wan, H.H.; Wang, S.F.; Meng, X.X.; Chen, W.Q.; Atia Tul, W.; Su, Y.; et al. Profiling of polyphenols for in-depth understanding of tartary buckwheat sprouts: Correlation between cultivars and active components, dynamic changes and the effects of ultraviolet B stress. *Food Chem. X* **2022**, *14*, 100295. [CrossRef]
28. Tsurunaga, Y.; Takahashi, T.; Katsube, T.; Kudo, A.; Kuramitsu, O.; Ishiwata, M.; Matsumoto, S. Effects of UV-B irradiation on the levels of anthocyanin, rutin and radical scavenging activity of buckwheat sprouts. *Food Chem.* **2013**, *141*, 552–556. [CrossRef]
29. Debski, H.; Wiczkowski, W.; Szawara-Nowak, D.; Horbowicz, M. Elicitation with sodium silicate and iron chelate affects the contents of phenolic compounds and minerals in buckwheat sprouts. *Pol. J. Food Nutr. Sci.* **2021**, *71*, 21–28. [CrossRef]
30. Jeong, H.; Sung, J.; Yang, J.; Kim, Y.; Jeong, H.S.; Lee, J. Effect of sucrose on the functional composition and antioxidant capacity of buckwheat (*Fagopyrum esculentum* M.) sprouts. *J. Funct. Foods* **2018**, *43*, 70–76. [CrossRef]
31. Qin, P.Y.; Wei, A.C.; Zhao, D.G.; Yao, Y.; Yang, X.X.; Dun, B.Q.; Ren, G.X. Low concentration of sodium bicarbonate improves the bioactive compound levels and antioxidant and α-glucosidase inhibitory activities of tartary buckwheat sprouts. *Food Chem.* **2017**, *224*, 124–130. [CrossRef]
32. Sim, U.; Sung, J.; Lee, H.; Heo, H.; Jeong, H.S.; Lee, J. Effect of calcium chloride and sucrose on the composition of bioactive compounds and antioxidant activities in buckwheat sprouts. *Food Chem.* **2020**, *312*, 126075. [CrossRef] [PubMed]
33. Peng, W.P.; Wang, N.; Wang, S.M.; Wang, J.Z.; Bian, Z.X. Effect of co-treatment of microwave and exogenous l-phenylalanine on the enrichment of flavonoids in Tartary buckwheat sprouts. *J. Sci. Food Agric.* **2023**, *103*, 2014–2022. [CrossRef] [PubMed]
34. Kang, L.T.; Luo, J.Q.; Su, Z.P.; Zhou, L.L.; Xie, Q.T.; Li, G.Y. Effect of sprouted buckwheat on glycemic index and quality of reconstituted rice. *Foods* **2024**, *13*, 1148. [CrossRef] [PubMed]
35. Qiao, Z.N.; Shi, Y.L.; Yi, J.J.; Zhu, J.Q.; Kang, Q.Z.; Qu, L.B.; Yang, R.; Lu, J.K.; Zhao, C.C. Low frequency ultrasound enhanced the antioxidant activity and isoflavones accumulation of soybean sprouts by inducing oxidant stress. *Food Biosci.* **2024**, *60*, 104360. [CrossRef]
36. Ren, S.C.; Sun, J.T. Changes in phenolic content, phenylalanine ammonia-lyase (PAL) activity, and antioxidant capacity of two buckwheat sprouts in relation to germination. *J. Funct. Foods* **2014**, *7*, 298–304. [CrossRef]
37. Chen, Y.; Zhu, Y.; Qin, L.K. The cause of germination increases the phenolic compound contents of tartary buckwheat (*Fagorpyrum tataricum*). *J. Future Foods* **2022**, *2*, 372–379. [CrossRef]
38. Li, W.X.; Liu, X.Y.; Ma, Y.; Huang, X.Q.; Hai, D.; Cheng, Y.X.; Bai, G.; Wang, Y.P.; Zhang, B.; Qiao, M.W.; et al. Changes in physio-biochemical metabolism, phenolics and antioxidant capacity during germination of different wheat varieties. *Food Chem. X* **2024**, *22*, 101429. [CrossRef]
39. Ma, H.; Xu, X.M.; Wang, S.M.; Wang, J.Z.; Peng, W.P. Effects of microwave irradiation on the expression of key flavonoid biosynthetic enzyme genes and the accumulation of flavonoid products in fagopyrum tataricum sprouts. *J. Cereal Sci.* **2021**, *101*, 103275. [CrossRef]
40. Rao, H.; Wang, Q.F.; Han, X.; Hu, G.S.; Zhao, D.D.; Hao, J.X. Quantitative proteomics reveals the mechanism of slightly acidic electrolyzed water-induced buckwheat sprouts growth and flavonoids enrichment. *Food Res. Int.* **2021**, *148*, 110634. [CrossRef]
41. Ke, J.; Ran, B.; Sun, P.Y.; Cheng, Y.Z.; Chen, Q.F.; Li, H.Y. An evaluation of the absolute content of flavonoids and the identification of their relationship with the flavonoid biosynthesis genes in tartary buckwheat seeds. *Agronomy* **2023**, *13*, 3006. [CrossRef]
42. Tan, Z.Y.; Deng, J.; Ye, Q.X.; Zhang, Z.F. The antibacterial activity of natural-derived flavonoids. *Curr. Top. Med. Chem.* **2022**, *22*, 1009–1019. [CrossRef] [PubMed]
43. Chon, S.U. Total polyphenols and bioactivity of seeds and sprouts in several legumes. *Curr. Pharm. Des.* **2013**, *19*, 6112–6124. [CrossRef] [PubMed]
44. Pajak, P.; Socha, R.; Galkowska, D.; Roznowski, J.; Fortuna, T. Phenolic profile and antioxidant activity in selected seeds and sprouts. *Food Chem.* **2014**, *143*, 300–306. [CrossRef] [PubMed]
45. Dong, Y.L.; Wang, N.; Wang, S.M.; Wang, J.Z.; Peng, W.P. A review: The nutrition components, active substances and flavonoid accumulation of Tartary buckwheat sprouts and innovative physical technology for seeds germinating. *Front. Nutr.* **2023**, *10*, 1168361. [CrossRef] [PubMed]

46. Lv, X.G.; Meng, G.L.; Li, W.N.; Fan, D.D.; Wang, X.; Espinoza-Pinochet, C.A.; Cespedes-Acuña, C.L. Sulforaphane and its antioxidative effects in broccoli seeds and sprouts of different cultivars. *Food Chem.* **2020**, *316*, 126216. [CrossRef] [PubMed]
47. Lee, L.S.; Choi, E.J.; Kim, C.H.; Sung, J.M.; Kim, Y.B.; Seo, D.H.; Choi, H.W.; Choi, Y.S.; Kum, J.S.; Park, J.D. Contribution of flavonoids to the antioxidant properties of common and tartary buckwheat. *J. Cereal Sci.* **2016**, *68*, 181–186. [CrossRef]
48. Shabbir, U.; Tyagi, A.; Ham, H.J.; Oh, D.H. Comprehensive profiling of bioactive compounds in germinated black soybeans via UHPLC-ESI-QTOF-MS/MS and their anti-Alzheimer's activity. *PLoS ONE* **2022**, *17*, e0263274. [CrossRef] [PubMed]
49. Zhang, Q.; Zheng, G.Y.; Wang, Q.; Zhu, J.X.; Zhou, Z.H.; Zhou, W.S.; Xu, J.J.; Sun, H.Y.; Zhong, J.W.; Gu, Y.H.; et al. Molecular mechanisms of flavonoid accumulation in germinating common bean (*Phaseolus vulgaris*) under salt stress. *Front. Nutr.* **2022**, *9*, 928805. [CrossRef]
50. Song, Z.H.; Yang, Q.; Dong, B.Y.; Li, N.; Wang, M.Y.; Du, T.T.; Liu, N.; Niu, L.L.; Jin, H.J.; Meng, D.; et al. Melatonin enhances stress tolerance in pigeon pea by promoting flavonoid enrichment, particularly luteolin in response to salt stress. *J. Exp. Bot.* **2022**, *73*, 5992–6008. [CrossRef]
51. Li, X.H.; Park, N.I.; Xu, H.; Woo, S.H.; Park, C.H.; Park, S.U. Differential expression of flavonoid biosynthesis genes and accumulation of phenolic compounds in common buckwheat (*Fagopyrum esculentum*). *J. Agric. Food Chem.* **2010**, *58*, 12176–12181. [CrossRef] [PubMed]
52. Song, Y.J.; Feng, J.C.; Liu, D.M.; Long, C.L. Different Phenylalanine Pathway Responses to Cold Stress Based on Metabolomics and Transcriptomics in Tartary Buckwheat Landraces. *J. Agr. Food Chem.* **2022**, *70*, 687–698. [CrossRef]
53. Tao, H.; Zhao, Y.; Li, L.Y.; He, Y.Q.; Zhang, X.Y.; Zhu, Y.; Hong, G.J. Comparative metabolomics of flavonoids in twenty vegetables reveal their nutritional diversity and potential health benefits. *Food Res. Int.* **2023**, *164*, 112384. [CrossRef] [PubMed]
54. Yin, Y.Q.; Liu, C.; Yang, Z.F.; Fang, W.M. Ethephon promotes isoflavone accumulation in germinating soybeans by its acceleration of isoflavone biosynthetic pathway. *Plant Physiol. Biochem.* **2023**, *201*, 107805. [CrossRef]
55. Zhang, L.; Dang, B.; Lan, Y.L.; Zheng, W.C.; Kuang, J.W.; Zhang, J.; Zhang, W.A. Metabolomics characterization of phenolic compounds in colored quinoa and their relationship with in vitro antioxidant and hypoglycemic activities. *Molecules* **2024**, *29*, 1509. [CrossRef] [PubMed]
56. Alfuraydi, A.A.; Aziz, I.M.; Almajhdi, F.N. Assessment of antioxidant, anticancer, and antibacterial activities of the rhizome of ginger (*Zingiber officinale*). *J. King Saud Univ. Sci.* **2024**, *36*, 103112. [CrossRef]
57. Zhang, H.H.; Gao, P.; Fang, H.W.; Zou, M.; Yin, J.J.; Zhong, W.; Luo, Z.; Hu, C.R.; He, D.P.; Wang, X.G. High-oleic rapeseed oil quality indicators and endogenous antioxidant substances under different processing methods. *Food Chem. X* **2023**, *19*, 100804. [CrossRef]
58. Yin, Y.Q.; Tian, X.; Yang, J.; Yang, Z.F.; Tao, J.; Fang, W.M.; Fang, W.M. Melatonin mediates isoflavone accumulation in germinated soybeans (*Glycine max* L.) under ultraviolet-B stress. *Plant Physiol. Biochem.* **2022**, *175*, 23–32. [CrossRef] [PubMed]
59. Zhang, M.C.; He, S.Y.; Qin, B.; Jin, X.J.; Wang, M.X.; Ren, C.Y.; Cao, L.; Zhang, Y.X. Exogenous melatonin reduces the inhibitory effect of osmotic stress on antioxidant properties and cell ultrastructure at germination stage of soybean. *PLoS ONE* **2020**, *15*, e0243537. [CrossRef]
60. Bin, T.; Shang-Zhong, X.U.; Xi-Ling, Z.; Yong-Lian, Z.; Fa-Zhan, Q. Changes of antioxidative enzymes and lipid peroxidation in leaves and roots of waterlogging-tolerant and waterlogging-sensitive maize genotypes at seedling stage. *Agric. Sci. China* **2010**, *9*, 651–661. [CrossRef]

plants

MDPI

Article

Changes in Metabolites Produced in Wheat Plants Against Water-Deficit Stress

Valentina Spanic [1,*], Jurica Duvnjak [1], Dubravka Hefer [1] and John C. D'Auria [2]

1 Agricultural Institute Osijek, Južno Predgrađe 17, 31000 Osijek, Croatia; jurica.duvnjak@poljinos.hr (J.D.); dubravka.hefer@poljinos.hr (D.H.)

2 Leibniz Institute of Plant Genetics and Crop Plant Research (IPK Gatersleben), OT Gatersleben, Corrensstraße 3, 06466 Seeland, Germany; dauria@ipk-gatersleben.de

* Correspondence: valentina.spanic@poljinos.hr

Abstract: Drought stress can adversely affect the seed germination and seedling growth of wheat plants. This study analyzed the effect of drought on seed germination and the morphological parameters of seedlings from ten winter wheat genotypes. The primary focus was to elucidate the effects of two drought intensities on metabolic status in wheat seedlings. The findings suggest that most wheat genotypes exhibited a significant reduction in germination and growth traits under severe drought, while the genotype Srpanjka exhibited less reduction under both drought conditions. Out of 668 metabolic features, 54 were altered under 10% PEG stress and 140 under 20% PEG stress, with 48 commonly shared between these two stress intensities. This study demonstrated that the metabolic response of shoots to 10% PEG stress contrasts with that of 20% PEG stress. Some growth metabolites, such as oxalic acid, sophorose, and turanose, showed the highest positive increase under both stresses, while butanoic acid, tropic acid, glycine, propionic acid, and phosphonoacetic acid decreased. It is suggested that the accumulation of amino acids, such as proline, contributed to the drought tolerance of the plants. Among all organic acids, succinic and aspartic acids particularly increased the plant response to mild and severe drought stress, respectively. Our results suggest that different metabolites in wheat seedlings enhance the potential ability of wheat to cope with drought stress in the early growth stages by activating a rapid and comprehensive tolerance mechanism. This discovery presents a new approach for enhancing wheat tolerance to abiotic stress, including water deficit.

Keywords: metabolic profiling; GC-MS; winter wheat; abiotic stress; drought stress

1. Introduction

Throughout the vegetative season, plants are exposed to various abiotic and biotic stresses. Abiotic stress, including drought, is one of the most significant problems affecting plant growth and development [1] and plant productivity [2]. Climate changes are increasing the average global temperature and leading to more frequent high-temperature extremes, such as heatwaves [3]. Alarmingly, it is predicted that up to 60% of the current wheat-growing area will be affected by severe water scarcity events by the end of the twenty-first century [4]. Therefore, drought-induced limitations on grain yields, causing economic losses, pose a threat to global food security. As the world population is projected to grow to an estimated 8.5 billion by 2030, it is essential to increase the grain yield of crops to meet the growing food demand [5]. Wheat (*Triticum aestivum* L.), one of the most important cereals, continues to play a crucial role in ensuring global food and nutrition security. Due to its widespread use in different regions, wheat frequently confronts drought stress as a result of climate changes. Developing and producing drought-tolerant wheat varieties with greater water-use efficiency is crucial, especially in the context of food sustainability [6]. Furthermore, it is expected that more drought-tolerant plants would have a stronger ability to initiate defense mechanisms against water deficit.

Drought can damage wheat plants at any point during the growing season, but specific stages, such as seedlings, are particularly influential in the further development of plants [7]. The first step in successful wheat production requires well-developed seedlings with a viable plant from each seed. Therefore, seed germination and seedling establishment are potentially the most critical stages in the life cycle of plants under drought conditions [8]. These two stages of wheat growth are significantly influenced by temperature and moisture conditions [9]. Water deficiency can greatly affect seed germination and early seedling growth by inducing oxidative damage and metabolic disruptions in plant cells [10]. Seeds lose vigor and become increasingly sensitive to stress upon germination, which occurs between imbibition and the emergence of the radicle from the seed [11]. Additionally, seeds are more tolerant to environmental stressors than seedlings. However, drought-induced reductions in traits and final grain yield depend on the severity and duration of the stress period. It is also important to note that wheat genotypes exhibiting drought tolerance at the germination stage tend to maintain this tolerance in later growth stages [12]. Previously, there was research conducted on after anthesis water deficit where it was evident that significant changes occurred at the metabolomic level in developing grains during the time from anthesis to maturity [13].

Drought stress causes morphological, physiological, biochemical, and molecular changes in plants as they attempt to mitigate the adversities of abiotic stresses. These changes, along with altered metabolic processes, result in a significant reduction in grain yield [14]. Kumar et al. [15] concluded that metabolic regulation is the key mechanism for maintaining cell osmotic potential under drought stress, as metabolites play a crucial role in plant growth and development through cell integrity, energy storage, cell signaling, membrane formation and scaffolding, and whole-plant resource allocation [16]. Thus, plants can modify their physiological mechanisms under drought stress to adapt through metabolic homeostasis. Metabolomics, transcriptomics, proteomics, and ionomics provide a broader scope of mechanisms underlying biological processes that can contribute to the identification and characterization of genes, proteins, metabolites, and ions involved in signaling pathways under drought stress [17]. Consequently, important metabolic pathways such as photosynthesis, sugar synthesis, the tricarboxylic acid cycle, glycolysis, and hormone synthesis are further affected [18]. Limited water availability and reduced CO_2 assimilation result in disturbances in photosynthesis in chloroplasts, leading to excessive generation of reactive oxygen species that disrupt various metabolic pathways. The accumulation of osmolytes and osmoprotectants affects both primary and secondary metabolism [4]. In wheat, increased concentrations of metabolites such as tricarboxylic acid intermediates, sugars, and amino acids have been implicated in response to drought [19]. Further, Saeidi et al. [13] reported that abscisic acid (ABA) and sucrose concentration and sucrose synthase activity were significantly higher 17 days after anthesis than 10 days after anthesis during drought in developing grains. It is also important to note that in drought outlearn varieties, the sum of the amino acid concentrations determined, i.e., asparagine (22%), aspartic acid (18%), glycine (48%), ACC (79%), and valine (133%), were significantly increased. Other research showed that drought implemented 22 days after sowing resulted in an increase in metabolites, such as some amino acids, most notably proline, some organic acids, and lipid classes PC 36:3 and TAG 56:9 [19]. In flag leaves of wheat varieties, it was found that the application of salicylic acid, trehalose, or their interaction induced a marked increase in growth vigor of root and shoot, water relations, and protein, as well as nucleic acids in flag leaves [20]. However, many aspects of metabolites in plant drought tolerance at different growth stages of wheat remain to be revealed, especially in seedlings stage when our research was conducted. Thus, the main objectives of the present study were to compare metabolic changes in wheat seedlings subjected to drought stress during germination and the seedling stage. Additionally, we aimed to identify the most important metabolites associated with drought stress.

2. Results

2.1. Morphological Effects of Two Drought Stress Treatments in Wheat

The seeds treated with different concentrations (10% and 20%) of PEG solution were compared with the control group. The germination energy of genotype Osk.3.286/1-19 significantly decreased with the 10% PEG treatment, while the 20% PEG concentration significantly reduced germination energy in all genotypes except Srpanjka (Figure 1A). The highest decrease was observed in Tika Taka (23%). However, the percentage of germination for Kraljica was significantly reduced even at the 10% PEG concentration, with a significant decrease in all genotypes at the 20% PEG concentration (Figure 1B). Under the 20% PEG treatment, Kraljica had the highest decrease in germination percentage (17%), whereas Srpanjka's germination percentage was not significantly affected by either PEG treatment.

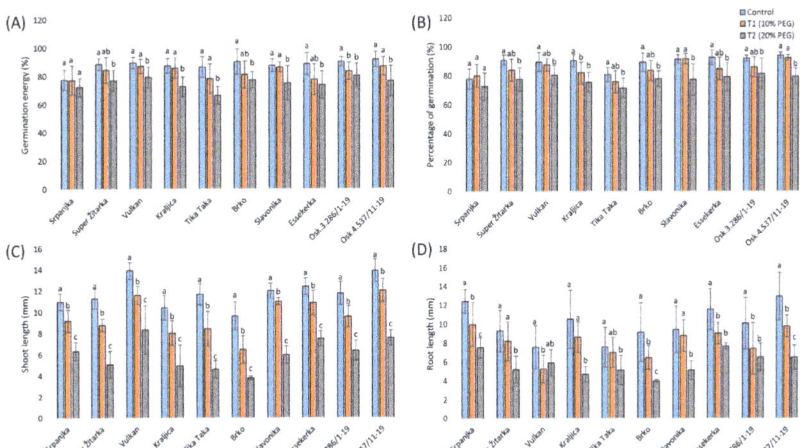

Figure 1. Germination energy (**A**), percentage of germination (**B**), shoot length (**C**), and root length (**D**) of ten winter wheat genotypes under control and two polyethylene glycol (PEG) treatments (T1-10% and T2-20%). Data are average values of ten biological replicates ± SD. Significant differences among treatments, for each genotype, separately, were assessed by the Fisher LSD test. Trait means with the same letter do not significantly differ at $p < 0.05$.

To further analyze the effect of different drought stresses on the investigated genotypes, seedling growth traits were recorded. Drought treatment with different concentrations of PEG affected shoot length, with all genotypes showing a significant decrease under both PEG treatments (Figure 1C). The highest decrease in shoot length was observed in Brko and Tika Taka (61% at 20% PEG concentration). The lowest significant reduction was observed in Essekerka, Vulkan, and Srpanjka.

The effect of the lower PEG concentration (10%) on root length was not significant for four genotypes (Super Žitarka, Kraljica, Tika Taka, and Slavonika), while only Vulkan showed a root length that was not significantly different from the control in the 20% PEG solution (Figure 1D). The highest decrease in root length was observed in Brko under the 20% PEG treatment (57%).

The effect of a lower concentration (10%) of PEG on the fresh weight of shoots was not significant for Vulkan, while the treatment with 20% PEG significantly reduced the fresh weight of shoots for all genotypes (Figure 2A). The dry weight of shoots in the 10% PEG treatment was not significantly different from the control in Essekerka and Osk.4.537/11-19 (Figure 2B). Under the 20% PEG treatment, the dry weight of shoots was significantly decreased in all genotypes compared with the control, except in Essekerka.

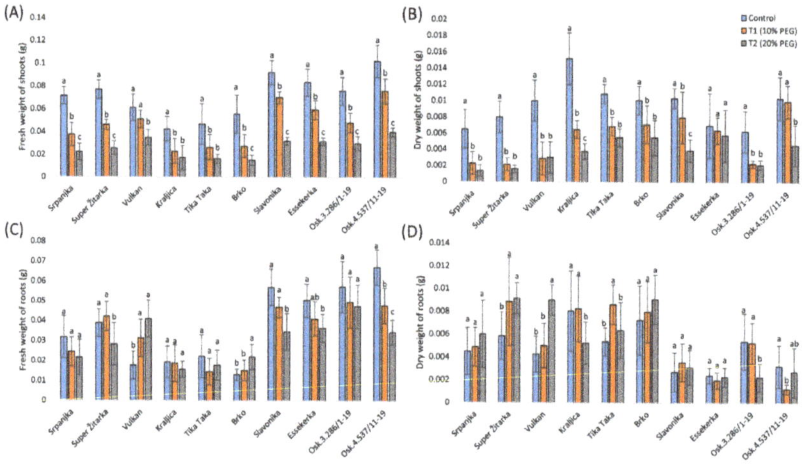

Figure 2. Fresh weight of shoots (**A**), dry weight of shoots (**B**), fresh weight of shoots (**C**), and dry weight of roots (**D**) of ten winter wheat genotypes under control and two polyethylene glycol (PEG) treatments (T1-10% and T2-20%). Data are average values of ten biological replicates ± SD. Significant differences among treatments for each genotype, separately, were assessed by the Fisher LSD test. Trait means with the same letter do not significantly differ at $p < 0.05$.

The fresh weight of roots was significantly decreased in the 10% PEG solution for Osk.4.537/11-19, while it was significantly increased in Vulkan (Figure 2C). Under the highest PEG treatment, the fresh weight of roots in four genotypes (Super Žitarka, Slavonika, Essekerka, and Osk.4.537/11-19) was significantly decreased, while it significantly increased in Vulkan and Brko. The effect of the lower concentration of PEG (10%) on the dry weight of roots was not significant for most genotypes, except for Super Žitarka, Tika Taka, and Osk.4.537/11-19 (Figure 2D). Under the highest PEG treatment (20%), the dry weight of roots significantly increased in Super Žitarka and Vulkan and significantly decreased in Kraljica and Osk.3.286/1-19.

2.2. Total Metabolite Profiling Under Two Drought Intensities in Wheat

Among the 668 metabolic features analyzed, 54 metabolites exhibited significant changes between the control and the treatment with 10% PEG ($p < 0.05$), and 140 metabolites showed significant changes between the control and the treatment with 20% PEG ($p < 0.05$). These were selected for further analyses (Supplementary Tables S1 and S2). The major groups of metabolites altered due to drought stress induced by the 10% PEG solution included organic acids (12), amino acids (10), carboxylic acids (7), sugars (2), hormones (3), sialic acid (neuraminic acid, N-acetyl), a flavonoid (taxifolin), a disaccharide (turanose

MP), a fatty acid (pentacosanoic acid methyl ester), a nucleic acid (uracil), nucleosides (adenosine alpha and inosine), a salt (glycyl-proline), and other organic compounds (12) (Supplementary Table S1, Figure 3A).

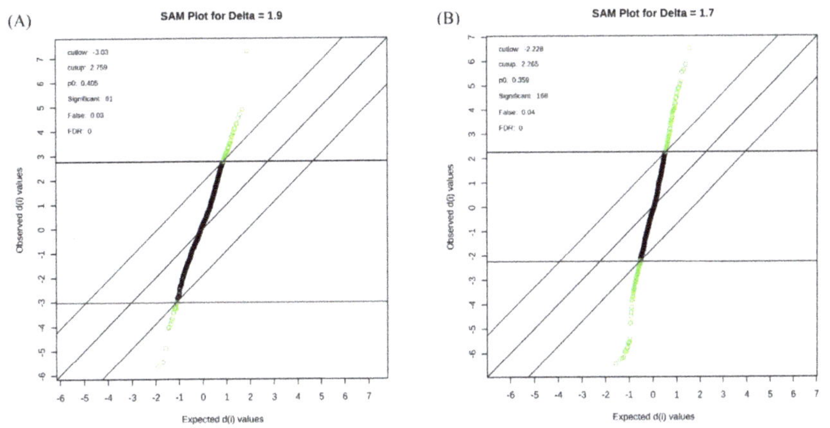

Figure 3. Significant analysis of metabolites (SAM) plot. The green points represent metabolite features that are differentially regulated between control and 10% PEG treatment (**A**) and 20% PEG treatment (**B**). The solid diagonal line represents "observed = expected". The more the variable deviates from the "observed = expected", the more likely it is to be significant.

Drought stress induced by 20% PEG treatment altered the content of 20 amino acids and 8 their derivatives or related compounds, 23 organic acids, 20 organic compounds, 16 carboxylic acids, 7 sugars, 6 sugar acids, 5 disaccharides, 4 alkaloids, 4 fatty acids, 2 amino alcohols, 2 monosaccharides, 2 plant metabolites, 2 benzoic acids, a lipid (dihydrosphingosine), a dipeptide (glycylglycine DL), a monoamine alkaloid (beta-D-Fructofuranosyl-(2,1)), a nucleoside (inosine), 3 phenolic acids (homogentisic acid), a carbohydrate (alpha-D-galactopyranosyl-(1,4)-D-galactopyranoside), a nucleotide base (guanine), a hydroxycinnamic acid (ferulic acid trans), a phosphodiesterase family enzyme (myo-Inositol-1-phosphate), a flavonoid (eriodictyol), hormones (gibberellin A3 and zeatin-trans), a salt (glycyl-proline), a phosphoric acid (serine O-phospho), a purine nucleoside (inosine 2-deoxy), an alcohol (ribitol), an acetoic acid (pyruvic acid 3-hydroxy), and a keto acid (pyruvic acid 4-hydroxyphenyl) (Supplementary Table S2, Figure 3B). Figure 4 shows the top 25 highly correlating and significant metabolites ($p < 0.05$) that change with increasing PEG concentration. These include metabolites that increased when PEG was added (shown in red; positive R) or decreased (highlighted in blue; negative R). It was observed that oxalic acid had the highest positive correlation between the control and 10% PEG, while tartronic acid had the highest negative correlation (Figure 4A). The highest positive correlation between the control and 20% PEG treatment was observed for turanose, and the highest negative correlation for phosphonoacetic acid (Figure 4B). The metabolites that were positively correlated between the control and both PEG treatments were turanose, sophorose, and oxalic acid. Tropic acid, glycine, butanoic acid, propionic acid, and phosphonoacetic acid were the most negatively correlated in the control and PEG treatments.

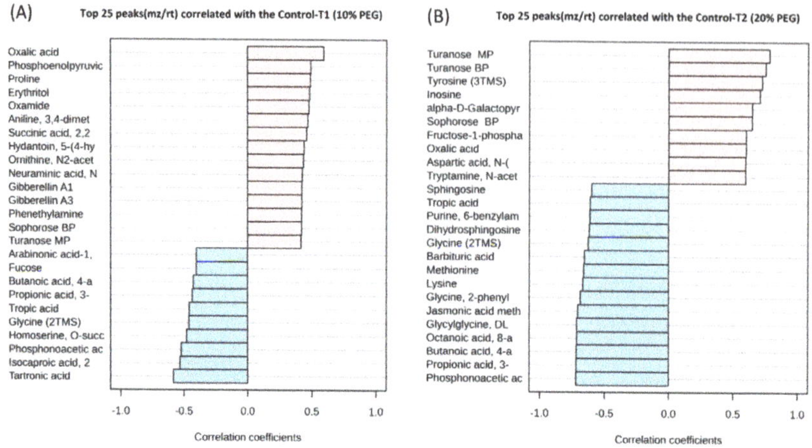

Figure 4. Top 25 metabolites with highest/lowest correlation coefficient from Pearson's correlation test between control and 10% polyethylene glycol (PEG) treatment (**A**) and control and 20% PEG treatment (**B**). Each row represents the most significant variable identified from the test ($p < 0.05$).

Subjecting the metabolite data to a PLS-DA revealed high variability in metabolite contributions to drought responses (Figure 5). PC1 explained 35.9% of the variability, while PC2 explained 9.6%. All control samples tended to cluster together, indicating similar metabolite profiles. Additionally, samples grown under severe drought conditions (20% PEG) clustered separately from the control samples. Samples grown under mild drought stress (10% PEG) were located in the intermediate region between control and severe drought, suggesting that mild drought stress represents a transition state. The PLS-DA biplot revealed that Vulkan in T2 was located on the opposite side of Vulkan in the control, indicating a direct negative relationship between the metabolites of these two treatments in this genotype. However, Srpanjka showed the least changes in metabolite profile across the three treatments, followed by Osk.4.537/11-19, which was closest to the origin. A group in the lower right quadrant consisted of four genotypes from the 20% PEG treatment (Super Žitarka, Brko, Tika Taka, and Kraljica) and six genotypes (Vulkan, Slavonika, Essekerka, Osk.3.286/1-19, Osk.4.537/11-19, and Srpanjka) in the upper right quadrant, which positively contributed to PC1. The group with the 10% PEG treatment (T1) that positively contributed to PC1 consisted of four genotypes (Srpanjka, Osk.3.286/1-19, Brko, and Kraljica) in the lower right quadrant.

Positive contributions to PC2 were observed in six genotypes under the 20% PEG treatment (Vulkan, Slavonika, Essekerka, Osk.3.286/1-19, Osk.4.537/11-19, and Srpanjka) in the upper right quadrant and a group of six genotypes (Slavonika, Essekerka, Super Žitarka, Vulkan, Tika Taka, and Osk.4.537/11-19) in the upper left quadrant under the T1 drought treatment, along with one genotype (Essekerka) under the control treatment.

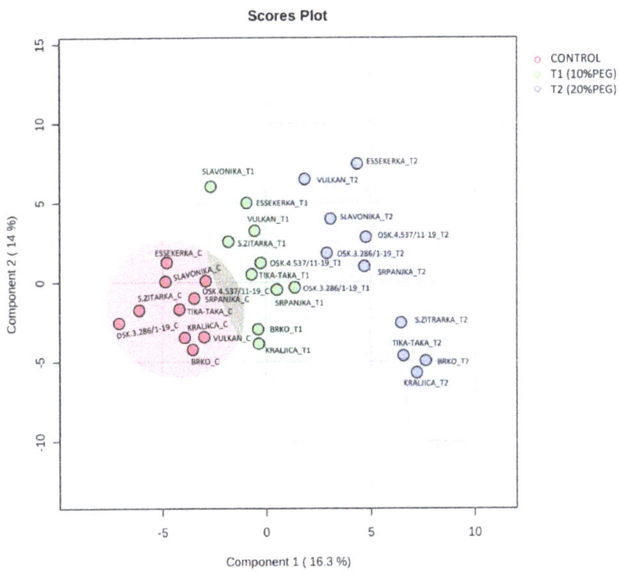

Figure 5. Partial least square discriminant analysis (PLS-DA) in seedlings of ten wheat genotypes under control and two drought conditions (T1 and T2). Samples at control and drought treatments did not overlap with each other indicating an altered state of metabolite levels in the wheat seedlings.

3. Discussion

New plant growth is greatly influenced by successful germination and seedling development. After seed germination, when the radicle protrudes, it is followed by the post-germinative growth of the seedling [21]. Furthermore, most of the future growth of a seedling depends on the cell divisions that occur in both the root and shoot meristems within the mature plant embryo [22]. At the germination and seedling stages, drought causes various physiological and biochemical changes, leading to dehydration of plant protoplasm and metabolic disorders, which affect normal plant growth [23]. Our current knowledge is limited in understanding the contrasting metabolic changes in wheat seedlings and the relationship between altered metabolite levels and performance under drought stress.

3.1. Morphological Changes in Wheat Seedlings from This Study

Previous research has shown that drought impairs germination and leads to poor stand establishment [24–26]. Although many seeds may remain viable, unfavorable seedling growth might occur. Previous studies report that germination, root, and leaf growth of wheat genotypes are considerably influenced by drought stress, depending on the drought tolerance of the genotypes [27]. The current results show that drought stress at 20% PEG has a strong inhibitory effect on the germination energy and percentage of germination in most genotypes. The reduction in germination with increased PEG levels could be associated with high seed nutrient imbalance, toxic ions, and reduced soluble osmotic potential [28]. Similar findings were recorded in rice genotypes, showing that the germination percentage, germination energy, germination rate, and germination index decreased as PEG concentrations increased [29]. Additionally, previous research by Duvnjak et al. [26] observed that PEG treatments negatively affected germination energy in all wheat varieties in a dose-dependent manner. However, in the current research, the genotype Srpanjka did not show significant changes in germination energy and percentage of germination under both drought treatments. Even higher reductions in wheat germination, up to 54% under PEG treatment, indicating that some genotypes better tolerated increasing osmotic potentials.

Drought differentially affects various plant parts, such as shoot and root length. For example, early seedling growth in rice subjected to drought stress results in a significant reduction in shoot and root length [30]. In the current study, the shoot length of ten genotypes was significantly reduced with increased PEG-induced drought levels. Similar findings, who showed that shoot length was reduced in wheat seedlings under PEG treatment by up to 54%. Under severe water deficiency, cell elongation in higher plants can be inhibited by the interruption of water flow from the xylem to the surrounding elongating cells [31]. Furthermore, Kerbauy [32] concluded that under drought stress conditions, seedling growth is affected due to reduced water uptake by the plants and lower cell turgor pressure. At the root level, drought reduces root length, density, and dry weight, which decreases water uptake capacity [16]. However, in the current research, it was evident that root growth appears to be less affected by milder drought than shoot growth. It should also be noted that root length is negatively correlated with shoot length and relative water content (RWC) [33], which could mean that a larger decrease in PEG treatments would result in higher RWC.

An unsolicited effect of drought on plants is the reduction in fresh and dry biomass [34]. In our study, the fresh and dry weights of shoots were significantly reduced in all genotypes due to 10% and 20% PEG stress treatments. However, the fresh and dry weights of roots indicated that slight drought stress increased these traits in most of the genotypes. Duvnjak et al. [26] revealed that the dry weight of roots in wheat seedlings was negatively correlated with germination energy, shoot length, fresh weight, and salicylic acid. Only the genotype Srpanjka showed no significant changes in the fresh and dry weights of roots. Under drought, plants may try to adapt to stress by altering their morphological structure, physiological, and biochemical functions to ensure the development of plant organs. As a result, leaf area may be reduced while root activity increases [27].

3.2. Metabolic Changes in Two Drought Intensities

Metabolomics is an effective tool for testing plant responses to different stresses, but the relationship between metabolites and their functions in response to drought stress remains uncertain. The study by Zhang et al. [35] demonstrated that drought stress increased levels of various small carbohydrates and soluble amino acids in the leaf, stem, root collar, and root. Different amino acids, as well as intermediates of the Krebs cycle and glycolysis, showed a reduction at the whole-plant level due to drought stress. In the current study, the 54 and 140 differential metabolites identified in the control and 10% PEG treatment and in the control and 20% PEG treatment, respectively, were clustered into carboxylic acids, amino acids, sugars, organic compounds, hormones, flavonoids, fatty acids, nucleic acids, phenolic acids, salts, nucleosides, cytokinins, and other compounds based on their comprehensive functions/pathways (Figure 3A,B). In the research by Guo et al. [18], mass spectrometry analysis under drought treatment enabled the identification of 56 metabolites, including amino acids, organic acids, carbohydrates, alkaloids, flavones, purines, and pyrimidines. Another study documented changes in soluble sugars, amino acids (especially proline), soluble proteins, betaine, and organic acids under drought stress [36]. Proline was the amino acid that significantly changed under the 10% PEG treatment in this study.

Metabolites can protect plants by exercising various physiological responses, such as strengthening membrane integrity, harmonizing enzymatic/antioxidant activity, and fulfilling water requirements [37]. Guo et al. [18] indicated that phenolics are important antioxidants that plants use to avoid oxidative damage caused by drought stress. Plants maintain turgor pressure and cell volume during drought through osmotic adjustment, accumulating and integrating compatible solutes like proline, sugars, and free amino acids to keep metabolic functions active [38]. More drought-tolerant wheat genotypes enhance physiological responses by upregulating regulatory genes and producing more sugars, organic acids, and important amino acids in shoots to maintain growth under drought stress [19]. Furthermore, plants can activate many enzymes, such as phenylalanine ammonia lyase, chalcone synthase, and phosphoenolpyruvate carboxylase, suggesting a shift from sucrose production to adaptation to drought [39]. In this experiment, phosphoenolpyruvic acid had one of the highest peaks in the 10% PEG treatment, indicating enhanced carboxylation of phosphoenolpyruvate to form oxaloacetate under mild stress. The metabolites of the control and two drought stresses were separated into three groups using PLS-DA. Among the 25 differential metabolites that showed the highest correlation coefficient between control and drought, three metabolites were upregulated by drought stress: oxalic acid, sophorose, and turanose. Four metabolites decreased under both drought treatments: propionic acid, tropic acid, glycine, and phosphonoacetic acid.

3.2.1. Organic Acids and Compounds

Organic acids play important roles in photosynthesis, nutrient uptake, tricarboxylic acid cycle metabolism, and drought tolerance in plants [40]. They are primarily formed in plants through the Krebs cycle, glyoxylate cycle, and photosynthesis. Moreover, organic acids such as succinic acid, malic acid, galacturonic acid, citric acid, glyceric acid, isocitric acid, and caffeic acid are involved in the plant response to drought [41]. Our study revealed that succinic and aspartic acids notably increased the plant response to mild and severe drought stress, respectively. In the study by Khan et al. [42], succinic acid, malic acid, and galacturonic acid also played roles in the response to long-term drought stress. It was also demonstrated that aspartic acid has an important role in plant resistance to drought stress [43].

Khosravi-Nejad et al. [44] showed that drought stress in wheat decreased malic acid, pyruvic acid, and citric acid content. Conversely, in our experiments, malic acid increased under both drought intensities compared with the control. It was also previously reported that malic acid increases under drought conditions, enhancing plant tolerance to water deficit by modulating osmotic potential [45]. Similarly, malic acid is the main organic acid secreted in maize roots, solubilizing phosphate for plant utilization [46]. The highest decrease under mild drought stress in this study was observed for arabinonic acid, and barbituric acid in the case of severe drought stress. Barbituric acid forms C-linked glycosides with many sugars [47], which could be an adaptation and response of plants to stress.

3.2.2. Amino Acids, Hormones, and Benzenoids

Amino acids are essential components of proteins and are involved in a wide variety of biochemical and physiological processes [48]. They play a crucial role in the biosynthesis of proteins, serve as building blocks for numerous biosynthetic pathways, act as precursors for various specialized metabolites, and are important in signaling processes in response to stress [49]. Increased amino acid biosynthetic capability and ROS scavenging ability, resulting from higher antioxidant activities and increased flavonoids, may be mechanisms underlying drought tolerance [36].

Increased proline in wheat plants helps overcome drought stress by supplying energy for growth and survival [50]. Rahman et al. [51] showed that amino acids increased under drought stress in wheat at the post-anthesis stage. In the current research, proline increased in drought-treated plants compared with the control, possibly to stabilize DNA, membranes, and protein complexes and provide carbon and nitrogen as an energy source to combat drought stress [52].

Sharma et al. [53] reported that drought stress decreases photosynthesis efficiency, increases ROS production and oxidative damage, increases abscisic acid accumulation and proline production, enhances antioxidant enzyme activity, and decreases cytokinin accumulation. In contrast, our research revealed that trans-zeatin was significantly elevated in both PEG treatments. The production of trans-zeatin is important for root and shoot meristematic activity, as it acts as a hormonal signaling transporter around its biosynthetic sites, playing a role in plant growth and drought response regulation [54]. Phenylalanine, tryptophan, and tyrosine are considered central molecules in plant metabolism, especially under environmental stresses [42]. Tryptophan significantly increased under both mild and severe drought during our experiment, while tyrosine and acetyl-5-hydroxy-tryptamine increased only under severe drought. Significant increases in proline, methionine, serine, asparagine, phenylalanine, aspartic acid, GABA, glycine, 5-hydroxynorvaline, threonine, and valine have been previously reported under drought [42]. Gibberellins are hormones that affect plant growth and development during germination and flowering, including seed germination [55]. In wheat, gibberellins play a crucial role in controlling and promoting germination [56]. From the current research, it can be suggested that under 20% PEG stress, the production of gibberellin A3 is inhibited, indicating that stronger drought stress is more rate-limiting in bioactive gibberellin biosynthesis. Gibberellins A1 and A3 were detected in germinating wheat grains under normal conditions [57,58]. It has been suggested that gibberellin content and metabolism are positively correlated with faster shoot growth rates in wheat hybrids or heterosis [59].

Two benzoic acids, salicylic and vanillic acid, were downregulated in the 20% PEG treatment. In some studies, vanillic acid increased due to abiotic stresses, including drought [60]. Salicylic acid plays a role in regulating important plant physiological processes, such as photosynthesis, nitrogen and proline metabolism, glycinebetaine formation, the antioxidant defense system, and plant–water relations under stress conditions [61].

3.2.3. Sugars, Amino Sugars, and Carboxylic Acids

Sugars play a crucial role in plant growth and development, influencing all stages of the plant's life cycle and interacting with other signaling molecules, including phytohormones [62]. In control and PEG treatments, turanose levels increased, with even higher levels reached during severe drought. Turanose is a disaccharide present naturally in plants and is implicated in altering plant physiology via the hormones ethylene and auxin [63]. In addition, turanose levels drastically increased in wheat accessions infected with the fungus *Tilletia caries* [64]. Increased levels of sugar help stress tolerance by increasing the osmotic potential in the plant [65]. The accumulation of many sugars, for example, fructan, trehalose, galactinol, and sugar alcohols such as mannitol and D-inositol, also plays a role in drought tolerance [42]. However, the type of sugars that accumulate vary depending on species, organ, and developmental stage. During mild drought stress, *N*-acylneuraminic acid significantly increased compared with controls. *N*-acetylneuraminic acid is abundant in terminal modifications of protein-linked glycans. It was demonstrated that *N*-acetylneuraminic acid was detected in tobacco cells, mung bean sprouts, apples, and bananas [66]. The concentration of total carbohydrates is limited under drought conditions [52]. Under 20% PEG stress, only alpha-D-galactopyranosyl-(1,4)-D-galactopyranoside was positively correlated with the control.

Regarding carboxylic acids, oxalic acid showed the highest positive correlation in the 10% PEG treatment among other metabolites and was also significant in the 20% PEG treatment. High concentrations of oxalic acid are known to induce programmed cell death, while low concentrations increase plant resistance to fungi [67]. Additionally, oxalic acid converts inorganic phosphorus to plant-available P in soil solution, enhancing P uptake [68]. Oxalates are involved in seed germination, calcium storage and regulation, ion balance, detoxification, structural strength, and insect repulsion in plants [69]. In the current study, it was evident that oxalic acid increased in seedling tissue under drought stress. This likely led to increased germin mRNA and oxalate oxidase activity, as previous research has shown that this occurs under both biotic and abiotic stresses [70].

3.2.4. Fatty Acids

Lipid content varies in plants when exposed to various stresses. These conditions induce fatty acid saturation, increasing membrane rigidity [19]. Fatty acids play specific roles in signaling events and metabolic processes [71]. 8-aminooctanoic acid, octadecanoic acid, and 2-oxovaleric acid showed negative correlations between the control and 20% PEG treatment. Maintaining high levels of fatty acids is important, as desaturation enables plants to stabilize membrane fluidity and reduce membrane damage caused by drought [72]. Gigon et al. [73] also reported an increase in unsaturated fatty acid levels during drought stress in *Arabidopsis thaliana*. Along with other metabolites, unsaturated fatty acids are crucial for drought tolerance in thyme [74]. Among the top 25 metabolites that correlated between the control and 20% PEG treatment, only 8-aminooctanoic acid was represented. This aligns with the research by Khosravi-Nejad et al. [44], which showed that organic acids exhibited the most variation in metabolic profiling in wheat, while fatty acids showed minimal change during the treatments. Previously, octanoic acid was also detected in disease-stressed shoots of wheat seedlings [75].

4. Materials and Methods

4.1. Wheat Material

Seeds of 10 winter wheat genotypes supplied by Agricultural Institute Osijek were used for this study (Table 1). Descriptions of recognized wheat varieties may be found in catalogs at https://www.poljinos.hr/katalozi/ (accessed on 3 October 2024).

Table 1. Origin and registration year of ten investigated wheat genotypes.

No.	Wheat Genotype	Origin	Registration Year
1	Srpanjka	AIO *, Croatia	1989
2	Super Žitarka	AIO, Croatia	1997
3	Vulkan	AIO, Croatia	2009
4	Kraljica	AIO, Croatia	2010
5	Tika Taka	AIO, Croatia	2014
6	Brko	AIO, Croatia	2020
7	Slavonika	AIO, Croatia	2023
8	Essekerka	AIO, Croatia	2023
9	Osk.3.286/1-19	AIO, Croatia	-
10	Osk.4.537/11-19	AIO, Croatia	-

* AIO—Agricultural Institute Osijek.

4.2. Experimental Layout

The experiment was conducted under controlled conditions in a plant growth chamber (Aralab, Rio de Mouro, Portugal) and designed as a completely randomized block design with eight replicates for each genotype per treatment. Prior to germination, wheat seeds were sterilized with 70% ethanol for 1 min and washed twice in deionized water (dH_2O). Germination assays were performed by evenly distributing 40 seeds in glass jars (H 9 × W 10 cm, V = 0.5 L) lined with filter paper. Sterile deionized water (10 mL) was added daily to the control jars, while the desired amounts of osmotic solutions were added to mimic drought stress. Seeds were germinated at 25/20 ± 2 °C day/night, with a relative humidity of 60% and a photoperiod of 16 h for 7 days. Polyethylene glycol (PEG) was added to the water at concentrations of 10% (treatment 1; T1) and 20% (treatment 2; T2) to reduce the water potential of the solution, thereby inducing drought stress in the plants. In our experiments, PEG-6000 (Sigma-Aldrich, Darmstadt, Germany) was used without causing toxicity to cells [23]. Seeds were checked regularly for germination and were considered germinated if the emerged radicles measured approximately 2 mm [76]. After 7 days of growth, the following germination attributes and morphological parameters were measured. For metabolomics analysis, the tissue of the seedlings was lyophilized.

4.3. Morphological Traits

To investigate the effects of drought stress, six morphological traits (germination energy, percentage of germination, shoot and root length, and fresh and dry weight of shoots and roots) were measured using seedling samples collected after seven days of drought stress induced by two PEG treatments. Germination energy was calculated in eight replicates by the formula: (number of seeds germinated on the fourth day/total number of seeds) × 100. The percentage of germination was calculated in eight replicates by the following formula: (number of seeds germinated on the seventh day/total number of seeds) × 100. Shoot and root lengths (mm) were measured in eight replicates using a ruler on the seventh day of the experiment. Fresh and dry weights of shoots and roots were obtained in eight replicates by weighing them separately, and again after 24 h of drying in a lab dryer at 105 °C. Shoot and root biomass was expressed in terms of fresh weight (FW) and dry weight (DW) per plant.

4.4. Metabolic Profiling

Prior to extraction for metabolic analyses, the seedlings were flash frozen in liquid nitrogen and ground in 10 mL plastic tubes together with a grinding ball for 2 min per sample in automatic Labman's cryogenic grinding system (Labman, Middlesbrough, UK). Polar and semipolar metabolites were extracted from 15 mg of deep-frozen homogenized plant material using polar metabolite extraction protocol according to Riewe et al. [77]. Extraction was proceeded by adding 1 mL of chilled extraction buffer (2.5:1:1 v/v $MeOH/CHCl_3/H_2O$) containing 1 µL of a 2 mg/mL stock solution of ^{13}C-sorbitol, and D4-alanine to the flash-frozen and pulverized tissue. Following 15 min incubation at 4 °C, 0.400 mL H_2O was added and centrifugated for 15 min at 4 °C and Vmax. Extraction was split into three batches and aliquots of 50 µL of polar phase into GC glass vials. Additionally, GC vials containing aliquoted samples were placed in a Speedvac overnight, crimped, and stored in sealed plastic bags with silica gel at −80 °C until analysis. Dried extracts were in-line derivatized directly prior to injection [78] using a Gerstel MPS2-XL autosampler (Gerstel, Mühlheim/Ruhr, Germany) with the front inlet temperature set at 200 °C, analyzed in split-less mode, using a LECO Pegasus BT time-of-flight mass spectrometer (LECO, St. Joseph, MI, USA, https://www.lecosoftware.com/chromatof, accessed on 10 March 2024) connected to an Agilent 8890 gas chromatograph and helium as the carrier gas at 1.0 mL min^{-1} flow and linear velocity as flow control mode. The capillary column used was an Agilent DB-35MS (30 m × 0.25 mm × 0.25 µm). Fatty acid methyl esters (FAME MIX: C9-C31) were used in the determination of the retention time index [79].

Metabolic features were identified by employing the LECO ChromaTOF software in conjunction with the Golm Metabolome Database (GMD, http://gmd.mpimp-golm.mpg.de/, accessed on 11 March 2024). Peak intensities were determined using the TargetSearch package [80] within the R software (4.3.1 version, R Core Team, 2023). For normalization, the content of the internal standards D4-alanine and ^{13}C Sorbitol were utilized. All metabolic class identifications were manually validated based on GOLM, KEGG (https://www.genome.jp/kegg/pathway.html, accessed on 13 March 2024), and HMDB (https://www.hmdb.ca/, accessed on 13 March 2024) metabolites databases.

4.5. Statistical Analysis

Samples were collected from each jar, and after data collection, they were tested for normality, which indicated that the data were normally distributed. Morphological parameters were expressed as the mean value of eight replicates ± standard deviation. Differences among treatments for each genotype were assessed by a one-way analysis of variance (ANOVA), followed by the Fisher test using Statistica software (version 14). Morphological parameters were analyzed as dependent variables, with genotype and treatment as fixed effects in a one-way ANOVA.

Partial least squares discriminant analysis (PLS-DA) and significant analysis of metabolites (SAM) were performed using MetaboAnalyst 6.0.

5. Conclusions

The PEG solution inhibited seed germination, but the percentage of germination remained high, indicating that all plant seeds had a certain level of drought tolerance, with Srpanjka showing the highest tolerance. There was a decrease in shoot and root-related traits with the rise in osmotic potential levels. Under mild stress, the highest negative correlation with the control was observed for organic acids, compounds, and carboxylic acids. However, under severe drought stress, the contents of carboxylic acids, organic compounds, and some amino acids showed the highest negative correlation with the control, indicating damage to the wheat seedlings. Among the top 25 metabolites that showed a positive correlation between the control and both PEG treatments were oxalic acid, sophorose, and turanose. In contrast, negative correlations were observed for butanoic acid, 3-ureidopropionic acid, tropic acid, glycine, and phosphoenolpyruvic acid. The drought tolerance of wheat genotypes is very complex, and further research is needed to explore the mechanisms of wheat seedlings' response to drought. By identifying and quantifying key metabolites, researchers can better understand the metabolic status of an organism and develop targeted interventions for various conditions.

Supplementary Materials: The following supporting information can be downloaded at https://www.mdpi.com/article/10.3390/plants14010010/s1. Table S1: Sam signatures between control and 10% PEG treatment; Table S2: Sam signatures between control and 20% PEG treatment.

Author Contributions: Conceptualization, V.S. and J.D.; methodology, V.S., J.D. and J.C.D.; formal analysis, J.D., D.H. and J.C.D.; investigation, V.S., J.D., D.H. and J.C.D.; resources, V.S. and J.C.D.; writing—original draft preparation, V.S.; writing—review and editing, J.D., D.H. and J.C.D.; visualization, J.D.; project administration, V.S. and J.C.D. All authors have read and agreed to the published version of the manuscript.

Funding: The work of the doctoral student Jurica Duvnjak was supported in part by the "Young researchers' career development project-training of doctoral students" of the Croatian Science Foundation. Financial support by the Leibniz Institute of Plant Genetics and Crop Plant Research (IPK) in Gatersleben by visiting programme for scientists from transitions countries is gratefully acknowledged.

Data Availability Statement: The data that support the findings are available on request from the corresponding author.

Conflicts of Interest: The authors declare no conflicts of interest.

References

1. Rajanna, G.A.; Suman, A.; Venkatesh, P. Mitigating drought stress effects in arid and semi-arid agro-ecosystems through bioirrigation strategies-A review. *Sustainability* **2023**, *15*, 3542. [CrossRef]
2. Seleiman, M.F.; Al-Suhaibani, N.; Ali, N.; Akmal, M.; Alotaibi, M.; Refay, Y.; Dindaroglu, T.; Abdul-Wajid, H.H.; Battaglia, M.L. Drought stress impacts on plants and different approaches to alleviate its adverse effects. *Plants* **2021**, *10*, 259. [CrossRef]
3. Sato, H.; Mizoi, J.; Shinozaki, K.; Yamaguchi-Shinozaki, K. Complex plant responses to drought and heat stress under climate change. *Plant J.* **2024**, *117*, 1873–1892. [CrossRef] [PubMed]
4. Szablińska-Piernik, J.; Lahuta, L.B. Polar metabolites profiling of wheat shoots (*Triticum aestivum* L.) under repeated short-term soil drought and rewatering. *Int. J. Mol. Sci.* **2023**, *24*, 8429. [CrossRef]
5. UN. World Population Prospects: Data Booklet. 2019. Available online: https://population.un.org/wpp/ (accessed on 23 July 2024).
6. Itam, M.; Mega, R.; Tadano, S.; Abdelrahman, M.; Matsunaga, S.; Yamasaki, Y.; Akashi, K.; Tsujimoto, H. Metabolic and physiological responses to progressive drought stress in bread wheat. *Sci. Rep.* **2020**, *10*, 17189. [CrossRef] [PubMed]
7. Fang, Y.; Xiong, L. General mechanisms of drought response and their application in drought resistance improvement in plants. *CMLS* **2015**, *72*, 673–689. [CrossRef]
8. Hu, X.W.; Fan, Y.; Baskin, C.C.; Baskin, J.M.; Wang, Y.R. Comparison of the effects of temperature and water potential on seed germination of *Fabaceae* species from desert and subalpine grassland. *Am. J. Bot.* **2015**, *102*, 649–660. [CrossRef]
9. Khaeim, H.; Kende, Z.; Jolánkai, M.; Kovács, G.P.; Gyuricza, C.; Tarnawa, Á. Impact of temperature and water on seed germination and seedling growth of maize (*Zea mays* L.). *Agronomy* **2022**, *12*, 397. [CrossRef]
10. Ma, L.; Wei, J.; Han, G.; Sun, X.; Yang, X. Seed osmopriming with polyethylene glycol (PEG) enhances seed germination and seedling physiological traits of *Coronilla varia* L. under water stress. *PLoS ONE* **2024**, *19*, e0303145. [CrossRef] [PubMed]
11. Reed, R.C.; Bradford, K.J.; Khanday, I. Seed germination and vigor: Ensuring crop sustainability in a changing climate. *Heredity* **2022**, *128*, 450–459. [CrossRef]
12. Khakwani, A.A.; Dennett, M.D.; Munir, M. Drought tolerance screening of wheat varieties by inducing water stress conditions. *SJST* **2011**, *33*, 135–142.
13. Saeidi, M.; Ahmadi, A.; Moradi, F.; Hajirezaei, M.-R. Comparative metabolome profiling of two contrasting wheat cultivars in late-season water deficit. *Front. Plant Physiol.* **2024**, *2*, 1386473. [CrossRef]
14. Zhao, J.; Zhang, S.; Yang, X.; Feng, K.; Wang, G.; Shi, Q.; Wang, X.; Yuan, X.; Ren, J. Hydrogen sulfide increases drought tolerance by modulating carbon and nitrogen metabolism in Foxtail Millet seedlings. *Agronomy* **2024**, *14*, 1080. [CrossRef]
15. Kumar, M.; Kumar Patel, M.; Kumar, N.; Bajpai, A.B.; Siddique, K.H.M. Metabolomics and molecular approaches reveal drought stress tolerance in plants. *Int. J. Mol. Sci.* **2021**, *24*, 9108. [CrossRef] [PubMed]
16. Kang, Z.; Babar, M.A.; Khan, N.; Guo, J.; Khan, J.; Islam, S.; Shrestha, S.; Shahi, D. Comparative metabolomic profiling in the roots and leaves in contrasting genotypes reveals complex mechanisms involved in post-anthesis drought tolerance in wheat. *PLoS ONE* **2019**, *11*, e0213502. [CrossRef]
17. You, J.; Zhang, Y.; Liu, A.; Li, D.; Wang, X.; Dossa, K.; Zhou, R.; Yu, J.; Zhang, Y.; Wang, L.; et al. Transcriptomic and metabolomic profiling of drought-tolerant and susceptible sesame genotypes in response to drought stress. *BMC Plant Biol.* **2019**, *19*, 267. [CrossRef] [PubMed]
18. Guo, X.; Xin, Z.; Yang, T.; Ma, X.; Zhang, Y.; Wang, Z.; Ren, Y.; Lin, T. Metabolomics response for drought stress tolerance in chinese wheat genotypes (*Triticum aestivum*). *Plants* **2020**, *9*, 520. [CrossRef]
19. Fadoul, H.E.; Martínez Rivas, F.J.; Neumann, K.; Balazadeh, S.; Fernie, A.R.; Alseekh, S. Comparative molecular and metabolic profiling of two contrasting wheat cultivars under drought stress. *Int. J. Mol. Sci.* **2021**, *22*, 13287. [CrossRef] [PubMed]
20. Aldesuquy, H.S.; Ibraheem, F.I.; Gahnem, H.E. Comparative morpho-biochemical responses of wheat cultivars sensitive and tolerant to water stress. *J. Stress Physiol. Biochem.* **2014**, *10*, 168–189.
21. Rajjou, L.; Duval, M.; Gallardo, K.; Catusse, J.; Bally, J.; Job, C.; Job, D. Seed germination and vigor. *Annu. Rev. Plant Biol.* **2012**, *63*, 507–533. [CrossRef]
22. Wolny, E.; Betekhtin, A.; Rojek, M.; Braszewska-Zalewska, A.; Lusinska, J.; Hasterok, R. Germination and the early stages of seedling development in *Brachypodium distachyon*. *Int. J. Mol. Sci.* **2018**, *19*, 2916. [CrossRef]
23. Lu, Y.; Liu, H.; Chen, Y.; Zhang, L.; Kudusi, K.; Song, J. Effects of drought and salt stress on seed germination of ephemeral plants in desert of northwest China. *Front. Ecol. Evol.* **2022**, *10*, 1026095. [CrossRef]

24. Harris, D.; Tripathi, R.S.; Joshi, A. On-farm seed priming to improve crop establishment and yield in dry direct-seeded rice. In *Direct Seeding: Research Strategies and Opportunities*; Pandey, S., Mortimer, M., Wade, L., Tuong, T.P., Lopes, K., Hardy, B., Eds.; International Research Institute: Manila, Philippines, 2002; pp. 231–240.
25. Vuković, R.; Štolfa Čamagajevac, I.; Vuković, A.; Šunić, K.; Begović, L.; Mlinarić, S.; Sekulić, R.; Sabo, N.; Španić, V. Physiological, biochemical and molecular response of different winter wheat varieties under drought stress at germination and seedling growth stage. *Antioxidants* **2022**, *11*, 693. [CrossRef]
26. Duvnjak, J.; Brkljačić, L.; Salopek Sondi, B.; Španić, V. Morpho-physiological and hormonal response of different winter wheat varieties to drought stress at germination and seedling stage. *Agron. Glas.* **2023**, *84*, 277–294. [CrossRef]
27. Duan, H.; Zhu, Y.; Li, J.; Ding, W.; Wang, H.; Jiang, L.; Zhou, Y. Effects of drought stress on growth and development of wheat seedlings. *Int. J. Agric. Biol.* **2017**, *19*, 1119–1124. [CrossRef]
28. Yang, Z.Y.; Zhou, B.Z.; Zhou, Y.; Ge, X.G.; Wang, X.M.; Cao, Y.H. Effects of drought stress simulated by peg on seed germination and growth physiological characteristics of *Phyllostachys edulis*. *For. Res.* **2018**, *31*, 47–54. [CrossRef]
29. Evamoni, F.Z.; Nulit, R.; Yap, C.K.; Ibrahim, M.F.; Sidek, N.B. Assessment of germination performance and early seedling growth of Malaysian indica rice genotypes under drought conditions for strategic cropping during water scarcity. *Chil. J. Agric. Res.* **2023**, *83*, 281–292. [CrossRef]
30. Zheng, M.; Tao, Y.; Hussain, S.; Jiang, Q.; Peng, S.; Huang, J.; Cui, K.; Nie, L. Seed priming in dry direct-seeded rice: Consequences for emergence, seedling growth and associated metabolic events under drought stress. *Plant Growth Regul.* **2016**, *78*, 167–178. [CrossRef]
31. Nonami, H. Plant water relations and control of cell elongation at low water potentials. *J. Plant Res.* **1998**, *111*, 373–382. [CrossRef]
32. Kerbauy, G.B. *Fisiologia Vegetal*, 2nd ed.; Guanabara Koogan: Rio de Janeiro, Brazil, 2012; 431p.
33. Ahmed, H.G.M.; Zeng, Y.; Shah, A.N.; Yar, M.M.; Ullah, A.; Ali, M. Conferring of drought tolerance in wheat (*Triticum aestivum* L.) genotypes using seedling indices. *Front. Plant Sci.* **2022**, *13*, 961049. [CrossRef]
34. Farooq, M.; Wahid, A.; Kobayashi, N.S.; Fujita, D.B.; Basra, S.M. Plant drought stress: Effects, mechanisms and management. *Sustain. Agric.* **2009**, *29*, 153–188. [CrossRef]
35. Zhang, J.; Chen, G.; Zhao, P.; Zhou, Q.; Zhao, X. The abundance of certain metabolites responds to drought stress in the highly drought tolerant plant *Caragana korshinskii*. *Acta Physiol. Plant.* **2017**, *39*, 116. [CrossRef]
36. Lv, L.; Chen, X.; Li, H.; Huang, J.; Liu, Y.; Zhao, A. Different adaptive patterns of wheat with different drought tolerance under drought stresses and rehydration revealed by integrated metabolomic and transcriptomic analysis. *Front. Plant Sci.* **2022**, *13*, 1008624. [CrossRef] [PubMed]
37. Ahmad, F.; Singh, A.; Kamal, A. Osmoprotective role of sugar in mitigating abiotic stress in plants. In *Protective Chemical Agents in the Amelioration of Plant Abiotic Stress. Biochemical and Molecular Perspectives*; Wiley: Hoboken, NJ, USA, 2020; pp. 53–70. [CrossRef]
38. Tatar, Ö.; Gevrek, M.N. Lipid peroxidation and water content of wheat. *Asian J. Plant Sci.* **2008**, *7*, 409–412. [CrossRef]
39. Ruiz, J.M.; Rivero, R.M.; Lopez-Cantarero, I.; Romero, L. Role of Ca^{2+} in the metabolism of phenolic compounds in tobacco leaves (*Nicotiana tabacum* L.). *Plant Growth Regul.* **2003**, *41*, 173–177. [CrossRef]
40. Kleiber, T.; Chadzinikolau, T.; Formela-Luboińska, M.; Lartey, J.L.; Kosiada, T. Enhancing lettuce drought tolerance: The role of organic acids in photosynthesis and oxidative defense. *Appl. Sci.* **2024**, *14*, 5119. [CrossRef]
41. Xie, H.; Bai, G.; Lu, P.; Li, H.; Fei, M.; Xiao, B.G.; Chen, X.J.; Tong, Z.J.; Wang, Z.Y.; Yang, D.H. Exogenous citric acid enhances drought tolerance in tobacco (*Nicotiana tabacum*). *Plant Biol.* **2022**, *24*, 333–343. [CrossRef] [PubMed]
42. Khan, N.; Ali, S.; Zandi, P.; Mehmood, A.; Ullah, S.; Ikram, M.; Ismail, I.; Babar, M.A. Role of sugars, amino acids and organic acids in improving plant abiotic stress tolerance. *Pak. J. Bot.* **2020**, *52*, 355–363. [CrossRef] [PubMed]
43. Lei, S.; Rossi, S.; Huang, B. Metabolic and physiological regulation of aspartic acid-mediated enhancement of heat stress tolerance in Perennial ryegrass. *Plants* **2022**, *13*, 199. [CrossRef]
44. Khosravi-nejad, F.; Khavari-nejad, R.A.; Moradi, F.; Najafi, F. Cytokinin and abscisic acid alleviate drought stress through changing organic acids profile, ion immolation, and fatty acid profile to improve yield of wheat (*Triticum aestivum* L.) cultivars. *Physiol. Mol. Biol. Plants* **2022**, *28*, 1119–1129. [CrossRef]
45. Alvarez, S.; Marsh, E.; Schroeder, S.; Schachtman, D. Metabolomic and proteomic changes in the xylem sap of maize under drought. *Plant Cell Environ.* **2008**, *31*, 325–340. [CrossRef]
46. Henry, A.; Doucette, W.; Norton, J.; Bugbee, B. Changes in crested wheatgrass root exudation caused by flood, drought, and nutrient stress. *J. Environ. Qual.* **2007**, *36*, 904–912. [CrossRef]
47. Mungi, C.V.; Singh, S.K.; Chugh, J.; Rajamani, S. Synthesis of barbituric acid containing nucleotides and their implications for the origin of primitive informational polymers. *Phys. Chem. Chem. Phys.* **2016**, *18*, 20144–20152. [CrossRef] [PubMed]
48. Ribarova, F. Amino acids: Carriers of nutritional and biological value foods. In *Food Processing for Increased Quality and Consumption*; Academic Press: London, UK, 2018; pp. 287–311. [CrossRef]
49. Hildebrandt, T.M.; Nesi, A.N.; Araújo, W.L.; Braun, H.P. Amino acid catabolism in plants. *Mol. Plant* **2015**, *8*, 1563–1579. [CrossRef] [PubMed]

50. Heshmat, A. Kernel biochemical aspects in stressed wheat in response to glycine betaine and salicylic scid. *Agric. Res. Technol.* **2017**, *7*, 555724. [CrossRef]
51. Rahman, M.; Akond, M.; Babar, M.A.; Beecher, C.; Erickson, J.; Thomason, K.; De Jong, F.A.; Mason, R.E. LC-HRMS Based non-targeted metabolomic profiling of wheat (*Triticum aestivum* L.) under post-anthesis drought stress. *Am. J. Plant Sci.* **2017**, *8*, 3024–3061. [CrossRef]
52. Eirini, S.; Paschalina, C.; Loannis, T.; Kortessa, D.T. Effect of drought and salinity on volatile organic compounds and other cecondary metabolites of Citrus aurantium leaves. *Nat. Prod. Commun.* **2017**, *12*, 193–196.
53. Sharma, B.; Yadav, L.; Shrestha, A.; Shrestha, S.; Subedi, M.; Subedi, S.; Shrestha, J. Drought stress and its management in wheat (*Triticum aestivum* L.): A review. *Agric. Sci. Technol.* **2022**, *14*, 3–14. [CrossRef]
54. Ali, E.F.; Aljarani, A.M.; Mohammed, F.A.; Desoky, E.M.; Mohamed, I.A.A.; El-Sharnouby, M.; Tammam, S.A.; Hassan, F.A.S.; Rady, M.M.; Shaaban, A. Exploring the potential enhancing effects of trans-zeatin and silymarin on the productivity and antioxidant defense capacity of cadmium-stressed wheat. *Biology* **2022**, *11*, 1173. [CrossRef]
55. Wang, T.; Li, J.; Jiang, Y.; Zhang, J.; Ni, Y.; Zhang, P.; Yao, Z.; Jiao, Z.; Li, H.; Li, L.; et al. Wheat gibberellin oxidase genes and their functions in regulating tillering. *PeerJ* **2023**, *11*, e15924. [CrossRef] [PubMed]
56. Gupta, R.; Chakrabarty, S.K. Gibberellic acid in plant. *Plant Signal. Behav.* **2013**, *8*, e25504. [CrossRef]
57. Pavlista, A.D.; Baltensperger, D.D.; Santra, D.K.; Hergert, G.W.; Knox, S. Gibberellic acid promotes early growth of winter wheat and rye. *Am. J. Plant Sci.* **2014**, *5*, 2984–2996. [CrossRef]
58. Lenton, J.R.; Appleford, N.E.J.; Croker, S.J. Gibberellins and αamylase gene expression in germinating wheat grains. *Plant Growth Regul.* **1994**, *15*, 261–270. [CrossRef]
59. Zhang, Y.; Ni, Z.; Yao, Y.; Nie, X.; Sun, Q. Gibberellins and heterosis of plant height in wheat (*Triticum aestivum* L.). *BMC Genet.* **2007**, *8*, 40. [CrossRef]
60. Šamec, D.; Karalija, E.; Šola, I.; Vujčić Bok, V.; Salopek-Sondi, B. The role of polyphenols in abiotic stress response: The influence of molecular structure. *Plants* **2021**, *10*, 118. [CrossRef] [PubMed]
61. Khan, M.I.; Fatma, M.; Per, T.S.; Anjum, N.A.; Khan, N.A. Salicylic acid-induced abiotic stress tolerance and underlying mechanisms in plants. *Front. Plant Sci.* **2015**, *6*, 462. [CrossRef]
62. Jeandet, P.; Formela-Luboińska, M.; Labudda, M.; Morkunas, I. The role of sugars in plant responses to stress and their regulatory function during development. *Int. J. Mol. Sci.* **2022**, *23*, 5161. [CrossRef] [PubMed]
63. Gonzali, S.; Novi, G.; Loreti, E.; Paolicchi, F.; Poggi, A.; Alpi, A.; Perata, P. A turanose-insensitive mutant suggests a role for WOX5 in auxin homeostasis in *Arabidopsis thaliana*. *Plant J.* **2005**, *44*, 633–645. [CrossRef] [PubMed]
64. Weed, R.A.; Savchenko, K.G.; Lessin, L.M.; Carris, L.M.; Gang, D.R. Untargeted metabolomic investigation of wheat infected with stinking smut *Tilletia caries*. *Phytopathology* **2021**, *111*, 2343–2354. [CrossRef]
65. Wind, J.; Smeekens, S.; Hanson, J. Sucrose: Metabolite and signaling molecule. *Phytochemistry* **2010**, *71*, 1610–1614. [CrossRef]
66. Zeleny, R.; Kolarich, D.; Strasser, R.; Altmann, F. Sialic acid concentrations in plants are in the range of inadvertent contamination. *Planta* **2006**, *224*, 222–227. [CrossRef]
67. Lehner, A.; Meimoun, P.; Errakhi, R.; Madiona, K.; Barakate, M.; Bouteau, F. Toxic and signaling effects of oxalic acid: Natural born killer or natural born protector? *Plant Signal. Behav.* **2008**, *3*, 746–748. [CrossRef] [PubMed]
68. Dotaniya, M.L.; Datta, S.C.; Biswas, D.R.; Meena, H.M.; Kumar, K. Production of oxalic acid as influenced by the application of organic residue and its effect on phosphorus uptake by wheat (*Triticum aestivum* L.) in an inceptisol of North India. *Natl. Acad. Sci. Lett.* **2014**, *37*, 401–405. [CrossRef]
69. Liu, E.E.; Luo, W.; Zhou, H.; Peng, X.X. Determination of oxalate in plant tissues with oxalate oxidase prepared from wheat. *Biol. Plant* **2009**, *53*, 129–132. [CrossRef]
70. Delisle, G.; Champoux, M.; Houde, M. Characterization of oxalate oxidase and cell death in al-sensitive and tolerant wheat roots. *Plant Cell Physiol.* **2001**, *42*, 324–333. [CrossRef]
71. Ullah, S.; Khan, M.N.; Lodhi, S.S.; Ahmed, I.; Tayyab, M.; Mehmood, T.; Din, I.U.; Khan, M.; Sohail, Q.; Akram, M. Targeted metabolomics reveals fatty acid abundance adjustments as playing a crucial role in drought-stress response and post-drought recovery in wheat. *Front. Genet.* **2022**, *13*, 972696. [CrossRef]
72. Sharma, P.; Lakra, N.; Goyal, A.; Ahlawat, Y.K.; Zaid, A.; Siddique, K.H.M. Drought and heat stress mediated activation of lipid signaling in plants: A critical review. *Front. Plant Sci.* **2023**, *14*, 1216835. [CrossRef] [PubMed]
73. Gigon, A.; Matos, A.; Laffray, D.; Zuily-Fodil, Y.; Pham-Thi, A. Effect of drought stress on lipid metabolism in the leaves of *Arabidopsis thaliana* (*Ecotype columbia*). *Ann. Bot.* **2004**, *94*, 345–351. [CrossRef]
74. Moradi, P.; Ford-Lloyd, B.; Pritchard, J. Metabolomic approach reveals the biochemical mechanisms underlying drought stress tolerance in thyme. *Anal. Biochem.* **2017**, *527*, 49–62. [CrossRef]
75. Sarsenbayev, K. *Determination of Signal Molecules, Proteins, and Low-Molecular-Weight Organic Compounds in Wheat Varieties Infected by leaf Rust Disease*; IntechOpen: London, UK, 2024. [CrossRef]

76. Saleem, M.H.; Ali, S.; Seleiman, M.F.; Rizwan, M.; Rehman, M.; Akram, N.A.; Liu, L.; Alotaibi, M.; Al-Ashkar, I.; Mubushar, M. Assessing the correlations between different traits in copper-sensitive and copper-resistant varieties of jute (*Corchorus capsularis* L.). *Plants* **2019**, *8*, 545. [CrossRef] [PubMed]
77. Riewe, D.; Jeon, H.J.; Lisec, J.; Heuermann, M.C.; Schmeichel, J.; Seyfarth, M.; Meyer, R.C.; Willmitzer, L.; Altmann, T. A naturally occurring promoter polymorphism of the *Arabidopsis* FUM2 gene causes expression variation, and is associated with metabolic and growth traits. *Plant J.* **2016**, *88*, 826–838. [CrossRef] [PubMed]
78. Erban, A.; Schauer, N.; Fernie, A.R.; Kopka, J. Nonsupervised construction and application of massspectral and retention time index libraries from time-of-flight gaschromatography–mass spectrometry metabolite profiles. *Methods Mol.* **2007**, *358*, 19–38. [CrossRef]
79. Kind, T.; Wohlgemuth, G.; Lee, D.Y.; Lu, Y.; Palazoglu, M.; Shahbaz, S.; Fiehn, O. FiehnLib: Mass spectral and retention index libraries for metabolomics based on quadrupole and time-of-flight gas chromatography/mass spectrometry. *Anal. Chem.* **2009**, *81*, 10038–10048. [CrossRef] [PubMed]
80. Cuadros-Inostroza, A.; Caldana, C.; Redestig, H.; Kusano, M.; Lisec, J.; Peña-Cortés, H.; Willmitzer, L.; Hannah, M.A. TargetSearch—A bioconductor package for the efficient preprocessing of GC-MS metabolite profiling data. *BMC Bioinform.* **2009**, *10*, 428. [CrossRef]

MDPI AG
Grosspeteranlage 5
4052 Basel
Switzerland
Tel.: +41 61 683 77 34

Plants Editorial Office
E-mail: plants@mdpi.com
www.mdpi.com/journal/plants

www.ingramcontent.com/pod-product-compliance
Lightning Source LLC
Chambersburg PA
CBHW061439210926
51850CB00008B/1557
9783725855490